Constraint Resolution Theories

Denis Berthier

Constraint Resolution Theories

Books by Denis Berthier:
Le Savoir et l'Ordinateur, Editions L'Harmattan, Paris, November 2002.
Méditations sur le Réel et le Virtuel, Editions L'Harmattan, Paris, May 2004.
The Hidden Logic of Sudoku (First Edition), Lulu.com, May 2007.
The Hidden Logic of Sudoku (Second Edition), Lulu.com, November 2007.
Constraint Resolution Theories (First Edition), Lulu.com, October 2011.

Keywords: Constraint Satisfaction Problem, Artificial Intelligence, Mathematical Logic, Constructive Logic, Logic Games, Sudoku.

9 8 7 6 5 4 3 2 1

Dépôt légal: Octobre 2011

ISBN: 978-1-4478-6888-0

Table of Contents

Foreword

The "Hidden Logic of Sudoku" heritage

The origins of the work reported in this book can be traced back to my choice of Sudoku as a topic of practical classes for an introductory course in Artificial Intelligence (AI) in early 2006. As I was formalising for myself the simplest classical techniques (Subset rules, xy-chains) before submitting them as exercises to my students, I had two ideas that kept me interested in this game longer than I had expected: logical symmetries between Subset rules and a simple extension of xy-chains (xyt-chains). As time passed, the short article I had planned to write grew to the size of a 430-page book: *The Hidden Logic of Sudoku – HLS* in the sequel, sometimes *HLS1* (first edition, May 2007) or *HLS2* (second edition, November 2007), in the rare cases where the distinction needs to be done.

The present book inherits many of the ideas I first introduced in *HLS* but it extends them to any finite Constraint Satisfaction Problem (CSP). *HLS* defined a conceptual framework for Sudoku solving. Based on the classical approach of *candidate elimination* (known as *domain restriction* in the CSP world), it provided a clear logical status for the notion of a candidate (which does not pertain to the original Sudoku problem formulation) and it introduced the notions of a *resolution rule* and a *resolution theory*. All the concepts were strictly formalised in Predicate Logic (FOL) – more precisely in Multi Sorted First Order Logic (MS-FOL) – which (surprisingly) was a new idea: previously, all the books and Web forums had always considered that Propositional Logic was enough. Indeed, *HLS* had to make a further step, because intuitionistic logic is necessary for the proper formalisation of the notion of a candidate.

The basic concepts appearing in the resolution rules introduced in *HLS* were straightforwardly grounded in the most elementary notions used to propose or solve a puzzle (numbers, rows, columns, blocks, ...); the more elaborate ones were progressively introduced and strictly defined from the basic ones. The *HLS* framework was thus totally player oriented from the start. It can be considered as a mere formalisation of what had always been looked for when it was said that a "pure logic solution" was wanted. It is almost unchanged in the present book for all that is related to Sudoku, but it is extended to any CSP, with the added technicalities in some definitions and proofs that often go with generality.

This book builds on *HLS* in the following way. In many places, the definitions of concepts and resolution rules, as well as the theorems for the general CSP are generalisations of those I first introduced in *HLS* for Sudoku; they are illustrated with their instantiation in the Sudoku example and, to a much lesser extent, with examples from another famous CSP, N-Queens. But *HLS* is not a pre-requisite.

On the practical puzzle solving side, *HLS1* introduced new resolution rules, based on natural generalisations of the famous xy-chains, such as xyt-, xyz- and zyzt- chains; contrary to those proposed in the current Sudoku literature, these were not based on Subsets (or almost locked sets – "ALS") and most of these chains were not "reversible"; the systematic clarification and exploitation of all the generalised symmetries of the game and the combination of my first two initial ideas had also led me to the "hidden" counterparts of the previous chains (hxy-, hxyt- hxyzt- chains). After this first edition, I found further generalisations (nrczt- chains and lassoes), pushing the idea of supersymmetry to its maximal extent and allowing to solve almost any puzzle with short chain patterns. Giving a more systematic presentation of these new "3D" chain rules was the main reason for the second edition.

Later, I introduced still other generalisations (that, in the simplified terminology of this book and in a form meaningful for any CSP, will appear as whips, braids, g-whips, S_p-whips, W_p-whips, ...). These may have justified a third edition of *HLS*, but I have just added a few pages to my *HLS* website instead. Since that time, I have been more concentrated on another type of generalisation.

It appeared to me that most of what I had done for Sudoku could be generalised to finite Constraint Satisfaction Problems and I published a few papers about these generalisations [Berthier 2008a, 2008b, 2009]. But, once more, as I found better formulations and/or further generalisations, and as the examination of additional CSPs with different characteristics was necessary to guarantee that my definitions were not too restrictive, the normal size of journal articles did not fit the purposes of a clear exposition; this is how this work grew into the present book.

The approach of the present book

As very efficient general purpose CSP solving algorithms (e.g. structured search) have been well known for a long time and improvements are regularly proposed, one may wonder why they would use the computationally much harder techniques that go with the approach adopted in *HLS* or in this book. The answer is, there is no reason at all if speed is the first or only criterion (as may be the case in such typical CSPs as scene labelling).

But one can easily imagine various additional requirements. Instead of just wanting a final result obtained by any available method, one may be concerned with how it was reached, i.e. by the *resolution path*. One may ask this path to be the "simplest" one or to be "understandable" or to be obtained by "pure logic" or to be built by "constructive" methods – whatever meaning they ascribe to the quoted words.

Starting from such requirements and elaborating a formal interpretation of them in Part I, we build a very general, pure logic resolution paradigm based on the notions of a *resolution rule* and a *resolution theory*; notwithstanding the necessarily more general formulation, this is very close to the framework developed in *HLS*.

As for the resolution rules themselves, starting from Part II, whereas *HLS* proceeded by successive generalisations of well-known elementary rules for Sudoku into more complex ones, in the present book we start from powerful rules meaningful in any CSP (whips, in chapter 5) equivalent (in the Sudoku case) to those that were only reached at the end of *HLS2* (nrczt- chains and lassoes).

As a result, in this book, patterns such as Subsets, with much less resolution power than whips and with more complex definitions in the general CSP than in Sudoku, come after bivalue-chains, whips and braids, and also after their "grouped" versions, g-whips and g-braids. Moreover, Subsets are introduced with purposes very different from those in *HLS*:
1) providing them with a definition meaningful in any CSP;
2) showing that whips subsume most cases of Subsets in any CSP;
3) illustrating by Sudoku examples how, in rare cases, such specific rules can nevertheless simplify the resolution paths obtained with whips;
4) defining in any CSP a "grouped" version of Subsets, g-Subsets; surprisingly, in the Sudoku case, g-Subsets do not lead to news rules, but they give a new perspective of the well-known Franken and Mutant Fish; this could be useful for the purposes of classification (classification of extended Fish patterns has always been the topic of much debate);
5) showing that, in any CSP, the basic principles according to which whips are built can be generalised to allow the insertion of Subsets into them, thus extending the resolution power of whips towards the exceptionally hard instances.

In relation with the purpose of "simplicity", there appears the question of rating and classifying the instances of a (fixed size) CSP according to their "inherent difficulty". This is a much more difficult topic than just solving them. The families of resolution rules introduced in this book go by couples (whips/braids); for each couple, we propose two associated rating:

– one based on braids, with a good theory and good computational properties; we devote much time to prove the confluence property of all our braid and extended braid resolution theories, because it justifies a "simplest first" resolution strategy, which allows to find the corresponding rating by trying only one resolution path;

– one based on whips, providing an easier to compute good approximation of the first (at least for Sudoku) when it is combined with the "simplest first" strategy.

We explain in which restricted sense all these ratings are compatible. But we also show that to each of them corresponds a different view of simplicity. In chapter 11, we also introduce the BB rating (that can easily be checked to be universal for Sudoku, in the sense that it assigns a finite rating to all the known puzzles, but not in the sense that it could provide a unique notion of simplicity).

Finally, in several places, we give results that are only valid for Sudoku (e.g. the whip classification results of minimal instances in chapter 6), for the purpose of illustrating with precise numerical data questions that cannot yet be tackled with such detail in other CSPs and that call for further studies, such as:

– the difficulty (much beyond what one may imagine) of finding uncorrelated unbiased samples of minimal instances of a CSP, a pre-requisite for any statistical analysis; the way we present it shows that it is likely to appear in many CSPs; (a related, well known, problem is that of finding hard instances of a CSP);

– the surprisingly high resolution power of short whips and braids.

1. Introduction

1.1. The general Constraint Satisfaction Problem (CSP)

Many real world problems, such as resource allocation, temporal reasoning, scheduling or scene labelling, naturally appear as Constraint Satisfaction Problems (CSP) [Guesguen et al. 1992, Tsang 1993]. Many theoretical problems and many logic games are also natural examples of CSPs: graph colouring, graph matching, cryptarithmetic, n-Queens, Latin Squares, Sudoku and its innumerable variants... In the past decades, the study of such problems has evolved into a main sub-area of Artificial Intelligence (AI) with its own specialised techniques.

1.1.1. Statement of the Constraint Satisfaction Problem

A CSP is defined by:

– a set of variables X_1, X_2, ... X_n, the "CSP variables", each with values in a given domain $Dom(X_1)$, $Dom(X_2)$, ... $Dom(X_n)$,

– a set of constraints (i.e. of relations) these variables must satisfy.

The problem consists of finding a value in the domain of each of these variables, such that these values satisfy all the constraints. Later (in Chapter 3), we shall show that a CSP can be re-written as a theory in First Order Logic.

As in many studies of CSPs, all the CSPs we shall consider in this book will be finite, i.e. the number of variables, each of their domains and the number of constraints will all be finite. When we write "CSP", it should therefore always be read as "finite CSP".

Also, we shall consider only CSPs with binary constraints. One can always tackle unary constraints by an appropriate choice of the domains. And, for $k > 2$, a k-ary constraint between a subset of k variables $(X_{n1}, .., X_{nk})$ can always be replaced by k binary constraints between each of these X_{ni} and an additional variable representing the original k-ary constraint; although this new variable has a large domain and this may not be the most efficient way of dealing with the given k-ary constraint, this is a very standard approach (for details, see [Tsang 1993]).

Moreover, a binary CSP can always be represented as a (generally large) labelled undirected graph: a node of this graph, called a *label*, is a possible value of some of the CSP variables (or, as we shall see, an equivalence class of such possibilities); given two nodes in this graph, each binary constraint not satisfied by this pair of labels (including the "strong" constraints induced by CSP variables, i.e. all the contradictions between different values for the same CSP variable) gives rise to an arc between them, labelled by the name of the constraint and representing it. We shall call this graph *the CSP graph*. (Notice that this is different from what is usually called the constraint graph.) The CSP graph expresses all the direct contradictions between any two labels (whereas the constraint graph usually considered expresses their compatibilities).

1.1.2. The Sudoku example

As explained in the foreword, Sudoku has been at the origin of our work on CSPs. In this book we shall keep it as our main example, even though we shall also refer to other CSPs in order to palliate its specificities.

Let us start with the usual formulation of the problem (with its specific vocabulary in italics): given a 9×9 *grid*, partially filled with *numbers* from 1 to 9 (the *givens* of the problem, also called the *clues* or the *entries*), *complete* it with numbers from 1 to 9 in such a way that in each of the nine *rows*, in each of the nine *columns* and in each of the nine disjoint *blocks* of 3×3 contiguous *cells*, the following property holds:

– there is at most one occurrence of each of these numbers.

Although this defining condition could be replaced by either of the following two, which are obviously equivalent to it, we shall stick to the first formulation, for reasons that will appear later:

– there is at least one occurrence of each of these numbers,

– there is exactly one occurrence of each of these numbers.

Figure 1.1 shows the standard presentations of a *problem grid* (also called a *puzzle*) and of a *solution grid* (also called a *complete Sudoku grid*).

							1	2
				3	5			
		6				7		
7					3			
		4			8			
1								
			1	2				
	8					4		
	5				6			

6	7	3	8	9	4	5	1	2
9	1	2	7	3	5	4	8	6
8	4	5	6	1	2	9	7	3
7	9	8	2	6	1	3	5	4
5	2	6	4	7	3	8	9	1
1	3	4	5	8	9	2	6	7
4	6	9	1	2	8	7	3	5
2	8	7	3	5	6	1	4	9
3	5	1	9	4	7	6	2	8

Figure 1.1. *A puzzle (Royle17#3) and its solution*

Since rows, columns and blocks play similar roles in the defining constraints, they will naturally appear to do so in many other places and a word that makes no difference between them is widely used in the Sudoku world: a *unit* is either a row or a column or a block. And one says that two cells *share a unit*, or that they *see* each other, if they are different and they are either in the same row or in the same column or in the same block (where "or" is non exclusive). We shall also say that these two cells are *linked*. It should be noticed that this (symmetric) relation between two different cells, whichever of the three equivalent names it is given, does not depend on the content of these cells but only on their place in the grid; it is therefore a straightforward and quasi physical notion.

As appears from the definition, a Sudoku grid is a special case of a Latin Square. Latin Squares must satisfy the same constraints as Sudoku, except the condition on blocks. The practical consequences of this relationship between Sudoku and Latin Squares will appear throughout this book and, following *HLS1*, the logical relationship between the two theories will be fully clarified in chapters 3 and 4.

What we need now is to see how the above natural language formulation of the Sudoku problem can be re-written as a CSP. In Chapter 2, the essential question of modelling in general and its practical implications on how to deal with a CSP will be raised and we shall see that the following formalisation is neither the only one nor the best one. But, for the time being, we only want to write the most straightforward one.

For each row r and each column c, introduce a variable X_{rc}, with domain the set $\{1, 2, 3, 4, 5, 6, 7, 8, 9\}$. Then the general Sudoku problem can be expressed as a CSP with the following set of (binary) constraints:
$X_{rc} \neq X_{r'c'}$ for all the pairs $\{rc, r'c'\}$ such that the cells rc and r'c' share a unit,
and a particular puzzle will add to this the set of unary constraints fixing the values of the X_{rc} corresponding to the givens.

Notice that the natural language phrase "complete the grid" in the original formulation has naturally been understood as "assign one and only one value to each of the cells" – which has then been translated into "assign a value to each of the X_{rc} variables" in the CSP formulation.

1.2. Paradigms of resolution

A CSP states the constraints a solution must satisfy, i.e. it says *what* is desired. But it does not say anything about *how* a solution can be obtained; this is the job of resolution methods, the choice of which will depend on the various purposes one may have in addition to merely finding a solution. A particular class of resolution methods, based on *resolution rules*, will be the main topic of this book.

1.2.1. Various purposes and methods

If the only goal is to get a solution by any available means, very efficient general-purpose algorithms have been known for a long time [Kumar 1992, Tsang 1993]; they guarantee that they will either find one solution or all the solutions (according to what is desired) or find a contradiction in the givens; they have lots of more recent variants and refinements. Most of these algorithms involve the combination of two very different techniques: some direct propagation of constraints between variables (in order to progressively reduce their sets of possible values) and some kind of structured search with "backtracking" (depth-first, breadth-first, ..., possibly with some forms of look-ahead); they consist of trying (recursively if necessary) a value for a variable and propagating (based on the constraints) the consequences of this tentative choice as restrictions on other variables; eventually, either a solution or a contradiction will be reached; the latter case allows to conclude that this value (or this combination of values simultaneously tried in the recursive case) is impossible and it restricts the possibilities for this (subset of) variables(s).

But, in some cases, such blind search is not possible for practical reasons (e.g. one is not in a simulator but in real life) or not allowed (for *a priori* theoretical or æsthetic reasons), or one wants to simulate human behaviour, or one wants to "understand" each step of the resolution process (as is the case with most Sudoku players), or one wants a "pure logic" or a "constructive" or the "simplest" solution, whatever meaning can be ascribed to the quoted words.

Contrary to the current CSP literature, this book will only be concerned with the latter cases, in which more attention is paid to the resolution path than to the final solution itself. Indeed, it can also be considered as an informal reflection on how notions such as "understandable proof of the solution", "pure logic", "constructive" and "simplest" can be defined (but we shall only be able to say more on this topic in the retrospective "final remarks" chapter). It does not mean that efficiency questions are not relevant to our approach, but they are not our primary goal, they are conditioned by such higher-level requirements. Without these additional requirements, there is no reason to use techniques computationally much harder (probably exponentially much harder) than the general-purpose algorithms.

In such situations, it is convenient to introduce the notion of a *candidate*, i.e. of a "still possible" value for a variable. As this intuitive notion does not pertain to the CSP problem itself, it must first be given a clear definition and a logical status. When this is done (in chapter 4), one can define the concepts of a *resolution rule* (a logical formula in the "condition $\Rightarrow$ action" form, which says what to do in some factual, observable situation described by the condition pattern), a *resolution theory* (a set of such rules), a *resolution strategy* (a particular way of using the rules in a resolution theory). One can then study the relationship between the original CSP and several of its resolution theories. One can also introduce several properties a resolution theory can have, such as confluence and completeness (contrary to general purpose algorithms, a resolution theory cannot in general solve all the instances of a given CSP; evaluating its scope is thus a new topic in its own; one can also study its statistical resolution power in specific examples). This "pattern-based" approach was first introduced in *HLS1*, in the limited context of Sudoku. It is the purpose of this book to show that it is indeed very general, but let us first illustrate how these ideas work concretely for Sudoku.

1.2.2. Candidates and candidate elimination in Sudoku

The process of solving a Sudoku puzzle "by hand" is generally initialised by defining the "candidates" for each cell. For later formalisation, one must be careful with this notion: if one analyses the natural way of using it, it appears that, *at any stage of the resolution process, a candidate for a cell is a number that has not yet been explicitly proven to be an impossible value for this cell.*

	c1	c2	c3	c4	c5	c6	c7	c8	c9	
r1	3,4,5,6,8,9	3,4,6,7,9	3,4,5,6,7,8,9	7,8,9	4,7,8,9	4,7,8,9	4,5,9	**1**	**2**	r1
r2	2,4,8,9	1,2,4,6,7,9	1,2,4,6,7,8,9	2,7,8,9	**3**	**5**	4,9	6,8,9	4,6,8,9	r2
r3	2,3,4,5,8,9	1,2,3,4,9	1,2,3,4,5,8,9	**6**	1,4,8,9	1,2,4,8,9	4,5,9	**7**	3,4,5,8,9	r3
r4	**7**	2,4,6,9	2,4,5,6,8,9	2,5,8,9	1,5,6,8,9	1,2,6,8,9	**3**	2,5,6,9	1,4,5,6,9	r4
r5	2,3,5,6,9	2,3,6,9	2,3,5,6,9	**4**	1,5,6,7,9	1,2,3,6,7,9	**8**	2,5,6,9	1,5,6,7,9	r5
r6	**1**	2,3,4,6,9	2,3,4,5,6,8,9	2,3,5,7,8,9	5,6,7,8,9	2,3,6,7,8,9	2,4,5,7,9	2,5,6,9	4,5,6,7,9	r6
r7	3,4,6,9	3,4,6,7,9	3,4,6,7,9	**1**	**2**	3,4,6,7,8,9	5,7,9	3,5,8,9	3,5,7,8,9	r7
r8	2,3,6,9	**8**	1,2,3,6,7,9	5,7,9	3,5,6,7,9	3,6,7,9	1,2,5,7,9	**4**	1,3,5,7,9	r8
r9	2,3,4,9	**5**	1,2,3,4,7,9	3,7,8,9	4,7,8,9	4,7,8,9	**6**	2,3,8,9	1,3,7,8,9	r9
	c1	c2	c3	c4	c5	c6	c7	c8	c9	

Figure 1.2. *Grid Royle17#3 of Figure 1.1, with the candidates remaining after the elementary constraints for the givens have been propagated*

Usually, candidates for a cell are displayed in the grid as smaller and/or clearer digits in this cell (as in Figure 1.2). Similarly, at any stage, a *decided value* is a number that has been explicitly proven to be the only possible value for this cell; it is written in big fonts, like the givens.

At the start of the game, one possibility is to consider that any cell with no input value admits all the numbers from 1 to 9 as candidates – but more subtle initialisations are possible (e.g. as shown in Figure 1.2) and a slightly different, more symmetric, view of candidates can be introduced (see chapter 2).

Then, according to the formalisation introduced in *HLS1*, a resolution process that corresponds to the vague requirement of a "pure logic" solution is a sequence of steps consisting of repeatedly applying "resolution rules" of the general condition-action type: if some *pattern* – i.e. configuration of cells, possible cell-values, links, decided values and candidates – defined by the condition part of the rule, is effectively present in the grid, then carry out the action(s) specified by the action part of the rule. Notice that any such pattern always has a purely "physical", invariant part (which may be called its "physical" or structural support), defined by conditions on possible cell-values and on links between them, and an additional part, related to the actual presence/absence of decided values and/or candidates in these cells in the current situation. (Again, this will be generalised in chapter 2 with the four "2D" views.)

Depending on the type of their action part, such resolution rules can be classified into two categories:

– either they assert a decided value for a cell (e.g. when it is proven that there is only one possibility left for it); there are very few rules of this type;

– or they eliminate some candidate(s) (which we call the *target(s)* of the pattern); as appears from a quick browsing of the available literature, almost all the classical Sudoku resolution rules are of this type; they express specific elaborated forms of constraints propagation; their general form is: if such pattern is present, then it is impossible for some number(s) to be in some cell(s) and the corresponding candidates must therefore be deleted; for the general CSP, it will appear that almost all the rules we shall meet in this book are also of this type.

The interpretation of the above resolution rules, whatever their type, should be clear: none of them claims that there is a solution with this value asserted or with this candidate deleted. Rather, it must be interpreted as saying: "from the current situation it can be asserted that any solution, if there is any, must satisfy the conclusion of this rule".

From both theoretical and practical points of view, it is also important to notice that, as one proceeds with resolution, candidates form a monotone decreasing set and decided values form a monotone increasing set. Whereas the notion of a candidate is the intuitive one for players, what is classical in logic is increasing monotonicity (what is known / what has been proven can only increase with time); but this is not a real problem, as it could easily be restored by considering non-candidates instead (i.e. what has been erased instead of what is still present).

For some very difficult puzzles, it seems necessary to (recursively) make a hypothesis on the value of a cell, to analyse its consequences and to eliminate it if it leads to a contradiction; techniques of this kind do not fit *a priori* the above condition-action form; they are proscribed by purists (for the main reason that they often make the game totally uninteresting) and they are assigned the infamous, though undefined, name of Trial-and-Error. As shown in *HLS* and in the statistics of chapter 6, they are needed in only extremely rare cases if one admits the kinds of chain rules (whips) that will be introduced in chapter 5.

1.2.3. Extension of this model of resolution to the general CSP

It appears that the above ideas can be generalised from Sudoku to any CSP. Candidate elimination corresponds to the now classical idea of domain restriction in CSPs. What has been called a candidate above is related to the notion of a *label* in the CSP world, a name coming from the domain of scene labelling, historically at the origin of the identification of the general Constraint Satisfaction Problem. However, contrary to labels that can be given a very simple set theoretic definition based on the data defining the CSP, the status of a candidate is *a priori* not clear from the point of view of mathematical logic, because this notion does not pertain *per se* to the CSP formulation, nor to its direct logic transcription.

In chapter 4, we shall show that a formal definition of a candidate must rely on intuitionistic logic and we shall introduce more formally our general model of resolution. Then we shall define the notion of a resolution theory and we shall show that, for each CSP, a Basic Resolution Theory can be defined. Even though this Basic Theory may not be very powerful, it will be the basis for defining more elaborate ones; it is therefore "basic" in the two meanings of the word.

1.3. Parameters and instances of a CSP; minimal instances; classification

Generally, rather than a single problem, a CSP defines a whole family of problems. Typically, there is an integer parameter that splits this family into subclasses. A good example of such a parameter is the size of the grid in n-Queens, Latin Squares or Sudoku. In the resource allocation problem, it could be some combination of the number of resources and the number of tasks competing for them. In graph colouring and graph matching, it can be the size of the graph (some combination of the number of nodes and the number of arcs).

1.3.1. Minimal instances

Typically also, once this main parameter has been fixed, there remains a whole family of instances of the CSP. In 9×9 Sudoku, an instance is defined by a set of givens. In n-Queens, although the usual presentation of the problem starts from an empty grid and asks for all the solutions, we shall adopt for our purposes another view of this CSP; it consists of setting a few initial entries and asking for a solution or a "readable" proof that there is none. In graph colouring, the possibilities are still more open: there may be lots of graphs of a given size and, once such a graph has been chosen, it may also be required to have predefined colours for some subsets of nodes. The same remarks apply to graph matching, where one may want to have predefined correspondences between some nodes of the two graphs.

In such cases, classifying all the instances of a CSP or doing statistics on the difficulty of solving them meets problems of two kinds. Firstly, lots of instances will have very easy solutions: if givens are progressively added to an instance, until only the values of few variables remain non given, the problem becomes easier and easier to solve. Conversely, if there are so few instances that the problem has several solutions, some of these may be much easier to find than others. These two types of situations make statistics on all the instances somewhat irrelevant. This is the motivation for the following definition (inherited from the Sudoku classics).

An instance of a CSP is called *minimal* if it has one and only one solution and any instance obtained from it by eliminating any of its givens has more than one solution.

For the above-mentioned reasons, all our statistical analyses of a CSP will be restricted to the set of its minimal instances.

1.3.2. Rating and the complexity distribution of instances

Classically, the complexity of a CSP is studied with respect to its main size parameter and one relies on a worst case (or more rarely on a mean case) analysis. It often reaches conclusions such as "this CSP is NP-complete" – as is the case for Sudoku(n) or Latin-Square(n), considered as depending on grid size.

The questions about complexity that we shall tackle in this book are of a very different kind; they will not be based on the main size parameter. Instead, they will be about the statistical complexity distribution of instances of a fixed size CSP.

This supposes that we define a measure of complexity for instances of a CSP. We shall therefore introduce several ratings (starting in chapter 5) that are meaningful for the general CSP. And we shall be able to give detailed results (in chapter 6) for the standard (i.e. 9×9) Sudoku case. In trying to do so, the problem arises of creating unbiased samples of minimal instances and it appears to be very much harder than one may expect. We shall be able to show this in full detail only for the particular Sudoku case, but our approach is sufficiently general to suggest that the same kind of problem is very likely to arise in any CSP.

Indeed, what we shall define is different measures of complexity for various families of resolution rules. For each of these families, the complexity of a CSP instance will be defined as the complexity of the hardest rule in this family necessary to solve it. This is *a priori* very different from a definition based on the full resolution path. Of course, there must be some relationship between a ranking based on the hardest step of the resolution paths and a ranking of the full resolution paths; but, given a fixed set of rules, Sudoku examples show that it can solve puzzles whose resolution paths vary largely in complexity (whatever intuitive notion of complexity one adopts for the paths). In this book however, we shall not deal with the very hard (and as yet untouched) problem of ranking the instances according to their full resolution path; we shall only consider the hardest step.

1.4. The basic and the more complex resolution theories of a CSP

Following the definition of the CSP graph in section 1.1.1, we say that two candidates are linked by a direct contradiction, or simply *linked*, if there is a constraint making them incompatible (including the "strong" constraints, usually not explicitly stated as such, that different values for a CSP variable are incompatible).

1.4.1. Universal elementary resolution rules and their limitations

Every CSP has a Basic Resolution Theory: BRT(CSP).

The simplest elimination rule (obviously valid for any CSP) is the direct translation of the initial problem formulation into operational rules for managing candidates. We call it the "elementary constraints propagation rule" (ECP):

– ECP: if a value is asserted for a CSP variable (as is the case for the givens), then remove any candidate that is linked to this value by a direct contradiction.

The simplest assertion rule (also obviously valid) is called Single (S):

– S: if a CSP variable has only one candidate left, then assert it as the only possible value of this variable.

There is also an obvious Contradiction Detection rule (CD):

– CD: if a CSP variable has no decided value and no candidate left, then conclude that the problem has no solution.

The "elementary rules" ECP, S and CD constitute the Basic Resolution Theory of the CSP, BRT(CSP).

In Sudoku, novice players may think that these three elementary rules express the whole problem and that applying them repeatedly is therefore enough to solve any puzzle. If such were the case, one would probably never have heard of Sudoku, because it would amount to mere paper scratching and it would soon become boring. Anyway, as they get stuck in situations in which they cannot apply any of these rules, they soon discover that, except for the simplest puzzles, this is very far from being sufficient. The puzzle in Figure 1.1 is a very simple illustration of how one gets stuck if one only knows and uses the elementary rules: the resulting situation is shown in Figure 1.2, in which none of these rules can be applied. For this puzzle, modelling considerations related to symmetry (chapter 2) lead to "Hidden Single" rules allowing to solve it, but even this is generally very far from being enough.

1.4.2. Derived constraints and more complex resolution theories

As we shall see later, there are lots of puzzles that need resolution rules of a much higher complexity than those in the Basic Sudoku Resolution Theory in order to be solved. And this is why this game has become so popular: all but the easiest puzzles require a particular combination of neuron-titillating techniques and they may even suggest the discovery of as yet unknown ones.

The general reason for the limited resolution power of the Basic Resolution Theory of any CSP can be explained as follows. Given a set of constraints, there are usually many "derived" or "implied" constraints not immediately obvious from the original ones. Resolution rules can be considered as a way of expliciting some of them. As we shall see that very complex resolution rules are needed to solve some instances of a CSP, this will show not only that derived constraints cannot be reduced to the elementary rules of the Basic Resolution Theory (which constitute a straightforward operationalization of the axioms) but also that they can be unimaginably more complex than the initial constraints.

With all our Sudoku examples being minimal instances, secondary questions about multiple or inexistent solutions are discarded. From an epistemological point of view, the gap between the *what* (the initial constraints) and the *how* (the resolution rules necessary to solve an instance) is thus exhibited in all its purity, in a concrete way understandable by anyone. I may be a very naive person, but I must confess that, in spite of my pure logic background and of my familiarity with all the well-known mathematical ideas more or less related to it (culminating in deterministic chaos), this gap has always been for me a subject of much wonder. It is undoubtedly one of the main reasons why I kept interested in the Sudoku CSP for much longer than I expected when I first chose it as a topic for practical classes in AI. All the families of resolution rules defined in this book can be seen as different ways of exploring this gap.

1.4.3. Resolution rules and resolution strategies; the confluence property

One last point can now be clarified: the difference between a resolution theory (a set of resolution rules) and a resolution strategy. Everywhere in this book, a *resolution strategy* must be understood in the following extra-logical sense:

 – a set of *resolution rules*, i.e. a *resolution theory*;

 – a *non-strict precedence ordering* of these rules. Non-strict means that two rules can have the same precedence (for instance, in Sudoku, there is no reason to give a rule higher precedence than that obtained from it by transposing rows and columns or by any of the generalised symmetries explained in chapter 2).

As a consequence of this definition, several resolution strategies can be based on the same resolution theory with different partial orderings and they may lead to different resolution paths for a given instance.

Moreover, with every resolution strategy one can associate several deterministic procedures for solving instances:
As a preamble (each of the following choices will generate a different procedure):
- list all the resolution rules in a way compatible with their precedence ordering (i.e. among the different possibilities of doing so, choose one);
- list all the labels in a predefined order;
Given an instance P, loop until a solution of P is found (or until all the solutions are found or until it is proven that P has no solution):
 | Do until a rule can effectively be applied:
 | | Take the first rule not yet tried in the list
 | | Do until its condition pattern is effectively active:
 | | | Try to apply all the possible mappings of the condition pattern of this rule
 | | | to subsets of labels, according to their order in the list of labels
 | | End do
 | End do
 | Apply the rule to the selected matching pattern
End loop

In this context, a natural question arises: given a resolution theory T, can different resolution procedures built on T lead to an instance being finally solved by some of them and unsolved by others? The answer lies in the *confluence property* of a resolution theory, to be explained in chapter 5; this fundamental property implies that the order in which the rules of T are applied is irrelevant as long as we are only interested in solving instances (but it can still be relevant when we also consider the efficiency of the procedure): all the resolution paths will lead to the same final state.

This apparently abstract confluence property (first introduced in *HLS1*) has very practical consequences when it holds in a resolution theory T. It allows any opportunistic strategy, such as applying a rule as soon as a pattern instantiating it is found (e.g. instead of waiting to have found all the potential instantiations of rules of the same precedence before choosing which should be applied first). Most importantly, it also allows to define a "simplest first" strategy that is guaranteed to produce a correct rating of an instance with respect to T after following a single resolution path (with the easy to imagine computational consequences).

1.5. The roles of logic, AI, Sudoku and other examples

As its organisation shows, this book about the general CSP problem has a large part (one third) dedicated to illustrating the abstract concepts with a detailed case study of Sudoku and (to a much lesser extent) with examples from n-Queens and n-SudoQueens. Nevertheless, it can also be considered as a long exercise in either logic or AI. Let us clarify the roles we grant each of these disciplines.

1.5.1. The role of logic

Throughout this book, the main function of logic will be to provide a rigorous framework for the precise definition of our basic concepts (such as a "candidate", a "resolution rule" and a "resolution theory"). Apart from the formalisation of the CSP itself, the simplest and most striking example is the formalisation (in section 4.3) of the CSP Basic Resolution Theory introduced above, and of all the forthcoming more complex resolution theories. Logic will also be used as a compact notation tool for expressing some resolution rules in a non-ambiguous way. In the Sudoku example, it will also be a very useful tool for expliciting the precise symmetry relationships between different "Subset rules" (in chapter 8).

For better readability, the rules we introduce are always formulated first in plain English and their validity is only established by elementary non-formal means. The non-mathematically oriented reader should thus not be discouraged by the logical formalism. Moreover, for all the types of chains we shall consider, the associated rules will always be represented in a very intuitive, almost graphical formalism.

As a fundamental and practical application of our strict logical foundations to the Sudoku CSP, its natural symmetry properties can be transposed into three formal meta-theorems allowing one to deduce systematically new rules from given ones (see chapter 2 and sections 3.6 and 4.7). In *HLS*, this allowed us to introduce chain rules of completely new types (e.g. "hidden chains"). It also allowed to state a clear relationship between Sudoku and Latin Squares.

Finally, the other role assigned to logic is that of a mediator between the intuitive formulation of the resolution rules and their implementation in an AI program (e.g. our SudoRules solver). This is a methodological point for AI (or software engineering in general): no program development should ever be started before precise definitions of its components are given (though not necessarily in strict logical form) – a common sense principle that is very often violated, especially by those who consider it as obvious (this is the teacher speaking!). Notice however that the logical formalism is only one among other preliminaries to implementation and that it does not dispense with the need for some design work (be it only for efficiency matters!).

1.5.2. The role of AI

The role we impart to AI in this book is mainly that of providing a quick testbed for the general ideas developed in the theoretical part. The main rules have been implemented in our SudoRules solver. It was initially designed for Sudoku only (whence the name) and its input and output functions remain dedicated to Sudoku (and Latin Squares), but its heart (the rules themselves) is fully general and could be applied unchanged to any CSP (except the "Subset rules" in chapter 8, for which we have kept a Sudoku specific implementation, for efficiency reasons).

One important facet of the rules introduced in this book is their resolution power. This can only be tested on specific examples and Sudoku has served as our testbed for this. The resolution of each puzzle by a human solver needs a significant amount of time and the number of puzzles that can be tested "by hand" against any resolution method is very limited. On the contrary, implementing our resolution rules in a solver allowed us to test about ten millions of puzzles (see chapter 6). This also gave us indications of the relative efficiency of different rules. It is not mere chance that the writing of *HLS* and of the present book occurred in parallel with successive versions of SudoRules. Abstract definitions of the relative complexities of rules were checked against our puzzle collections for their resolution times and for their memory requirements (in terms of the number of partial chains generated).

This book can also be considered as the basis for a long exercise in AI. Many computer science departments in universities have used Sudoku for various projects. According to our personal experience, it is a most welcome topic for student projects in computer science or AI. Trying to implement some rules, even the "simple" Subset rules of chapter 8, shows how re-ordering the conditions can drastically change the behaviour of a knowledge-based system: without care, Quads easily lead to memory overflow problems. (We give detailed formulations for Subset rules in Sudoku, so that they can be used for such exercises without too long preliminaries.) Trying to implement S_p-whips or W_p-whips is a real challenge.

1.5.3. The role of Sudoku

Because some parts of this book related to the general CSP are rather abstract (e.g. chapters 3 and 4) or technical (e.g. chapters 9 and 10), a detailed case study was needed to show how the general concepts work in practice. It is also necessary to show how the general theory can easily be adapted, in the most important initial modelling phase, for dealing more efficiently or more naturally with each specific case. Choosing Sudoku for these purposes was for us a natural consequence of the historic development of the techniques described here, both the general approach and all the types of resolution rules. But there are many other reasons why it is an excellent example for the general CSP.

As can be seen from a fast browsing of this book, examples from the Sudoku CSP appear in many chapters (generally at the end, in order not to overload the main text with long resolution paths) and we keep our *HLS* constraint that all of them should originate in a real minimal puzzle – instead of an artificially created partial situation (resolution state).

But it should be clear for the readers of *HLS* that the purpose here is very different: we have no goal of illustrating with a Sudoku example each of the rules we introduce (for this, there is *HLS* or our website). Instead, each example satisfies a precise function with respect to the general Constraint Satisfaction Problem, such as providing a counter-example to some conjecture. As a result, most of our Sudoku examples will be exceptional cases, with very long resolution paths – which could give (without this warning) a very bad idea of how difficult the resolution paths look for the vast majority of puzzles; the statistics in chapter 6 will give a better idea: most of the time, the chains used and the paths are short.

1.5.3.1. Why Sudoku is a good example

Sudoku is known to be NP-complete [Gary & al. 1979]; more precisely, the CSP family Sudoku(n) on square grids of sizes n×n for all n is NP-complete. As we fix n = 9, this should not have any impact on our analyses. But the Sudoku case will exemplify very clearly (in chapter 6) that, for fixed n, the instances of an NP-complete problem often have a broad spectrum of complexity. It also shows that standard analyses, only based on worst case (worst instances), can be very far from reflecting the realities of a CSP.

For fixed n = 9, Sudoku is much easier to study than other readily formalised problems such as Chess or Go or any "real world" example. But it keeps enough structure so that it is not obvious.

Sudoku is a particular case of Latin Squares. Latin Squares are more elegant (and somehow more respectable) from a mathematical point of view, because they enjoy a complete symmetry of all the variables: numbers, rows, columns. In Sudoku, the constraint on blocks introduces some apparently mild complexity that makes it more exciting for players. But this lack of full symmetry also makes it more interesting from a theoretical point of view. In particular, it allows to introduce the notion of a grouped label and to identify a kind of general pattern based on it which is not present in Latin Squares: g-labels and the corresponding g-whips and g-braids (see chapter 7).

There are millions of Sudoku players all around the world and many forums where the rules defined in *HLS* have been the topic of much debate. A huge amount of invaluable experience has been cumulated and is readily available – including generators of random (but biased) puzzles, collections of puzzles with very specific properties (fish patterns, symmetry properties, ...) and other collections of extremely hard puzzles. The lack of similar collections and of generators of minimal instances is a strong limitation for the detailed analysis of other CSPs.

1.5.3.2. Origin of our Sudoku examples

Most of our Sudoku examples rely on the following sets of minimal puzzles:

– the Sudogen0 collection consists of 1,000,000 puzzles randomly generated with the top-down suexg generator (http://magictour.free.fr/suexco.txt), with seed 0 for the random numbers generator; puzzle number n is named Sudogen0#n;

– the cb collection consists of 5,926,343 puzzles produced by a new kind of generator, the controlled-bias generator (we first introduced it on the Sudoku Player's Forum; see also [Berthier 2009] and chapter 6 below); it is still biased (though less than the previous ones) but in a known way, so that it allows to compute unbiased statistics; puzzle number n is named cb#n;

– the Magictour collection of 1465 puzzles considered to be the hardest (at the time of its publishing); puzzle number n is named Magictour-top1465#n;

– the gsf collection of 8152 puzzles considered to be the hardest (at the time of its publishing); puzzle number n is named gsf-top8152 #n.

Occasionally, we shall also use other puzzles from collections with particular properties, as specified in the text.

1.5.4. The role of non Sudoku examples

Although Sudoku is a very good CSP example, it has a few specificities, such as (the major one of) having all its constraints defined by CSP variables. The role of our other examples (Latin Squares, n-Queens, n-SudoQueens) is limited to reviewing a few of these specificities and to showing that they have no negative impact on our general theory, in particular that our main resolution rules (for whips, g-whips, Subsets, S_p-whips, W-whips, braids, ...) can effectively be applied to other CSPs and how different these patterns may look in these cases.

We are aware that more examples should be given as much consideration as we have granted Sudoku and it is a shortcoming of this book that we did not do it; but time is not extensible and we hope that the approach described in this book will motivate more research for applications to other CSPs.

1.5.5. Uniform presentation of all the examples

If we displayed the full resolution path of an instance, it would generally take several pages, most of which would describe obvious or uninteresting steps. We shall skip most of these steps, with the following conventions (same as in *HLS*):

– elementary constraint propagation rules (ECP) will never be displayed;

– as the final rules that apply to any instance are always ECP and Singles (at least when these rules are given higher priority than more complex ones), they will be omitted from the end of the path.

All our examples respect the following uniform format. After an introductory text explaining the purpose of the example, the resolution theory T applied to it and/or comments on some particular point, a row of two (sometimes three) grids is displayed: the original puzzle (sometimes an intermediate state) and its solution. Then comes *the resolution path, a proof of the solution within theory T, where "proof" is meant in the strict sense of constructive logic.*

Each line in the resolution path consists of the name of the rule applied, followed by: the description of how the rule is "instantiated" (i.e. how the condition part is satisfied), the "==>" sign, the conclusion allowed by the "action" part. The conclusion is always either that a candidate can be eliminated (symbolically written as r4c8 $\neq$ 6 in Sudoku) or that a value must be asserted (symbolically written as r4c8 = 5). When the same rule instantiation justifies several conclusions, they are written on the same line, separated by commas: e.g. r4c8 $\neq$ 8, r5c8 $\neq$ 8. Occasionally, the detailed situation at some point in the resolution path (the "resolution state") is displayed so that the presence of the pattern under discussion can be directly checked, but, due to place constraints, this cannot be systematic.

For Sudoku, all the resolution paths given in this book were obtained with versions 13 or 15b of our SudoRules solver (with some occasional hand editing for a shorter and cleaner appearance), using the CLIPS inference engine (release 6.30), on a MacPro™ 2006 running at 2.66 GHz.

1.6. Notations

Throughout this book, we consider an arbitrary, but fixed, finite Constraint Satisfaction Problem. We call it CSP, generically. BRT(CSP) refers to its Basic Resolution Theory, RT to any of its resolution theories, W_n [respectively B_n, gW_n, gB_n, S_pW_n, S_pB_n, ...] to its nth whip [respectively braid, g-whip, g-braid, S_p-whip, S_p-braid, ...] resolution theory. The same letters, with no index, are used for the associated ratings. We use BSRT as an abbreviation for BRT(Sudoku) and BLSRT for BRT(LatinSquare).

Part One

LOGICAL FOUNDATIONS

2. The role of modelling, illustrated with Sudoku

Before we start with the logical formalisation of a general CSP, the main purpose of this chapter is to show in detail, using the Sudoku example, how some initial modelling choices and/or associated mental or graphical representations can radically change our view of a CSP. Together with consequences of several modelling choices that will appear throughout this book, it will also illustrate the general epistemological principle that changing our representations of a problem can drastically change its apparent complexity. Almost all of the material here was first introduced in *HLS1*.

It may seem strange to start a part on the "logical foundations" with a chapter on modelling that is almost only about Sudoku. But we mean to insist that, in CSP as in any other domain, modelling choices are the starting point of any good application of any general theory. And most of such choices can only be application specific.

Complementary considerations on modelling will appear in section 5.11, when we introduce the n-Queens and the n-SudoQueens CSPs, after we have defined our general logical framework and our first resolution rules.

2.1. Symmetries, analogies and supersymmetries

2.1.1. Symmetries

Throughout this book, the word "symmetry" is used in the general abstract mathematical sense. A Sudoku symmetry, or symmetry for short, is a transformation that, when applied to *any* valid Sudoku grid, produces a valid Sudoku grid. Any combination of symmetries is a symmetry, there is a null symmetry (that does not change anything) and every symmetry has a reverse; therefore symmetries form a group.

Two grids (completed or not) that are related by some symmetry are said to be essentially equivalent. The reason is that when the first is solved, its solution and its resolution path can be transposed by the same symmetry to a solution and a resolution path for the second. The abstract notions above become very concrete and intuitive as soon as a set of generators for the whole group of symmetries is given. By definition, any symmetry is then composed of a finite sequence of these generating ones. The simplest set of generators one can consider is composed of two different types of obvious symmetries (see e.g. [Russell 2005]):

– permutations of the numbers: the numerical values of the numbers used to fill the grid are totally irrelevant; they could indeed be replaced by arbitrary symbols; any permutation of the digits (which is just a relabeling of the entries) defines a symmetry of the game; there are obviously 9! = 362,880 such symmetries.

– "geometrical" symmetries of the grid:
 - permutations of individual rows 1, 2, 3;
 - permutations of individual rows 4, 5, 6;
 - permutations of individual rows 7, 8, 9;
 - permutations of triplets of rows ("floors") 1-2-3, 4-5-6 and 7-8-9;
 - symmetry relative to the first diagonal (row-column symmetry).

From these primary geometrical symmetries, others can be deduced:
 - permutations of individual columns 1, 2, 3;
 - permutations of individual columns 4, 5, 6;
 - permutations of individual columns 7, 8, 9;
 - permutations of triplets of columns ("towers") 1-2-3, 4-5-6 and 7-8-9;
 - reflection (left-right symmetry);
 - up-down symmetry;
 - symmetry relative to the second diagonal;
 - ± 90° rotation,
 - and, more generally, any combination of symmetries in the generating set.

As of the writing of *HLS1*, the above-mentioned symmetries had been used mainly to count the number of essentially non-equivalent grids. Expressed in terms of elementary symmetries, two grids (completed or not) are essentially equivalent if there is a sequence of elementary symmetries such that the second is obtained from the first by application of this sequence.

Thus, it has been shown in [Russell 2005] that the number of non essentially equivalent complete Sudoku grids is 5,472,730,538 – much less than the *a priori* possibly different 6,670,903,752,021,072,936,960 complete grids. But the number of essentially different minimal puzzles is still much greater, its exact value being still unknown (however, see our estimate in chapter 6: 2.55×10^{25}). The point is that each complete grid is, in the mean, the solution for 4.67×10^{15} minimal puzzles.

Later we shall formulate axioms for Sudoku in a logical language and in a way that exhibits all the previous symmetries. In turn, such symmetries in the axioms will lead to symmetries in the logical formulation of our resolution rules. But all the types of symmetries will not be expressed in the same way in these axioms or rules.

Primary symmetries other than row-column will be totally transparent, in that they will make use of variable names (for numbers, rows, columns…) but they will refer to no specific values of these entities.

As for row-column symmetry, in elementary resolution rules, our formalisation will stick to their classical formulation and it will be expressed by the presence of two similar axioms or rules, each of which can be obtained from the other by a simple permutation of the words "row" and "column". As a consequence of this symmetry in the axioms, there will be a meta-symmetry in the theorems and the resolution rules, as expressed by the following intuitively obvious

meta-theorem 2.1 (informal): for any valid Sudoku resolution rule, the rule deduced from it by permuting systematically the words "row" and "column" is valid and it has obviously the same logical complexity as the original. We shall express this as: the set of valid Sudoku resolution rules is closed under row-column symmetry.

In more evolved resolution rules, in particular in chain rules, we shall show that a more powerful approach consists of building them only on primary predicates that already take all the symmetries into account.

2.1.2. The two canonical coordinate systems on a grid

Let the nine rows be numbered 1, 2, …, 9 from top to bottom. Let the nine columns be numbered 1, 2, …, 9 from left to right. Let the nine blocks and the nine squares inside any fixed block be numbered according to the same scheme, as follows:

$$1\ 2\ 3$$
$$4\ 5\ 6$$
$$7\ 8\ 9$$

Any cell, in "natural" row-column space, can be unambiguously located on the grid via either of its two pairs of coordinates (row, column) or [block, square]. One can therefore consider two coordinate systems on the grid. We call them the two canonical coordinate systems and we write the coordinates of a cell in each of them as (r, c) or as [b, s], respectively.

Change of coordinates F: (r, c) → [b, s] is defined by the following formulæ:
$$b = block\ (r, c) = 1 + 3*IP((r - 1)/3) + IP((c - 1)/3);$$
$$s = square(r, c) = 1 + 3*mod((r + 2), 3) + mod((c + 2), 3).$$

Conversely, change of coordinates [b, s] → (r, c) is defined by:
$$r = row(b, s) = 1 + 3*IP((b - 1)/3) + IP((s - 1)/3);$$
$$c = column(b, s) = 1 + 3*mod((b + 2), 3) + mod((s + 2), 3),$$
where "IP" stands for "integer part" and "mod" for "modulo".

Notice that transformation F: (r, c) → [b, s]: is involutive, i.e. $F^{-1} = F$ or $F \bullet F = Id$ (the identity), where "F^{-1}" denotes as usual the inverse of F and "$\bullet$" denotes function composition.

2.1.3. Coordinates and names

Coordinates should not be confused with the various names that can be given to the rows, columns, blocks, squares and cells for displaying purposes. Various displaying conventions can be used (e.g. the chess convention: A1, A2, ... G7, G8), but we shall systematically stick to the following one, which we have found the most convenient:

– rows are named: r1, r2, r3, r4, r5, r6, r7, r8, r9;

– columns are named: c1, c2, c3, c4, c5, c6, c7, c8, c9;

– cells in natural rc-space are named accordingly, in the obvious way: r1c1, r1c2, ..., r9c9;

– blocks are named: b1, b2, b3, b4, b5, b6, b7, b8, b9;

– squares in a block are named: s1, s2, s3, s4, s5, s6, s7, s8, s9;

– as a result, cells in rc-space can also be named: b1s1, b1s2, ..., b9s9;

– when needed, numbers are named n1, n2, n3, n4, n5, n6, n7, n8, n9; this will be useful in the next sections when we consider "abstract spaces": row-number, column-number and block-number and we want to name cells in these spaces: r1n1, r1n2... in rn-space; c1n1, c1n2,... in cn-space; b1n1, b1n2,... in bn-space; the reason is that r11, r12... or c11, c12... would be rather obscure and confusing.

Notice that the same lower case letters as for constants will be used for naming variables, but with subscripts, e.g. r_1, b_3, ...; these close conventions should not lead to any confusion between variables and constants. In any case, the risk of confusion is very limited: no variable symbol can appear in the description of any real fact on a real grid and no constant symbol will ever appear in an axiom (except of course in the axioms corresponding to the givens of the puzzle) or a resolution rule.

2.1.4. Supersymmetries

Up to now, symmetries relative to the entries (numbers) and "geometrical" symmetries relative to the grid have been considered separately. One of the results of *HLS1* was the elicitation of other symmetries (named *supersymmetries*) that mix numbers, rows and columns. It showed how they translate into relationships between some of the constraints propagation rules, how they entail a new logical classification of these rules, how this allows clearer definitions of the rules themselves and how this leads to introduce new types of chains ("hidden" chains and "supersymmetric" chains) and associated rules.

The main reason for our interest in supersymmetry is the following:

meta-theorem 2.2 (informal): for any valid Sudoku resolution rule mentioning only numbers, rows and columns (i.e. neither blocks nor squares nor any property referring to such objects), any rule deduced from it by any systematic permutation of the words "number", "row" and "column" is valid and it has obviously the same logical complexity as the original. We shall express this as: the set of valid Sudoku resolution rules is closed under supersymmetry.

Meta-theorem 2.2 is not intuitively as obvious as meta-theorem 2.1. From a logical point of view, it is nevertheless a straightforward consequence of the subsequent logical formulation of the problem in Multi-Sorted First Order Logic (more on this in chapters 3 and 4). And, from a practical point of view, subtle correspondences between subset rules become explicit (see chapter 8). If we consider the LatinSquare CSP, the above theorem has a much simpler formulation: *for any valid LatinSquare resolution rule, any rule deduced from it by a systematic permutation of the words "number", "row" and "column" is valid.*

2.1.5. Analogies

Analogies should not be confused with symmetries. There are analogies between rows and blocks (or between columns and blocks) but there is no real symmetry.

This is related to the fact that the two canonical coordinate systems do not share the same properties with respect to the rules of Sudoku. There is a symmetry between the coordinates in the first system (rows and columns) and, relying explicitly on this symmetry, many axioms and rules exist by pairs; but there is no symmetry between the coordinates in the second system (blocks and squares) so that transposing rules from the first system to the second would be meaningless.

There is nevertheless a partial analogy between rows (or columns) and blocks, captured by the following informal

meta-theorem 2.3 (informal): for any valid Sudoku resolution rule mentioning only numbers, rows and columns (i.e. neither blocks nor squares nor any property referring to such objects), if this rule displays a systematic symmetry between rows and columns but it can be proved without using the axiom on columns, then the rule deduced from it by systematically replacing the word "row" by "block" and the word "column" by "square" is valid and it has obviously the same logical complexity as the original one. We shall express this as: the set of valid Sudoku resolution rules is closed under analogy.

What the phrases "systematic symmetry between rows and columns" and "proved without using the axiom on columns" mean will be defined precisely in chapter 3.

2.2. Introducing the four 2D spaces: rc, rn, cn and bn

To better visualise the symmetries, analogies and supersymmetries defined in the previous section, we introduce three new 2D spaces and their 2D graphical representations. The latter can be grouped with the usual one to form an extended Sudoku board. These representations were first introduced in *HLS1*. How to build and use them was explained in detail in *HLS2*; we do not repeat it here.

In the Subset rules of chapter 8, they will be used to illustrate how apparently complex familiar rules (such as X-wing, Swordfish or Jellyfish) are no more than the supersymmetric versions of obvious ones (Naked-Pairs, Naked-Triplets and Naked-Quads, respectively); all this was in *HLS1*, where they have also been the basis for the notion of hidden chains and associated resolution rules.

In this book, however, the main role of these new spaces and representations will be to justify intuitively the introduction of additional CSP variables.

2.2.1. Additional graphical representations of a puzzle

In addition to the standard "natural" row-column space (or rc-space), we consider three new "abstract" spaces: row-number, column-number and block-number. In the sequel, these four spaces will also be called respectively rc-space, rn-space, cn-space and bn-space and "cells" in these four spaces will be called rc-cells, rn-cells, cn-cells and bn-cells. As for their graphical representations, when they are displayed together, they are aligned so that rows in the first two coincide and columns in the first and the third coincide (cn space is thus displayed as nc).

When it comes to candidates, the reason for considering rn-cell with coordinates (r, n) in rn-space is that it will contain all the possibilities (all the possible columns) for the unique instance of number n that must occur in row r; similarly, the reason for considering cn-cell with coordinates (c, n) in cn-space is that it will contain all the possibilities (all the possible rows) for the unique instance of number n that must occur in column c; finally, the reason for considering bn-cell with coordinates (b, n) in bn-space is that it will contain all the possibilities (all the possible squares) for the unique instance of number n that must occur in block b.

At any point in the resolution process, all the data in the grid (values and candidates) can be displayed in any of these four representations. We insist that each of them displays exactly the same logical information content – or, to say it more formally: they correspond to the same underlying set of ground atomic formulæ in the (basically 3D) logical language that will be introduced later. They should be considered only as different visual supports for symmetry, supersymmetry and analogy, in the sense that it is easier to detect some patterns in some representations than in others, as illustrated by several chapters in this book and in *HLS*.

The correspondences are straightforward and given by the equivalences:

– Boolean symbol True is present in nrc-cell (n, r, c), (3D view, to be discussed in section 2.4),

   – number n is present in rc-cell (r, c), (standard view),

   – column c is present in rn-cell (r, n),

   – row r is in present in cn-cell (c, n),

   – square s is in present in bn-cell (b, n), where (r, c) = [b, s].

Notice that pseudo blocks (i.e. groups of 3×3 rn, cn or bn cells) have no meaning in the new rn, cn or bn representations (this is why we do not mark them with thick borders): only constraints valid for Latin Squares can be directly propagated in rn or cn spaces (as will be proved in chapter 3). Moreover, links in bn-space cannot use the number coordinate.

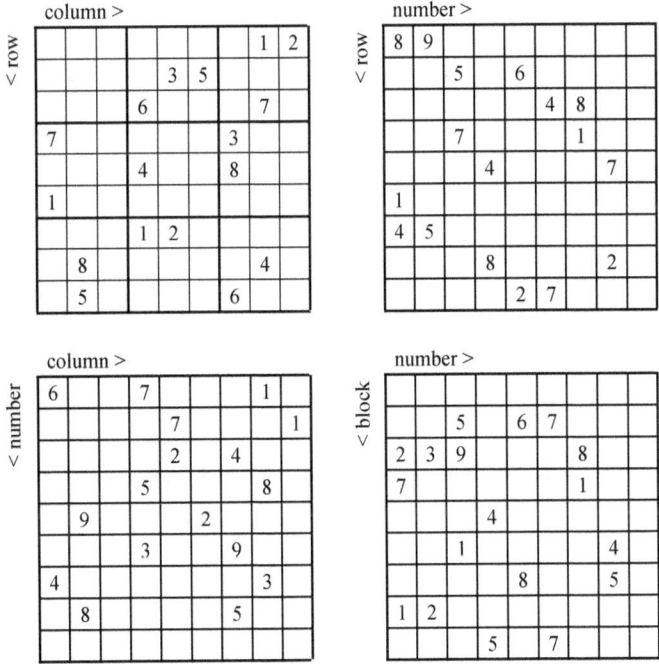

Figure 2.1. *Same puzzle Royle17#3 as in Figure 1.1, but viewed in the four different representation spaces (rc, rn, cn, bn)*

Generating these new grid representations by hand is easy as long as we consider only values, as in Figure 2.1, but it is tedious when it comes to the candidates.

Nevertheless, with some practice, it is relatively simple to apply the above stated equivalences (see *HLS*). Moreover, programming a spreadsheet computing the three new grids and their candidates automatically from the first is an easy exercise.

Let us illustrate these new representations with the example given in Figure 1.1 (puzzle Royle17#3). Starting from the standard form of the puzzle, we can first display its entries in the standard grid and in the three new grids of Figure 2.1. After applying all the elementary constraints propagation rules in rc-space, we get the usual representation of the resolution state in rc-space (Figure 1.2).

	c1	*c2*	*c3*	*c4*	*c5*	*c6*	*c7*	*c8*	*c9*	
n1	r6	r2 r3	r2 r3 r8 r9	r7	r3 r4 r5	r3 r4 r5	**r8**	r1	r4 r5 r8 r9	*n1*
n2	r2 r3 r5 r8 r9	r2 r3 r4 r5 r6	r2 r3 r4 r5 r6 r8 r9	r2 r4 r6	r7	r3 r4 r5 r6	r6 r8	r4 r5 r6 r9	r1	*n2*
n3	r1 r3 r5 r7 r8 r9	r1 r3 r5 r6 r7	r1 r3 r5 r6 r7 r8 r9	r6 r8 r9	r2	r5 r6 r7 r8 r9	r4	r7 r9	r3 r7 r8 r9	*n3*
n4	r1 r2 r3 r7 r9	r1 r2 r3 r4 r6 r7	r1 r2 r3 r4 r6 r7 r9	r5	r1 r3 r9	r1 r3 r7 r9	r1 r2 r3 r6	r8	r2 r3 r4 r6	*n4*
n5	r1 r3 r5	r9	r1 r3 r4 r5 r6	r4 r6 r8	r4 r5 r6 r8	r2	r1 r3 r6 r7 r8	r4 r5 r6 r7	r3 r4 r5 r6 r7 r8	*n5*
n6	r1 r2 r5 r7 r8	r1 r2 r4 r5 r6 r7	r1 r2 r4 r5 r6 r7 r8	r3	r4 r5 r6 r8	r4 r5 r6 r7 r8	r9	r2 r4 r5 r6	r2 r4 r5 r6	*n6*
n7	r4	r1 r2 r7	r1 r2 r7 r8 r9	r1 r2 r6 r8 r9	r1 r5 r6 r8 r9	r1 r5 r6 r7 r8 r9	r6 r7 r8	r3	r5 r6 r7 r8 r9	*n7*
n8	r1 r2 r3	r8	r1 r2 r3 r4 r6	r1 r2 r4 r6 r9	r1 r3 r4 r6 r9	r1 r3 r4 r6 r7 r9	r5	r2 r7 r9	r2 r3 r7 r9	*n8*
n9	r1 r2 r3 r5 r7 r8 r9	r1 r2 r3 r4 r5 r6 r7	r1 r2 r3 r4 r5 r6 r7 r8 r9	r1 r2 r4 r6 r8 r9	r1 r3 r4 r5 r6 r8 r9	r1 r3 r4 r5 r6 r7 r8 r9	r1 r2 r3 r6 r7 r8	r2 r4 r5 r6 r7 r9	r2 r3 r4 r5 r6 r7 r8 r9	*n9*
	c1	*c2*	*c3*	*c4*	*c5*	*c6*	*c7*	*c8*	*c9*	

Figure 2.2. *Same puzzle Royle17#3 as in Figure 1.2, but viewed in cn-space*

Now, suppose we generate the full rn, cn and bn representations with candidates. For our puzzle, there is nothing particularly appealing in the rn and bn representations, so we skip them. But a surprise is awaiting us with its cn representation (Figure 2.2). It makes it obvious that there is a cn-cell (c7n1) with only one possibility left: the unique instance of number 1 that must appear somewhere in column 7 is in fact confined to row 8 (i.e. cn-cell c7n1 has only one row candidate: r8).

As an example that the groups of 3×3 contiguous cn-cells have no meaning, we can see that there are many of these pseudo-blocks in which the same candidate (row) appears two or more times.

Now, it appears that, if we had considered more attentively the standard rc representation with candidates (Figure 1.2 of the Introduction), we could have seen that, in column c7, there is only one row (row r8) having number 1 among its candidates. Therefore, the unique instance of number 1 that must be found somewhere in column c7 has only one possibility left of finding its place in this column and that is in row r8. But the difference is, this cannot be seen in rc-space by looking only at one rc-cell (namely r8c7) since it still has five candidates: 1, 2, 5, 7 and 9. What the representation in cn-space provides is the possibility of *detecting locally* this forced value by looking at a single cn-cell, while in "natural" rc-space we must examine all the nine rc-cells of column c7. This is a very elementary example of how rn, cn or bn spaces can be used in practice.

This is our first example of a "Hidden-Single" (HS) in a column. Notice that the phrase "hidden single in a column" suggests properly that, in column c7, cell r8c7 has a single possible value but that this fact is hidden, i.e. is not visible by looking only at the candidates for this cell in the usual rc-representation. Of course, one can also find Hidden-Singles in rows or in blocks. Actually, this Royle17#3 puzzle can be solved using only these types of Hidden-Singles (in addition, of course, to Naked Single and the elementary constraints propagation rules).

Graphically, in the standard rc representation, spotting a Hidden-Single-in-a-row [respectively in-a-column, in-a-block] for some Number n supposes that one checks that the other eight cells in this row [resp. this column, this block] do not contain n among their candidates. In the new rn [resp. cn, bn] representation, all that is needed is checking that one cell has a single possibility left. Thus, even in very elementary cases, the new representations simplify the detection job.

Now, a few comments about these new graphical representations are in order. Should one consider them as a practical basis for human solving? There will probably never be any general agreement on this point. Our personal opinion is that, given the additional paperwork needed for building and maintaining the four representations in parallel, they are not very useful for easy puzzles; but, one can easily imagine a computerised interface that maintains the coherency between the four grids (any time a candidate is eliminated from one of them or a value is asserted in one of them, this information is transferred to the others). Moreover, there are many difficult puzzles that become easier to solve if we use such representations (and rules based on them): see *HLS*, a significant part of which was based on symmetries, supersymmetries and "hidden" structures.

Anyway, in the present book, they will mainly be considered as a step towards the introduction of new CSP variables and as a representation system for them.

	c1	c2	c3	c4	c5	c6	c7	c8	c9	
r1	n1 n2 n3 n4 n5 n6 n7 n8 n9	n1 n2 n3 n4 n5 n6 n7 n8 n9	n1 n2 n3 n4 n5 n6 n7 n8 n9	n1 n2 n3 n4 n5 n6 n7 n8 n9	n1 n2 n3 n4 n5 n6 n7 n8 n9	n1 n2 n3 n4 n5 n6 n7 n8 n9	n1 n2 n3 n4 n5 n6 n7 n8 n9	n1 n2 n3 n4 n5 n6 n7 n8 n9	n1 n2 n3 n4 n5 n6 n7 n8 n9	r1
r2	n1 n2 n3 n4 n5 n6 n7 n8 n9	n1 n2 n3 n4 n5 n6 n7 n8 n9	n1 n2 n3 n4 n5 n6 n7 n8 n9	n1 n2 n3 n4 n5 n6 n7 n8 n9	n1 n2 n3 n4 n5 n6 n7 n8 n9	n1 n2 n3 n4 n5 n6 n7 n8 n9	n1 n2 n3 n4 n5 n6 n7 n8 n9	n1 n2 n3 n4 n5 n6 n7 n8 n9	n1 n2 n3 n4 n5 n6 n7 n8 n9	r2
r3	n1 n2 n3 n4 n5 n6 n7 n8 n9	n1 n2 n3 n4 n5 n6 n7 n8 n9	n1 n2 n3 n4 n5 n6 n7 n8 n9	n1 n2 n3 n4 n5 n6 n7 n8 n9	n1 n2 n3 n4 n5 n6 n7 n8 n9	n1 n2 n3 n4 n5 n6 n7 n8 n9	n1 n2 n3 n4 n5 n6 n7 n8 n9	n1 n2 n3 n4 n5 n6 n7 n8 n9	n1 n2 n3 n4 n5 n6 n7 n8 n9	r3
r4	n1 n2 n3 n4 n5 n6 n7 n8 n9	n1 n2 n3 n4 n5 n6 n7 n8 n9	n1 n2 n3 n4 n5 n6 n7 n8 n9	n1 n2 n3 n4 n5 n6 n7 n8 n9	n1 n2 n3 n4 n5 n6 n7 n8 n9	n1 n2 n3 n4 n5 n6 n7 n8 n9	n1 n2 n3 n4 n5 n6 n7 n8 n9	n1 n2 n3 n4 n5 n6 n7 n8 n9	n1 n2 n3 n4 n5 n6 n7 n8 n9	r4
r5	n1 n2 n3 n4 n5 n6 n7 n8 n9	n1 n2 n3 n4 n5 n6 n7 n8 n9	n1 n2 n3 n4 n5 n6 n7 n8 n9	n1 n2 n3 n4 n5 n6 n7 n8 n9	n1 n2 n3 n4 n5 n6 n7 n8 n9	n1 n2 n3 n4 n5 n6 n7 n8 n9	n1 n2 n3 n4 n5 n6 n7 n8 n9	n1 n2 n3 n4 n5 n6 n7 n8 n9	n1 n2 n3 n4 n5 n6 n7 n8 n9	r5
r6	n1 n2 n3 n4 n5 n6 n7 n8 n9	n1 n2 n3 n4 n5 n6 n7 n8 n9	n1 n2 n3 n4 n5 n6 n7 n8 n9	n1 n2 n3 n4 n5 n6 n7 n8 n9	n1 n2 n3 n4 n5 n6 n7 n8 n9	n1 n2 n3 n4 n5 n6 n7 n8 n9	n1 n2 n3 n4 n5 n6 n7 n8 n9	n1 n2 n3 n4 n5 n6 n7 n8 n9	n1 n2 n3 n4 n5 n6 n7 n8 n9	r6
r7	n1 n2 n3 n4 n5 n6 n7 n8 n9	n1 n2 n3 n4 n5 n6 n7 n8 n9	n1 n2 n3 n4 n5 n6 n7 n8 n9	n1 n2 n3 n4 n5 n6 n7 n8 n9	n1 n2 n3 n4 n5 n6 n7 n8 n9	n1 n2 n3 n4 n5 n6 n7 n8 n9	n1 n2 n3 n4 n5 n6 n7 n8 n9	n1 n2 n3 n4 n5 n6 n7 n8 n9	n1 n2 n3 n4 n5 n6 n7 n8 n9	r7
r8	n1 n2 n3 n4 n5 n6 n7 n8 n9	n1 n2 n3 n4 n5 n6 n7 n8 n9	n1 n2 n3 n4 n5 n6 n7 n8 n9	n1 n2 n3 n4 n5 n6 n7 n8 n9	n1 n2 n3 n4 n5 n6 n7 n8 n9	n1 n2 n3 n4 n5 n6 n7 n8 n9	n1 n2 n3 n4 n5 n6 n7 n8 n9	n1 n2 n3 n4 n5 n6 n7 n8 n9	n1 n2 n3 n4 n5 n6 n7 n8 n9	r8
r9	n1 n2 n3 n4 n5 n6 n7 n8 n9	n1 n2 n3 n4 n5 n6 n7 n8 n9	n1 n2 n3 n4 n5 n6 n7 n8 n9	n1 n2 n3 n4 n5 n6 n7 n8 n9	n1 n2 n3 n4 n5 n6 n7 n8 n9	n1 n2 n3 n4 n5 n6 n7 n8 n9	n1 n2 n3 n4 n5 n6 n7 n8 n9	n1 n2 n3 n4 n5 n6 n7 n8 n9	n1 n2 n3 n4 n5 n6 n7 n8 n9	r9

	c1	c2	c3	c4	c5	c6	c7	c8	c9	
n1	r1 r2 r3 r4 r5 r6 r7 r8 r9	r1 r2 r3 r4 r5 r6 r7 r8 r9	r1 r2 r3 r4 r5 r6 r7 r8 r9	r1 r2 r3 r4 r5 r6 r7 r8 r9	r1 r2 r3 r4 r5 r6 r7 r8 r9	r1 r2 r3 r4 r5 r6 r7 r8 r9	r1 r2 r3 r4 r5 r6 r7 r8 r9	r1 r2 r3 r4 r5 r6 r7 r8 r9	r1 r2 r3 r4 r5 r6 r7 r8 r9	n1
n2	r1 r2 r3 r4 r5 r6 r7 r8 r9	r1 r2 r3 r4 r5 r6 r7 r8 r9	r1 r2 r3 r4 r5 r6 r7 r8 r9	r1 r2 r3 r4 r5 r6 r7 r8 r9	r1 r2 r3 r4 r5 r6 r7 r8 r9	r1 r2 r3 r4 r5 r6 r7 r8 r9	r1 r2 r3 r4 r5 r6 r7 r8 r9	r1 r2 r3 r4 r5 r6 r7 r8 r9	r1 r2 r3 r4 r5 r6 r7 r8 r9	n2
n3	r1 r2 r3 r4 r5 r6 r7 r8 r9	r1 r2 r3 r4 r5 r6 r7 r8 r9	r1 r2 r3 r4 r5 r6 r7 r8 r9	r1 r2 r3 r4 r5 r6 r7 r8 r9	r1 r2 r3 r4 r5 r6 r7 r8 r9	r1 r2 r3 r4 r5 r6 r7 r8 r9	r1 r2 r3 r4 r5 r6 r7 r8 r9	r1 r2 r3 r4 r5 r6 r7 r8 r9	r1 r2 r3 r4 r5 r6 r7 r8 r9	n3
n4	r1 r2 r3 r4 r5 r6 r7 r8 r9	r1 r2 r3 r4 r5 r6 r7 r8 r9	r1 r2 r3 r4 r5 r6 r7 r8 r9	r1 r2 r3 r4 r5 r6 r7 r8 r9	r1 r2 r3 r4 r5 r6 r7 r8 r9	r1 r2 r3 r4 r5 r6 r7 r8 r9	r1 r2 r3 r4 r5 r6 r7 r8 r9	r1 r2 r3 r4 r5 r6 r7 r8 r9	r1 r2 r3 r4 r5 r6 r7 r8 r9	n4
n5	r1 r2 r3 r4 r5 r6 r7 r8 r9	r1 r2 r3 r4 r5 r6 r7 r8 r9	r1 r2 r3 r4 r5 r6 r7 r8 r9	r1 r2 r3 r4 r5 r6 r7 r8 r9	r1 r2 r3 r4 r5 r6 r7 r8 r9	r1 r2 r3 r4 r5 r6 r7 r8 r9	r1 r2 r3 r4 r5 r6 r7 r8 r9	r1 r2 r3 r4 r5 r6 r7 r8 r9	r1 r2 r3 r4 r5 r6 r7 r8 r9	n5
n6	r1 r2 r3 r4 r5 r6 r7 r8 r9	r1 r2 r3 r4 r5 r6 r7 r8 r9	r1 r2 r3 r4 r5 r6 r7 r8 r9	r1 r2 r3 r4 r5 r6 r7 r8 r9	r1 r2 r3 r4 r5 r6 r7 r8 r9	r1 r2 r3 r4 r5 r6 r7 r8 r9	r1 r2 r3 r4 r5 r6 r7 r8 r9	r1 r2 r3 r4 r5 r6 r7 r8 r9	r1 r2 r3 r4 r5 r6 r7 r8 r9	n6
n7	r1 r2 r3 r4 r5 r6 r7 r8 r9	r1 r2 r3 r4 r5 r6 r7 r8 r9	r1 r2 r3 r4 r5 r6 r7 r8 r9	r1 r2 r3 r4 r5 r6 r7 r8 r9	r1 r2 r3 r4 r5 r6 r7 r8 r9	r1 r2 r3 r4 r5 r6 r7 r8 r9	r1 r2 r3 r4 r5 r6 r7 r8 r9	r1 r2 r3 r4 r5 r6 r7 r8 r9	r1 r2 r3 r4 r5 r6 r7 r8 r9	n7
n8	r1 r2 r3 r4 r5 r6 r7 r8 r9	r1 r2 r3 r4 r5 r6 r7 r8 r9	r1 r2 r3 r4 r5 r6 r7 r8 r9	r1 r2 r3 r4 r5 r6 r7 r8 r9	r1 r2 r3 r4 r5 r6 r7 r8 r9	r1 r2 r3 r4 r5 r6 r7 r8 r9	r1 r2 r3 r4 r5 r6 r7 r8 r9	r1 r2 r3 r4 r5 r6 r7 r8 r9	r1 r2 r3 r4 r5 r6 r7 r8 r9	n8
n9	r1 r2 r3 r4 r5 r6 r7 r8 r9	r1 r2 r3 r4 r5 r6 r7 r8 r9	r1 r2 r3 r4 r5 r6 r7 r8 r9	r1 r2 r3 r4 r5 r6 r7 r8 r9	r1 r2 r3 r4 r5 r6 r7 r8 r9	r1 r2 r3 r4 r5 r6 r7 r8 r9	r1 r2 r3 r4 r5 r6 r7 r8 r9	r1 r2 r3 r4 r5 r6 r7 r8 r9	r1 r2 r3 r4 r5 r6 r7 r8 r9	n9
	c1	c2	c3	c4	c5	c6	c7	c8	c9	

Figure 2.3. *The Extended Sudoku Board, with the four rc, rn, cn and bn spaces; each cell in this Extended Board represents a CSP variable of the extended list.*

	n1	n2	n3	n4	n5	n6	n7	n8	n9	
r1	c1 c2 c3 c4 c5 c6 c7 c8 c9	c1 c2 c3 c4 c5 c6 c7 c8 c9	c1 c2 c3 c4 c5 c6 c7 c8 c9	c1 c2 c3 c4 c5 c6 c7 c8 c9	c1 c2 c3 c4 c5 c6 c7 c8 c9	c1 c2 c3 c4 c5 c6 c7 c8 c9	c1 c2 c3 c4 c5 c6 c7 c8 c9	c1 c2 c3 c4 c5 c6 c7 c8 c9	c1 c2 c3 c4 c5 c6 c7 c8 c9	r1
r2	c1 c2 c3 c4 c5 c6 c7 c8 c9	c1 c2 c3 c4 c5 c6 c7 c8 c9	c1 c2 c3 c4 c5 c6 c7 c8 c9	c1 c2 c3 c4 c5 c6 c7 c8 c9	c1 c2 c3 c4 c5 c6 c7 c8 c9	c1 c2 c3 c4 c5 c6 c7 c8 c9	c1 c2 c3 c4 c5 c6 c7 c8 c9	c1 c2 c3 c4 c5 c6 c7 c8 c9	c1 c2 c3 c4 c5 c6 c7 c8 c9	r2
r3	c1 c2 c3 c4 c5 c6 c7 c8 c9	c1 c2 c3 c4 c5 c6 c7 c8 c9	c1 c2 c3 c4 c5 c6 c7 c8 c9	c1 c2 c3 c4 c5 c6 c7 c8 c9	c1 c2 c3 c4 c5 c6 c7 c8 c9	c1 c2 c3 c4 c5 c6 c7 c8 c9	c1 c2 c3 c4 c5 c6 c7 c8 c9	c1 c2 c3 c4 c5 c6 c7 c8 c9	c1 c2 c3 c4 c5 c6 c7 c8 c9	r3
r4	c1 c2 c3 c4 c5 c6 c7 c8 c9	c1 c2 c3 c4 c5 c6 c7 c8 c9	c1 c2 c3 c4 c5 c6 c7 c8 c9	c1 c2 c3 c4 c5 c6 c7 c8 c9	c1 c2 c3 c4 c5 c6 c7 c8 c9	c1 c2 c3 c4 c5 c6 c7 c8 c9	c1 c2 c3 c4 c5 c6 c7 c8 c9	c1 c2 c3 c4 c5 c6 c7 c8 c9	c1 c2 c3 c4 c5 c6 c7 c8 c9	r4
r5	c1 c2 c3 c4 c5 c6 c7 c8 c9	c1 c2 c3 c4 c5 c6 c7 c8 c9	c1 c2 c3 c4 c5 c6 c7 c8 c9	c1 c2 c3 c4 c5 c6 c7 c8 c9	c1 c2 c3 c4 c5 c6 c7 c8 c9	c1 c2 c3 c4 c5 c6 c7 c8 c9	c1 c2 c3 c4 c5 c6 c7 c8 c9	c1 c2 c3 c4 c5 c6 c7 c8 c9	c1 c2 c3 c4 c5 c6 c7 c8 c9	r5
r6	c1 c2 c3 c4 c5 c6 c7 c8 c9	c1 c2 c3 c4 c5 c6 c7 c8 c9	c1 c2 c3 c4 c5 c6 c7 c8 c9	c1 c2 c3 c4 c5 c6 c7 c8 c9	c1 c2 c3 c4 c5 c6 c7 c8 c9	c1 c2 c3 c4 c5 c6 c7 c8 c9	c1 c2 c3 c4 c5 c6 c7 c8 c9	c1 c2 c3 c4 c5 c6 c7 c8 c9	c1 c2 c3 c4 c5 c6 c7 c8 c9	r6
r7	c1 c2 c3 c4 c5 c6 c7 c8 c9	c1 c2 c3 c4 c5 c6 c7 c8 c9	c1 c2 c3 c4 c5 c6 c7 c8 c9	c1 c2 c3 c4 c5 c6 c7 c8 c9	c1 c2 c3 c4 c5 c6 c7 c8 c9	c1 c2 c3 c4 c5 c6 c7 c8 c9	c1 c2 c3 c4 c5 c6 c7 c8 c9	c1 c2 c3 c4 c5 c6 c7 c8 c9	c1 c2 c3 c4 c5 c6 c7 c8 c9	r7
r8	c1 c2 c3 c4 c5 c6 c7 c8 c9	c1 c2 c3 c4 c5 c6 c7 c8 c9	c1 c2 c3 c4 c5 c6 c7 c8 c9	c1 c2 c3 c4 c5 c6 c7 c8 c9	c1 c2 c3 c4 c5 c6 c7 c8 c9	c1 c2 c3 c4 c5 c6 c7 c8 c9	c1 c2 c3 c4 c5 c6 c7 c8 c9	c1 c2 c3 c4 c5 c6 c7 c8 c9	c1 c2 c3 c4 c5 c6 c7 c8 c9	r8
r9	c1 c2 c3 c4 c5 c6 c7 c8 c9	c1 c2 c3 c4 c5 c6 c7 c8 c9	c1 c2 c3 c4 c5 c6 c7 c8 c9	c1 c2 c3 c4 c5 c6 c7 c8 c9	c1 c2 c3 c4 c5 c6 c7 c8 c9	c1 c2 c3 c4 c5 c6 c7 c8 c9	c1 c2 c3 c4 c5 c6 c7 c8 c9	c1 c2 c3 c4 c5 c6 c7 c8 c9	c1 c2 c3 c4 c5 c6 c7 c8 c9	r9
	n1	n2	n3	n4	n5	n6	n7	n8	n9	

	n1	n2	n3	n4	n5	n6	n7	n8	n9	
b1	s1 s2 s3 s4 s5 s6 s7 s8 s9	s1 s2 s3 s4 s5 s6 s7 s8 s9	s1 s2 s3 s4 s5 s6 s7 s8 s9	s1 s2 s3 s4 s5 s6 s7 s8 s9	s1 s2 s3 s4 s5 s6 s7 s8 s9	s1 s2 s3 s4 s5 s6 s7 s8 s9	s1 s2 s3 s4 s5 s6 s7 s8 s9	s1 s2 s3 s4 s5 s6 s7 s8 s9	s1 s2 s3 s4 s5 s6 s7 s8 s9	b1
b2	s1 s2 s3 s4 s5 s6 s7 s8 s9	s1 s2 s3 s4 s5 s6 s7 s8 s9	s1 s2 s3 s4 s5 s6 s7 s8 s9	s1 s2 s3 s4 s5 s6 s7 s8 s9	s1 s2 s3 s4 s5 s6 s7 s8 s9	s1 s2 s3 s4 s5 s6 s7 s8 s9	s1 s2 s3 s4 s5 s6 s7 s8 s9	s1 s2 s3 s4 s5 s6 s7 s8 s9	s1 s2 s3 s4 s5 s6 s7 s8 s9	b2
b3	s1 s2 s3 s4 s5 s6 s7 s8 s9	s1 s2 s3 s4 s5 s6 s7 s8 s9	s1 s2 s3 s4 s5 s6 s7 s8 s9	s1 s2 s3 s4 s5 s6 s7 s8 s9	s1 s2 s3 s4 s5 s6 s7 s8 s9	s1 s2 s3 s4 s5 s6 s7 s8 s9	s1 s2 s3 s4 s5 s6 s7 s8 s9	s1 s2 s3 s4 s5 s6 s7 s8 s9	s1 s2 s3 s4 s5 s6 s7 s8 s9	b3
b4	s1 s2 s3 s4 s5 s6 s7 s8 s9	s1 s2 s3 s4 s5 s6 s7 s8 s9	s1 s2 s3 s4 s5 s6 s7 s8 s9	s1 s2 s3 s4 s5 s6 s7 s8 s9	s1 s2 s3 s4 s5 s6 s7 s8 s9	s1 s2 s3 s4 s5 s6 s7 s8 s9	s1 s2 s3 s4 s5 s6 s7 s8 s9	s1 s2 s3 s4 s5 s6 s7 s8 s9	s1 s2 s3 s4 s5 s6 s7 s8 s9	b4
b5	s1 s2 s3 s4 s5 s6 s7 s8 s9	s1 s2 s3 s4 s5 s6 s7 s8 s9	s1 s2 s3 s4 s5 s6 s7 s8 s9	s1 s2 s3 s4 s5 s6 s7 s8 s9	s1 s2 s3 s4 s5 s6 s7 s8 s9	s1 s2 s3 s4 s5 s6 s7 s8 s9	s1 s2 s3 s4 s5 s6 s7 s8 s9	s1 s2 s3 s4 s5 s6 s7 s8 s9	s1 s2 s3 s4 s5 s6 s7 s8 s9	b5
b6	s1 s2 s3 s4 s5 s6 s7 s8 s9	s1 s2 s3 s4 s5 s6 s7 s8 s9	s1 s2 s3 s4 s5 s6 s7 s8 s9	s1 s2 s3 s4 s5 s6 s7 s8 s9	s1 s2 s3 s4 s5 s6 s7 s8 s9	s1 s2 s3 s4 s5 s6 s7 s8 s9	s1 s2 s3 s4 s5 s6 s7 s8 s9	s1 s2 s3 s4 s5 s6 s7 s8 s9	s1 s2 s3 s4 s5 s6 s7 s8 s9	b6
b7	s1 s2 s3 s4 s5 s6 s7 s8 s9	s1 s2 s3 s4 s5 s6 s7 s8 s9	s1 s2 s3 s4 s5 s6 s7 s8 s9	s1 s2 s3 s4 s5 s6 s7 s8 s9	s1 s2 s3 s4 s5 s6 s7 s8 s9	s1 s2 s3 s4 s5 s6 s7 s8 s9	s1 s2 s3 s4 s5 s6 s7 s8 s9	s1 s2 s3 s4 s5 s6 s7 s8 s9	s1 s2 s3 s4 s5 s6 s7 s8 s9	b7
b8	s1 s2 s3 s4 s5 s6 s7 s8 s9	s1 s2 s3 s4 s5 s6 s7 s8 s9	s1 s2 s3 s4 s5 s6 s7 s8 s9	s1 s2 s3 s4 s5 s6 s7 s8 s9	s1 s2 s3 s4 s5 s6 s7 s8 s9	s1 s2 s3 s4 s5 s6 s7 s8 s9	s1 s2 s3 s4 s5 s6 s7 s8 s9	s1 s2 s3 s4 s5 s6 s7 s8 s9	s1 s2 s3 s4 s5 s6 s7 s8 s9	b8
b9	s1 s2 s3 s4 s5 s6 s7 s8 s9	s1 s2 s3 s4 s5 s6 s7 s8 s9	s1 s2 s3 s4 s5 s6 s7 s8 s9	s1 s2 s3 s4 s5 s6 s7 s8 s9	s1 s2 s3 s4 s5 s6 s7 s8 s9	s1 s2 s3 s4 s5 s6 s7 s8 s9	s1 s2 s3 s4 s5 s6 s7 s8 s9	s1 s2 s3 s4 s5 s6 s7 s8 s9	s1 s2 s3 s4 s5 s6 s7 s8 s9	b9
	n1	n2	n3	n4	n5	n6	n7	n8	n9	

2.2.2. Extended Sudoku Board

As several examples in *HLS* have shown, especially when we deal with chains, the rn, cn and bn spaces allow to describe simple "hidden" patterns and rules that would need much more complex descriptions in the standard rc-space. In order to facilitate their use, the rn, cn and bn representations can be grouped with the standard one into the Extended Sudoku Board of Figure 2.3. Notice that these representations do not replace the standard one; they are added to it, so that the four representations, when placed in the proper relative positions, form an extended board. In order to avoid confusion between numbers, rows and columns, in this extended board we tend to use systematically their full names: n1, n2, ...; r1, r2, ...; c1, c2, ... But, when an example uses only the rc-space, we may be lax on this.

2.3. CSP variables associated with the rc, rn, cn and bn cells

What is more important for the present book is that, ***corresponding to the new 2D views, one can define an extended set of CSP variables*** (with cardinality 324 instead of 81): in addition to all the $Xr°c°$ as before, one can now introduce all the $Xr°n°$, $Xc°n°$ and $Xb°n°$ for n° in {n1, n2, n3, n4, n5, n6, n7, n8, n9}, r° in {r1, r2, r3, r4, r5, r6, r7, r8, r9}, c° in {c1, c2, c3, c4, c5, c6, c7, c8, c9} and b° in {b1, b2, b3, b4, b5, b6, b7, b8, b9}. And one has the following interpretation:

The Extended Sudoku Board represents the extended set of CSP variables for Sudoku, the content of each cell representing the set of still possible values (the candidates) for the corresponding CSP variable.

The original CSP can now be reformulated in a very different way: find a value for each of these 324 CSP variables such that, for each n°, r°, c°, b°, s° with $(r°, c°) = [b°, s°]$, one has: $Xr°c° = n° \Leftrightarrow Xr°n° = c° \Leftrightarrow Xc°n° = r° \Leftrightarrow Xb°n° = s°$.

From a logical point of view, there is nothing really new, only obvious rewritings of the initial natural language constraints with redundant CSP variables. One may therefore wonder whether introducing such new variables and constraints can be of any practical use. All this book will show that it is, but part of the answer is already given, at a very intuitive and elementary level, by our analysis of the Hidden Single rule in the example of Figure 2.1: written with the new variables, this rule appears as a mere Naked Single rule. Thus, a very straightforward extension of the original set of CSP variables is enough to suggest new resolution rules or to extend the scope of the existing ones.

Moreover, ***this apparently innocuous method is indeed very powerful***, even at this basic level: no (known) minimal Sudoku puzzle can be solved using only Elementary Constraints Propagation and Naked Singles; but 29% of the minimal puzzles (in unbiased statistics) can be solved if we add Hidden Singles (for detailed statistics, see *HLS* or chapter 6 of this book for a better version).

2.4. Introducing the 3D nrc-space

Can one go further? Could the above 2D representations be a mere stage towards a more abstract, more synthetic, 3D representation? Instead of considering the four 2D spaces, one could consider a 3D space, with coordinates n, r, c. In the nrc-cell with coordinates (n, r, c), one would put the Boolean True (or a 1, or a dot, or any arbitrarily chosen sign) if n is present in rc-cell (r, c). The 2D spaces would then appear as the 2D projections of 3D nrc-space.

Corresponding to this 3D view, there would be a still larger set (of cardinality $2 \times 9^3 = 1458$) of possible CSP variables: all the $Xn°r°c°$ and $Xn°b°s°$ for all the constants $n°, r°, c°, b°, s°$ as above. Each of these CSP variables would take Boolean values (True or False). The constraints would then have to be re-written in a different, more complex way:
$Xn°r°c° \wedge Xn°'r°'c°' = $ False, for all the pairs $\{n°r°c°, n°'r°'c°'\}$ such that

  – either $n° = n°'$ and the rc-cells $r°c°$ and $r°'c°'$ share a unit;

  – or $n° \neq n°'$ and $r°c° = r°'c°'$;
together with similar constraints for the $Xn°b°s°$. Moreover, obvious relationships could be written between these "3D" CSP variables and the "2D" CSP variables of the previous section: $Xn°r°c° = $ True $\Leftrightarrow Xr°c° = n° \Leftrightarrow Xr°n° = c° \Leftrightarrow Xc°n° = r° \Leftrightarrow Xb°n° = s°$ whenever $(r°, c°) = [b°, s°]$.

Considered as CSP variables, these 3D variables would not bring anything new (with respect to the four sets of "2D" CSP variables), because all the CSP constraints can already be written in the four 2D sets. Actually, Sudoku has no 3D diagonal constraints. Invoking Occam's razor principle, we shall not adopt them.

Nevertheless, the 3D view will not be completely forgotten: each of these non-CSP-variables will reappear later as a "label" (see section 3.2.1), i.e. as a name $n°r°c°$ or $(n°, r°, c°)$ for the set of four equivalent possibilities: $\{Xr°c° = n°, Xr°n° = c°, Xc°n° = r°, Xb°n° = s°\}$. And 3D space will reappear as a representation of the set of labels.

3. The logical formalisation of a CSP

Although this book may be used as a support for exercises in Logic or AI and it must therefore adopt a clear and non ambiguous formalism, it is not intended to be an introductory textbook on these disciplines and it also aims at defining resolution techniques readable with no pre-requisite in them. The non mathematically oriented reader should not be discouraged by the formalism introduced in this chapter: apart from the proof (in chapter 4) of meta-theorems 2.1, 2.2 and 2.3 and some local remarks, it will be used mainly as a general background for our resolution paradigm. On the practical side of things, starting with Part II, the resolution rules will always be formulated in plain English, so that it will be possible to skip the logical version, if it is ever written. Moreover, most of the resolution rules (and, in particular, the chain rules of the various types considered in this book) will also be displayed in very simple, intuitive, quasi-graphical representations. As for the Sudoku example, the Sudoku Grid Theory (SGT) and Sudoku Theory (ST) introduced in section 3.5 below can be considered as completely obvious from an intuitive point of view (so that this chapter and the next can be skipped or kept for later reading).

3.1. A quick introduction to Multi-Sorted First Order Logic (MS-FOL)

In order to have a logical formalism as concrete and intuitive as possible, we want our formulæ to be simple and compact; we shall therefore use Multi-Sorted First Order Logic with equality (MS-FOL). A theory in formal logic always deals with some limited topic and it does this in a well-defined language adapted to its purpose. The distinctive characteristic of MS-FOL is that it assumes the topic of interest has different types of objects, called sorts.

From a theoretical point of view, such logic is known to be formally equivalent to standard First Order Logic with equality (FOL): formulæ, theories and proofs in MS-FOL translate easily to and from formulæ, theories and proofs in FOL. But, for practical purposes, the natural expressive power of MS-FOL is much greater, i.e. things are generally much easier to write. For a more extensive introduction to MS-FOL and an easy but technical proof of its equivalence with FOL, see e.g. [Meinke et al. 1993].

In most of the real world applications of logic and in computer science (where modern languages are typed – and even object oriented), MS-FOL rather than FOL is the natural reference, whether or not any kind of variant or extension (intuitionistic, modal, temporal, dynamic and so on) is required. This is not to suggest that the specific sorts needed for an application are in any way "natural"; they can only be the result of a modelling process, as shown in the previous chapter.

Our introduction to MS-FOL follows the standard lines of any introduction to logic. It is here only for purposes of (almost) self-containment of this book. It also introduces a few unusual but intuitive and useful abbreviations.

3.1.1. The language of a theory in MS-FOL

Every theory in FOL or MS-FOL is defined by a specific language reflecting the concepts and only the concepts pertaining to the underlying domain (its "vocabulary"); but the syntax or "grammar" of all these specific languages is built according to universal principles.

3.1.1.1 Specific sorts, constants and variables

First is given a set Sort of *sorts*; these are just abstract symbols (generally written as Greek letters or with a capital first letter), naming the various types of objects of the application. Attached to each sort σ, there are two disjoint sets of symbols: $ct(\sigma)$ for naming *constants* of this sort and $var(\sigma)$ for naming *variables* of this sort. Moreover, the sets attached to two different sorts are disjoint (unless one sort is a sub-sort of the other). When a variable appears anywhere (e.g. after a quantifier), its sort does not have to be further specified: it is known from its name.

3.1.1.2 Specific predicates and functions

In FOL, *predicate symbols* (also called relation symbols) are names used to express either properties of objects or relations between objects they relate. A predicate symbol has an "arity": an integer number defining the number of arguments it takes. In MS-FOL, it also has a "signature": a sequence of sorts, the length of its arity, specifying that each of the arguments of this predicate must be of the sort corresponding to the place it occupies in it.

One generally considers theories with equality. In this case, for each sort σ, there is an equality predicate: "$=_\sigma$" (= with subscript σ) expressing equality between objects of the same sort σ. "$=_\sigma$" has arity 2 and signature (σ, σ). We shall also use $\neq_\sigma$ to express non-equality: if x_1 and x_2 are variables of sort σ, then $x_1 \neq_\sigma x_2$ is an abbreviation for $\neg(x_1 =_\sigma x_2)$. As sorts are known from the names of the variables, a loose notation with = instead of $=_\sigma$ is generally used.

Similarly, a *function symbol* is a name used to refer to a function. In MS-FOL, it has a sort (the sort of the result), an arity and a signature (specifying respectively the number and the sequence of sorts of its arguments).

3.1.1.3 Terms and atomic formulæ

From now on, we describe general principles for building formulæ from the above-defined specific "vocabulary" – the grammar or syntax of MS-FOL.

Terms of sort σ are defined recursively:

  – if "a" is a symbol for a constant of sort σ, then it is a term of sort σ;

  – if "x" is a symbol for a variable of sort σ, then it is a term of sort σ;

  – if f is a symbol for a function of sort σ, arity n and signature $(\sigma_1, ..., \sigma_n)$, and if $t_1, ..., t_n$ are terms of respective sorts $\sigma_1, ..., \sigma_n$, then $f(t_1, ..., t_n)$ is a term of sort σ.

An *atomic formula* is the standard means for expressing elementary relations between its arguments. Atomic formulæ are defined as follows:

  – if R is a symbol for a predicate of arity n and signature $(\sigma_1, ..., \sigma_n)$, and if $t_1, ..., t_n$ are terms of respective sorts $\sigma_1, ..., \sigma_n$, then $R(t_1, ..., t_n)$ is an atomic formula.

An atomic formula $R(t_1, ..., t_n)$ is said to be *ground* if for every i from 1 to n, t_i is a constant symbol. It thus expresses a relation between constants.

3.1.1.4 Logical connectives (or logical operators)

The language of MS-FOL has the standard logical connectives of FOL:

  – "$\wedge$", "&" or "and" are used indifferently to express conjunction;

  – "$\vee$" or "or" are used indifferently to express disjunction;

  – "$\neg$" or "not" are used indifferently to express negation;

  – "$\Rightarrow$" expresses logical implication;

  – "$\forall x$" expresses universal quantification over objects of the sort of x;

  – "$\exists x$" expresses existential quantification over objects of the sort of x.

We shall also make an extensive use of the following (not all very standard) abbreviations (especially for the formal expression of the chain rules in chapter 5 and of the Subset rules in chapter 8), where F is any formula:

  – "$\exists!xF(x)$" expresses that "there exists one and only one x such that F(x)";

  – "$\forall x{\neq}x_1,x_2,...,x_nF$" expresses a single quantification over x; by definition, it means: $\forall x[x=x_1 \vee x=x_2 \vee ... \vee x=x_n \vee F]$;

  – "$\forall{\neq}(x_1,x_2,...,x_n)F$" expresses n universal quantifications for n different objects of the same sort; it should not be confused with the previous abbreviation; by definition, it means:
$\forall x_1 \forall x_2 ... \forall x_n[x_2=x_1 \vee x_3=x_1 \vee x_3=x_2 \vee ... \vee x_n=x_1 \vee x_n=x_2 \vee ... \vee x_n=x_{n-1} \vee F]$;

  – "$\forall x \in \{x_1,x_2,...x_n\}F(x)$" does not surreptitiously introduce set theory; it merely expresses the conjunction of n non quantified formulæ: $F(x_1) \wedge F(x_2) \wedge ... \wedge F(x_n)$;

– similarly, "$\exists x \in \{x_1, x_2, ..., x_n\} F(x)$" merely expresses the disjunction of n non-quantified formulæ: $F(x_1) \vee F(x_2) \vee ... \vee F(x_n)$.

3.1.1.4 Formulæ

Formulæ of an MS-FOL theory are defined recursively:

– if $R(t_1, ..., t_n)$ is an atomic formula, then it is a formula;

– Boolean combinations of formulæ are formulæ: if F and G are formulæ, then ¬F (also written "not F"), F ∧ G (also written "F & G") , F ∨ G (also written "F or G") and F ⇒ G are formulæ;

  – if F is a formula and x is a variable of any sort, then $\forall x F$ and $\exists x F$ are formulæ.

A variable x appearing in a formula is called free if it is not in the scope of a $\forall x$ or $\exists x$ quantifier. A formula with no free variables is called closed (all its variables are quantified); otherwise, the formula is called open. An open formula may have quantifiers (when only some but not all of its variables are quantified).

3.1.2. General logic axioms and inference rules

Notice that, up to this point, no notion of truth has been introduced: a formula is only a syntactic construct. Provability (rather than truth) will be defined via axioms and rules of inference. As we shall need classical logic to formulate the CSP problem in the rest of this chapter and intuitionistic logic to define the CSP resolution theories in chapter 4, we shall introduce these axioms in a way that allows a clear separation between classical and intuitionistic logic.

3.1.2.1 Gentzen's "natural logic"

There are two main formulations of logic. Hilbert's is probably the most familiar one (it is the one we adopted in HLS). Here, we shall prefer Gentzen's "natural logic" [Gentzen 1934], for three reasons:

– it makes no formal distinction between an axiom (such as: A ∧ B ⇒ A) and a rule of inference (such as Modus Ponens: from A and A ⇒ B, infer B);

– each logical connective is defined in itself by two complementary and very intuitive rules of elimination and introduction (whereas some of Hilbert's axioms mix several connectives and they can have many equivalent formulations);

– in many occasions, proofs can be made recursively by following the definition of a formula; a separate rule for each axiom makes this easier; in particular, our three meta-theorems will be shown to be obvious.

Gentzen's formulation is a set of rules in the form: $\dfrac{\text{premises}}{\text{conclusion}}$ (name of the rule), more precisely:

$$\frac{\Gamma_1 \mathbin{|\!\!-} \phi_1, \quad \Gamma_2 \mathbin{|\!\!-} \phi_2, \quad \Gamma_3 \mathbin{|\!\!-} \phi_3, \dots}{\Delta \mathbin{|\!\!-} \psi} \qquad \text{(name of the rule)}$$

$\Gamma \mathbin{|\!\!-} \phi$ is interpreted as: ϕ can be deduced from Γ;
the whole rule is interpreted as: if ϕ_i can be deduced from Γ_i, for $i = 1, 2, 3, \dots$, then ψ can be deduced from Δ;
here ϕ_i and ψ are formulæ, Γ_i and Δ are finite sets of formulæ (sets, not sequences – the order of their elements is irrelevant).

This formalism is the same for classical and intuitionistic logic, but the intended meaning of "can be deduced from" is stronger in intuitionistic logic: it means that there is an effective, constructive proof (e.g. not only a proof by contradiction). Whereas the classical interpretations are in terms of True and False (i.e. ϕ means that ϕ is True), the intuitionistic ones are in terms of Provable and Contradictory (i.e. ϕ means that ϕ is provable; $\phi_1 \wedge \phi_2$ means that ϕ_1 is provable and ϕ_2 is provable; $\phi_1 \vee \phi_2$ means that ϕ_1 is provable or ϕ_2 is provable).

3.1.2.2 Propositional axioms common to intuitionistic and classical logic

Most of the rules for the various connectives go by pairs. We use the standard abbreviations such as: Γ, ϕ_1, ϕ_2 for $\Gamma \cup \{\phi_1, \phi_2\}$; we also use the symbol $\perp$ for the absurd, considered as a proposition always false.

– *Implication*:

$$\frac{\Gamma \mathbin{|\!\!-} \phi \Rightarrow \psi \quad \Gamma \mathbin{|\!\!-} \phi}{\Gamma \mathbin{|\!\!-} \psi} \; (\Rightarrow E) \qquad\qquad \frac{\Gamma, \phi \mathbin{|\!\!-} \psi}{\Gamma \mathbin{|\!\!-} \phi \Rightarrow \psi} \; (\Rightarrow I)$$

$(\Rightarrow E)$ is the way *Modus Ponens* is expressed in Gentzen's natural logic.

– *Conjunction* (there are two elimination rules, one for each conjunct):

$$\frac{\Gamma \mathbin{|\!\!-} \phi_1 \wedge \phi_2}{\Gamma \mathbin{|\!\!-} \phi_i} \; (\wedge E_i) \qquad\qquad \frac{\Gamma \mathbin{|\!\!-} \phi_1 \quad \Gamma \mathbin{|\!\!-} \phi_2}{\Gamma \mathbin{|\!\!-} \phi_1 \wedge \phi_2} \; (\wedge I)$$

– *Disjunction* (there are two introduction rules, one for each disjunct):

$$\frac{\Gamma \mathbin{|\!\!-} \phi_1 \vee \phi_2 \quad \Gamma, \phi_1 \mathbin{|\!\!-} \psi \quad \Gamma, \phi_2 \mathbin{|\!\!-} \psi}{\Gamma \mathbin{|\!\!-} \psi} \; (\vee E) \qquad \frac{\Gamma \mathbin{|\!\!-} \phi_i}{\Gamma \mathbin{|\!\!-} \phi_1 \vee \phi_2} \; (\vee I_i)$$

– *Negation*: there is no rule for negation, $\neg\phi$ is considered as an abbreviation for $\phi \Rightarrow \bot$. Instead there is an elimination rule for the absurd:

  – *Absurd*:

$$\frac{\Gamma \mid\!\!- \bot}{\Gamma \mid\!\!- \phi} \quad (\bot\,E)$$

The meaning of rule ($\bot$ E) is that anything can be deduced from the absurd. Contrary to the other connectives, there is (fortunately) no rule ($\bot$ I) for introducing the absurd.

3.1.2.3 Propositional axioms specific to classical logic: "the excluded middle"

These are four *intuitionistically equivalent* forms of the only law specific to classical logic, the "law of the excluded middle":
  – Excluded middle: $\mid\!\!- A \vee \neg A$
  – *Reductio ad absurdum* (reduction to the absurd): $\mid\!\!- \neg\neg A \Rightarrow A$
  – Contraposition: $\mid\!\!- (\neg B \Rightarrow \neg A) \Rightarrow (A \Rightarrow B)$
  – Material implication: $\mid\!\!- (A \Rightarrow B) \Leftrightarrow (\neg A \vee B)$

3.1.2.4 Axioms on quantifiers

They can also be written as natural deductions:

  – *Universal quantification*:

$$\frac{\Gamma, \phi[t/x] \mid\!\!- \psi}{\Gamma, \forall x\phi \mid\!\!- \psi} \quad (\forall\,E) \qquad\qquad \frac{\Gamma \mid\!\!- \phi}{\Gamma \mid\!\!- \forall x\phi} \quad (\forall\,I)$$

  – *Existential quantification*:

$$\frac{\Gamma, \phi \mid\!\!- \psi}{\Gamma, \exists x\phi \mid\!\!- \psi} \quad (\exists\,E) \qquad\qquad \frac{\Gamma \mid\!\!- \phi[t/x]}{\Gamma \mid\!\!- \exists x\phi} \quad (\exists\,I)$$

In these rules, $\phi[t/x]$ is the formula obtained by replacing every free occurrence of variable x in $\phi(x)$ by term t. Notice that, in intuitionistic logic, contrary to classical logic, $\exists x$ is not equivalent to $\neg\forall x\neg$. This is usually interpreted by saying that proofs of existence by the absurd are not allowed; proofs of existence must be constructive; they must explicitly exhibit the object whose existence is asserted.

3.1.3. Theory specific axioms, proofs and theorems in an MS-FOL theory

In any logic, an axiom is defined as a closed formula and a theory as a set of axioms including the general logic axioms. In Gentzen's natural logic, an axiom appears as a rule with no premise and with empty set Γ. In short notation, it can be written, as: $| - A$ (as we did in section 3.1.2.3).

A proof is a sequence of expressions of the form $\Gamma \ | - \phi$, each of which is either an axiom or the conclusion of a logic rule with premises equal to previous expressions in the sequence. A theorem is the last expression of a proof, with empty set Γ.

3.1.4. Model theory, consistency and completeness theorems

In this section, we shall consider classical logic only. Models of intuitionistic logic will be introduced in chapter 4.

Definition: an *interpretation* of a theory T is a set of sets, one for each sort (it is thus a particular kind of function i from Sort to Set, the category of sets), together with:

– for each sort σ, an application from $ct(\sigma)$ into $i(\sigma)$;

– for each n-ary function symbol f with sort σ and signature $(\sigma_1,... \sigma_n)$, a function $i(f)$: $i(\sigma_1)$ x....x $i(\sigma_n) \rightarrow i(\sigma)$;

– for each n-ary predicate symbol R with signature $(\sigma_1,... \sigma_n)$, a subset $i(R)$ of $i(\sigma_1)$ x....x $i(\sigma_n)$.

An interpretation i of T can be extended to any formula of T in an obvious way following the recursive definition of formulæ. If i is an interpretation of T and F is a formula, we introduce the symbol "$|=$" (read satisfies) and the expression $i \ |= F$ to mean that i satisfies F.

Definition: a *model* of T is an interpretation i of T such that its extension satisfies all the axioms of T.

The most basic theorems of logic are Gödel's consistency and completeness theorems. They establish the correspondence between syntax and semantics, i.e. between formal proof and set theoretic interpretations:

– *Consistency theorem: a formula provable in T is valid in any model of T;*

– *Completeness theorem: a formula valid in any model of T is provable in T.*

3.1.5. Non uniqueness of models of an MS-FOL theory

In FOL or MS-FOL, there is no general means of specifying that a theory has a unique model. For theories with an infinite model, it is even the contrary that is true:

due to the "compactness" theorem, there are always infinitely many models and there are models of arbitrarily large infinite cardinality.

3.2. The formalisation of a CSP in MS-FOL: T(CSP)

The CSP axioms can generally be classified into four general categories: CSP sorts axioms (defining the domain of the variables, e.g. rows, columns, …), CSP background axioms (expliciting general structural properties of the problem, e.g. the structure of the Sudoku grid), CSP constraints axioms (the core content of the CSP, e.g. the famous four Sudoku axioms), CSP instance axioms (relative to each instance of the CSP, e.g. the entries of a puzzle).

3.2.1. Sorts and predicates of the CSP

There are many ways a CSP could be expressed as a logical theory T(CSP). Some of them may be simpler than the one proposed here, but our universal formalisation is mainly intended to be a step towards the introduction of CSP resolution theories.

Our approach will be based on the following two remarks. Firstly, as mentioned in the Introduction, any constraint (including the implicit constraints between different values for the same variable) is supposed to be re-written as a set of binary constraints and we can thus suppose that our CSP is binary.

Secondly, the notion of a *label* will play a central role. Labels will be the basis for a proper definition of candidates in chapter 4. Our non standard definition of a label (as an equivalence class of pre-labels) may seem a little convoluted, but it provides for the possibility of having multiple representations of the same basic facts without confusing the underlying CSP variables. As shown in chapter 2 with the four "2D" spaces in Sudoku, multiple representations are very useful in practice.

From a set theoretic point of view, a binary constraint c between two CSP variables X_1 and X_2 (which may be the same one) is the subset of pairs in $Dom(X_1) \times Dom(X_2)$ satisfying this constraint; equivalently, it is also a symmetric subset of
$[\{X_1\} \times Dom(X_1) \oplus \{X_2\} \times Dom(X_2)] \times [\{X_1\} \times Dom(X_1) \oplus \{X_2\} \times Dom(X_2)])$, which is itself a symmetric subset of $P \times P$ (where P is the set of pre-labels, defined below).

The complement of this set in $P \times P$ is a symmetric subset DC(c) of $P \times P$; it is obviously equivalent to a set of pairwise c-links between pre-labels, if we say that there is a c-link between two pre-labels p_1 and p_2 if and only if $(p_1, p_2) \in DC(c)$, i.e. if they are contradictory with respect to constraint c. The following definitions make this more formal.

Definition: in a CSP, a *pre-label* is a <variable, value> pair, i.e. a pair $\langle X^\circ, x^\circ \rangle$, where X° is a CSP variable and $x^\circ \in \text{Dom}(X^\circ)$. The set P of pre-labels is thus the disjoint union (the "direct sum", the $\oplus$) of the domains of the variables. Informally, this can also be viewed as the union of all the elements of all the domains, after each element has been subscripted by the name of the variable.

Definition: in a CSP, two pre-labels $\langle X^\circ, x^\circ \rangle$ and $\langle X^{\circ\prime}, x^{\circ\prime} \rangle$ are equivalent if, with respect to the CSP, $X^\circ = x^\circ$ is equivalent to $X^{\circ\prime} = x^{\circ\prime}$. In the present context, this entails that they are related to any other pre-labels by exactly the same constraints. This is a first test of correctness for the modelling.

Definition: a *label* is a name for an equivalence class of pre-labels (with respect to the above defined equivalence relation). If l° is a label and $\langle X^\circ, x^\circ \rangle$ is an element of this class, i.e. if $\langle X^\circ, x^\circ \rangle \in l^\circ$, we often use $\langle X^\circ, x^\circ \rangle$ to mean l°, by abuse of language. It should be noted that, given a CSP variable X° and a value x° in its domain, there is a unique label associated with the $\langle X^\circ, x^\circ \rangle$ pair. But, conversely, due to our approach of introducing several redundant representations in the modelling process, given a label, there will generally be several elements in its equivalence class.

Given a label l° and a CSP variable X°, there are only two possibilities: either there is one and only one value x° in $\text{Dom}(X^\circ)$ such that $\langle X^\circ, x^\circ \rangle \in l^\circ$ (in which case we say that $\langle X^\circ, x^\circ \rangle$ *is a representative of* l° and that l° *is a label for* X°) or there is no such x°.

Definition: two different labels l_1 and l_2 are *linked by constraint c* if there are representatives $p_1 = \langle X_1, x_1 \rangle$ of l_1 and $p_2 = \langle X_2, x_2 \rangle$ of l_2 such that $(p_1, p_2) \in \text{DC}(c)$. "linked-by c" is a symmetric (but neither reflexive nor transitive) relation. This definition entails that $(p_1, p_2) \in \text{DC}(c)$ for any representatives p_1 of l_1 and p_2 of l_2. By abuse of language, we sometimes write that $(l_1, l_2) \in \text{DC}(c)$.

Definition: two different labels l_1 and l_2 are *linked by some constraint* or simply *linked* if $(l_1, l_2) \in \text{DC}(c)$ for some c. "linked" is a symmetric (but nor reflexive and not transitive) relation.

Pre-labels are used as a technical tool for the definition of labels. From now on, we shall meet mainly CSP variables, values and labels.

We can now define the logical language of T(CSP). Basically, it has the following *sorts*, sort constants and sort variables:

– for each CSP variable X, a sort $\underline{X}$; for CSP variable X, for each element in $\text{Dom}(X)$, there is a constant symbol of sort $\underline{X}$ (considered as a name for this possible value of X); variables of sort $\underline{X}$ are: $x, x', x_1, x_2, \ldots$;

– a sort Label; for each element in the set of labels, there is a constant symbol of sort Label (the name of this label); variables of sort Label are: $l, l', l_1, l_2, \ldots$ but also (because it will be convenient when we define chains) z and r, r', r_1, r_2, ...; we shall also use capital letters for labels;

– a sort Constraint; for each constraint in the CSP, there is a constant symbol of sort Constraint (the name of this constraint); variables of sort Constraint are c, c', c_1, c_2, ...; [alternatively or additionally, as in the Sudoku or the n-Queens cases, one may have a sort Constraint-Type, when each constraint can be defined in a unique way by a label and a constraint type; modifying accordingly the general theory and all the resolution rules defined later in this book is straightforward];

– a sort CSP-Variable; for each CSP variable X, there is a constant symbol X of sort CSP-Variable (CSP variables are considered to be their own name); variables of sort CSP-Variable are V, V', V_1, V_2, ...; *CSP-Variable is considered as a sub-sort of Constraint*; [one could also have CSP-Variable-Type, a sub-sort of Constraint-Type];

– a sort Value; for each value in the (ordinary, set theoretic) union of the domains of the CSP variables, there is a constant symbol; variables of sort Value are v, v', v_1, v_2, ...

The logical language of the CSP has only the following four *predicates*:

– a unary predicate: value, with signature (Label); the intended meaning of value(l) is that, if <X, x> is any representative of l, then x is the value of variable X;

– a ternary predicate: linked-by, with signature (Label, Label, Constraint); the intended meaning is that the first two arguments, labels l_1 and l_2, are linked by the constraint given in the third argument, i.e. they are incompatible for this constraint;

– a binary predicate: linked, with signature (Label, Label); the intended meaning is that the two arguments, labels l_1 and l_2, are linked by some of the constraints.

For technical reasons, it also has the following *predicate*:

– a ternary predicate: label, with signature (Label, CSP-Variable, Value); the intended meaning of label(l, X, x) is that l is the label of the <variable, value> pair <X, x>.

Notice that, contrary to the sorts Label, Constraint [and/or Constraint-Type] and "CSP-Variable" [and/or "CSP-Variable-Type"] that will play a major theoretical role in the sequel, sort "Value" and associated predicate "label" are here mainly for the technical purpose of specifying the correspondence between labels and <variable, value> pairs (see axiom "meaning of labels" below) and for formulating the completeness of the solution (see the eponym axiom below). In applications, there may be simpler, perhaps implicit ways of specifying this correspondence and of writing this axiom (see section 3.5).

Optionally, the language of the CSP may include additional sorts useful for formulating certain types of rules or for interacting with the outer world in natural terms; in some cases, the general sorts above may be defined from these additional sorts. For details about this, see the Sudoku example (section 3.5).

What is most important here is that:

– the universal language necessary to formulate the general CSP theory is very restricted;

– with the mere addition of a single predicate "candidate" in the CSP resolution theories, this language will be enough to define very general and powerful resolution rules valid for any CSP.

3.2.2. Implicit CSP sort axioms

In MS-FOL, sort axioms do not have to be written explicitly, as would be the case in FOL, because they are considered as part of the definition of sorts. For each sort X, implicit sort axioms for a finite CSP will be of two kinds: exhaustiveness of domain constants (the domain of X has no other value than those corresponding to constants of this sort) and unique names assumption (two different constants for X name two different objects of sort X). Notice that, contrary to constants, there is no unique names assumption on variables: two variables (of same sort) can designate the same object (of this sort); when we want to specify that they refer to different objects, this must be stated explicitly.

3.2.3. CSP background axioms

Until now, we have defined sorts, predicates and functions and we have given their intended meaning. But we have written nothing that would formally ensure that they really have this meaning. The role of the following background axioms is to express the fixed structure of the problem and its translation into a graph of labels, independently of any values; they deal with correspondences between original <variable, value> pairs and labels, and with the re-writing of the original constraints into symmetric links between labels:

meaning of labels: for each CSP variable $X°$, for each $x°$ in $Dom(X°)$, if $l°$ is the (unique) label of $<X°, x°>$, the axiom defined by the ground atomic formula: label($l°, X°, x°$);

re-writing of each constraint as a set of links: for each constraint $c°$, for each pair of labels $l°_1$ and $l°_2$ such that $(l°_1, l°_2) \in DC(c°)$, the axiom defined by the ground atomic formula: linked-by($l°_1, l°_2, c°$);

symmetry of links: $\forall c\ \forall l_1\ \forall l_2$ {linked-by(l_1, l_2, c) $\Leftrightarrow$ linked-by(l_2, l_1, c)}; (this is normally useless, because it should be ensured by the modelling process);

exhaustiveness of constraints: $\forall l_1 \forall l_2$ {linked(l_1, l_2) $\Leftrightarrow$ $\exists c$ linked-by(l_1, l_2, c)}.

This is the general, slightly artificial, formulation of background axioms for any CSP. In each particular CSP, the concrete expression of these axioms may be adapted to the specificities of the problem. They may even be partly implicit in the definition of the "technical sorts". This will appear clearly in the Sudoku example.

3.2.4. CSP constraints axioms

It is not enough to associate a link with each constraint; the fact that these links really stand for constraints must also be written. We can now state what could be called the "core" CSP axioms (the background ones being only technicalities):

Meaning of links as constraints: $\forall l_1 \forall l_2 \{value(l_1) \wedge linked(l_1, l_2) \Rightarrow \neg value(l_2)\}$;

Completeness of solution: $\forall V \exists! v \exists l [label(l, V, v) \wedge value(l)]$.

We have written the first axiom in an asymmetrical way that will make the transition to CSP resolution theories more natural. As for the second axiom, it can be read as: each CSP variable has a unique value. Notice that this does not mean that the CSP has a unique solution; it only means that, in any solution, there is a unique value for each CSP variable.

3.2.5. Logical theory of the CSP: T(CSP)

Finally, define the Theory of the CSP, **T(CSP)**, as the MS-FOL theory written in the above defined language and consisting of (the implicit sort axioms,) the CSP background axioms and the CSP constraints axioms.

3.2.6. CSP instance axioms

A given corresponds to the assertion of a value for a label: $value(l^0)$. An instance P of the CSP is specified by a set of n givens $l^0_1, ..., l^0_n$ (where all the l^0_i are meta-symbols for – i.e. they stand for – constant label symbols) and it thus corresponds to the conjunction:

$value(l^0_1) \wedge ... \wedge value(l^0_n)$. We name it indifferently E(P) or E_P (E for "entries").

Finally, we have the obvious *theorem: there is a natural correspondence between a solution of the original CSP instance P and a model of its logical theory T(CSP) $\cup E_P$.*

Consequence: as a logical theory can only prove properties that are true in all its models, the CSP Theory for a given instance can only prove values that are common to all the solutions of this instance, if there is at least one (it can prove anything if there is no solution, i.e. if the instance axioms are inconsistent).

3.3. Remarks on the existence and uniqueness of a solution

Notice that, given any instance P, the axioms of T(CSP) together with E_P *a priori* imply neither the existence nor the uniqueness of a solution for P. Concerning the existence, this may seem to contradict the axiom of completeness, but this axiom only puts a condition on a solution, it does not assert that there is a solution (i.e. that E_P is consistent with T(CSP)). Indeed, any axiom that would assert the existence of a solution for any P would be trivially inconsistent. Let us consider the Sudoku example (see section 3.5 for the specific notations).

In this case, no set of *a priori* conditions on the entries of an instance P is known that would ensure that P has a solution (at least one). Obviously, some trivial necessary conditions for existence can be written (such as not having the same entry twice in a row, a column or a block) but they are very far from being sufficient.

As for uniqueness, for any puzzle P and corresponding axiom E_P, one may think that it could be expressed by the following additional axiom:

– ST-U: there is at most one solution:

$$\forall r \forall c \forall n_{rc} \forall n'_{rc} \, [\text{value}(n_{rc}, r, c) \wedge \text{value}(n'_{rc}, r, c) \Rightarrow n_{rc} = n'_{rc}].$$

But this is not true: such an axiom for uniqueness cannot imply that the solution is unique. It can only imply that, if the solution is not unique, then E_P contradicts this axiom; i.e. theory $ST \cup ST\text{-}U \cup \{E_P\}$ is inconsistent. This is why we prefer to speak of the *assumption* rather than the *axiom* of uniqueness. Whereas the Sudoku axioms are constraints the player must satisfy, the assumption of uniqueness puts a constraint on the puzzle creator; a player may choose to believe it or not; if he does, it amounts to accepting an oracle.

Uniqueness of a solution is a very delicate question (see also section 3.1.5). As was the case for existence, some trivial necessary conditions on the givens can be written for uniqueness (such as having entries for at least eight different numbers – otherwise, given any solution, one could get a different one by merely permuting two of the remaining numbers) but, again, they are very far from being sufficient.

Uniqueness of the solution (i.e. of a model of the puzzle theory) can only be a consequence of the givens. But is it possible to write a formula U(P) that would be equivalent to the uniqueness of the solution if the set of givens of P satisfies it? It is likely that this problem is much more difficult than solving the puzzle.

There are famous examples of puzzles that have been proposed and asserted as having a unique solution and that have indeed several. Many of the resolution rules that have been proposed to take uniqueness into account have been used inconsistently to *conclude* that some puzzle has a unique solution. Moreover, the uniqueness of a solution for a given puzzle can be asserted only if it has already been proven – which supposes that there exists some means for proving it. In our approach, unless explicitly stated otherwise, we shall never take the uniqueness of a solution as granted and we therefore do not adopt this assumption for any CSP.

3.4. Operationalizing the axioms of a CSP Theory

From a logical point of view, the above-defined theory T(CSP) is necessary and sufficient to define the CSP: given any instance P (with axiom E_P corresponding to its entries) and any complete solution G of P, the following are equivalent:

– G is a solution (in the intuitive sense) of instance P of the CSP;

– G is a *model* of T(CSP) $\cup$ {E_P} (in the standard sense of mathematical logic introduced in section 3.1);

– G satisfies the axioms of T(CSP) $\cup$ {E_P}.

T(CSP) is therefore theoretically perfect: for any instance of the CSP, its formal and intuitive meanings coincide. The only problem with it is practical: it does not give any indication on *how* to build a solution.

From an operational point of view, the "meaning of links as constraints" axioms could be considered as contradiction detection rules. For instance, they could be re-written in the following operational form: if, at some point in the resolution process of an instance, we reach a situation in which two different values should be assigned to the same variable, then we can conclude that this instance has no solution (the entries of this instance are contradictory with the axioms). This is, somehow, an operational form of these axioms. But do these forms express all the operational consequences of the original formulæ? Actually, the developments in chapter 4 will show that they do not (and they are indeed very far from doing so). The situation for the "completeness of a solution" axiom is still worse, since it does not tell anything about how it can be used in practice.

Vague as this may remain, let us define the aim we shall pursue with CSP Resolution Theories: replace the above axioms by another set of axioms that could be easily interpreted as (or transformed into) a set of operational rules for building a solution. And, since most known resolution rules in the Sudoku case are based on the notion of a candidate and on the progressive elimination of candidates, and since this idea corresponds to the common one of domain restriction in the general CSP, we want to write rules explicitly designed for this purpose. The problem is that, unless one admits recursive Trial and Error (which is not a rule), no theory of this kind is known that would be equivalent to T(CSP).

This book can thus be considered as being about the operationalization of the axioms of a CSP Theory – or about its replacement by a set of axioms that can be used in a constructive way.

3.5. Example: Sudoku Theory, T(Sudoku) or ST

The rest of this chapter illustrates the abstract general theory with the Sudoku case. T(Sudoku) is written ST for short. With the detailed Sudoku example, our goal is to illustrate both the above formalism and the ways of taking some liberty with it in order to simplify it in any specific case. For this purpose, we start with the "natural" formalisation of Sudoku and we show how it can be made compliant with the above general approach. For the most part, at the cost of some redundancy, the following sections are designed in such a way that they can be read independently of the previous ones or before them, for readers who do not like the abstract technicalities of formal logic.

3.5.1 Sudoku background axioms: Sudoku Grid Theory, SGT

The minimal underlying framework of Sudoku – the minimal support necessary for the representation of any Sudoku puzzle and any intermediate state in the resolution process – is a 9×9 grid composed of nine disjoint square blocks of 3×3 contiguous cells. Therefore, whichever formulation one chooses for the constraints (in rows, columns and blocks) defining the game, any theory of Sudoku must include an appropriate theory of such a grid. In the sequel, (our version of) this theory will be called 9-Sudoku Grid Theory (or simply Sudoku Grid Theory or SGT); it will contain all the general and "static" or "structural" knowledge about grids and only this knowledge, i.e. all the knowledge that does not depend on any particular entries for a puzzle and that does not change throughout the resolution process.

3.5.1.1. Sorts

In the limited world of SGT (and of ST in the next section), we shall consider the following sorts:

– Number: "Number" is the type of the objects intended to fill up the rc-cells of a grid; when, outside of the formal ST world, we need to refer to other kinds of numbers, we shall use their standard specific mathematical type: for instance, integers from 0 to infinity are simply called integers; the subscripts appearing in variables of any sort are integers, not Numbers; we have chosen to introduce the sort Number, because Sudoku is generally expressed in terms of digits, but one could introduce instead a sort Symbol, with nine arbitrary constant symbols;
  - constant symbols: n1, n2, n3, n4, n5, n6, n7, n8, n9;
  - variable symbols: n, n', n'', n_0, n_1, n_2, ...;
– Row:
  - constant symbols: r1, r2, r3, r4, r5, r6, r7, r8, r9;
  - variable symbols: r, r', r'', r_0, r_1, r_2, ...;

  – Column:

    - constant symbols: c1, c2, c3, c4, c5, c6, c7, c8, c9;

    - variable symbols: c, c', c'', c_1, c_2, ...;

  – Block:

    - constant symbols: b1, b2, b3, b4, b5, b6, b7, b8, b9;

    - variable symbols: b, b', b'', b_0, b_1, b_2, ...;

  – Square:

    - constant symbols: s1, s2, s3, s4, s5, s6, s7, s8, s9;

    - variable symbols: s, s', s'', s_0, s_1, s_2, ...;

  – Label: we define Label as a sort with domain the 729 elements $(n°, r°, c°)$ such that $n°$ is a Number constant, $r°$ is a Row constant and $c°$ is a Column constant; each label $(n°, r°, c°)$ will be the label for four different <variable, value> pairs, one associated with each of the four groups of CSP-Variables, namely: $(n°, r°, c°)$ = {<$Xr°c°$, $n°$>, <$Xr°n°$, $c°$>, <$Xc°n°$, $r°$>, <$Xb°n°$, $s°$>}, where [$b°$, $s°$] = $(r°, c°)$; labels can be assimilated with cells in 3D space; we sometimes use a loose notation $n°r°c°$ for $(n°, r°, c°)$;

    - constant symbols: (n1, r1, c1), ... (n9, r9, c9); sometimes also written in a loose notation: n1r1c1, ... n9r9c9;

    - variable symbols: l, l', ..., r, r', ..., z;

  – Constraint-Type:

    - constant symbols: rc, rn, cn, bn; notice that we use only four symbols corresponding to the four original types of constraints (*a* number in *a* cell, *a* row, *a* column or *a* block), not to specific constraints (e.g. a given number in a given row);

    - variable symbols: lk, lk', lk'', lk_0, lk_1, lk_2, ... ("lk" instead of "c" in the general theory, because symbol "c" is used for columns in Sudoku; we choose the "lk" symbol because constraint types are used to *link* candidates).

As the variable symbols explicitly carry their sort with the first letter of their name, they can be used straightforwardly in quantifiers or in equality with no further specification. For instance:

  – $\forall r$ always means "for all rows r",

  – $\forall c$ always means "for all columns c",

  – $\exists n$ always means "there exists a number n",

  – = can only be used with objects of the same sort, so that writing r = c is not allowed; to be more formal, the = sign should also be subscripted according to the type of objects it relates; for instance, to assert that two rows r_1 and r_2 are equal, we should use a specific equality symbol $=_r$ and write $r_1 =_r r_2$ (but we shall be lax on this notation also, since no confusion can arise from it).

Here is a very simple example of how MS-FOL simplifies formulæ: one can write ∀rF instead of what could only be written in FOL with an additional "row" predicate, something like ∀r[row(r) ⇒ F]. In longer formulæ, this may lead to drastic simplifications.

Remark on Constraint versus Constraint-Type: while the four elements of Constraint-Type correspond to the four 2D-spaces, the elements of Constraint (if we used this sort instead of Constraint-Type) would be represented by the 324 2D-cells of these four 2D spaces. Given any label l = (n°, r°, c°) and any constraint type lk, there is one and only one constraint of type lk "passing through l".

3.5.1.2. Function and predicate symbols

The SGT language has the "label" predicate necessary to specify all the correspondences between each label n°r°c° and its four <Xr°c°, n°>, <Xr°n°, c°>, <Xc°n°, r°>, <Xb°n°, s°> representatives. It also has the following functions: block and square [both with signature (Row, Column) and with respective sorts Block and Square], row and column [both with signature (Block, Square) and with respective sorts Row and Column], establishing the correspondences between the two coordinate systems: (r, c) and [b, s]. See sections 2.3 and 2.4 for details.

3.5.1.3. Background axioms (Axioms of Sudoku Grid Theory: SGT)

SGT has all the axioms asserting the equivalences stated in section 2.3.5, but they are now written in the form specified by the general theory (meaning of labels), i.e. for each Number constant n°, for each Row constant r°, for each Column constant c°, for each Block constant b° and for each Square constant s° such that [b°, s°] = (r°, c°), the following four ground atomic formulæ are axioms of SGT: label(n°r°c°, r°c°, n°), label(n°r°c°, r°n°, c°), label(n°r°c°, c°n°, r°), label(n°r°c°, b°n°, s°).

3.5.1.4. Block-free Grid Theory, LatinSquare Grid Theory (LSGT)

The Sudoku Grid Theory defined above can be simplified according to the following principles:

– forget the sorts Block and Square,

– forget all the functions and predicates referring to the above sorts.

Then we obtain a theory of grids that does not mention blocks and that is appropriate for Latin Squares: LSGT.

Theorem 3.1: There is a one-to-one correspondence between the models of SGT and the models of LSGT with added functions defining the proper correspondence between the two coordinate systems.

Proof: the proof involves some easy but tedious technicalities concerning the correspondence between theories in MS-FOL and in FOL (along the lines of [Meinke & al. 93]). Given a model of SGT, just forget anything about blocks and squares to get a model of LSGT. Conversely, given a model of LSGT, the key is that the added functions can be used to define new predicates for blocks and squares and that these predicates can, in turn, be used to introduce the new sorts Block and Square. Details of the proof are left as an exercise for the motivated reader.

3.5.2. Sudoku axioms, Sudoku Theory (ST)

With a proper choice of the sorts, Sudoku Theory (ST) can be axiomatised as a mere transliteration of the naive problem formulation. ST is an extension of Sudoku Grid Theory (SGT).

3.5.2.1. The sorts, functions and predicates of Sudoku Theory

ST has the same sorts, functions and axioms as SGT.

In addition, in conformance with the general theory, ST also has a predicate **value** with signature (Number, Row, Column). We define an auxiliary predicate **value'** with signature (Number, Block, Square) by the change-of-coordinates axiom:

CC: $\forall n \forall b \forall s$ {value'[n, b, s] $\Leftrightarrow$ value(n, row(b, s), column(b, s))}.

3.5.2.2. The axioms of Sudoku Theory

The only point in stating the axioms of ST is that we must be careful if we want to guarantee the best possible proximity with the resolution theories to be defined later. For instance, if we write that there must be one value for each cell (*in fine* an inescapable condition of the problem), this precludes all intermediate states from satisfying this axiom; we therefore try to limit the number of such assertions: indeed it will appear in only one axiom (ST-C). All the other general conditions in the statement of the problem can be expressed as "single occupancy" or "mutual exclusion" axioms – this is why, anticipating on the present formalisation, we adopted the first presentation of the game in the Introduction.

ST is defined as the specialisation of SGT (i.e. it has all the axioms of SGT) with CC and the following additional five axioms.

The first four axioms, "meaning of links as constraints axioms" are the quasi direct transliteration of the English formulation of the problem, as given in the Introduction:

– **ST-rc**: in natural rc-space, every rc-cell has at most one number as its value (i.e. given any rc-cell, it can have at most one value):

$$\forall r \forall c \forall n_1 \forall n_2 \text{ {value}}(n_1, r, c) \land n_1 \neq n_2 \Rightarrow \neg \text{value}(n_2, r, c)\};$$

notice that the condition linked-by(l_1, l_2, rc) of the general theory is here written more explicitly by giving the same values to the r and c components of both labels $l_1 = n_1rc$ and $l_2 = n_2rc$ and different values to their n components; the same remark applies to the next three axioms;

– **ST-rn**: in abstract rn-space, every rn-cell has at most one column as its value (i.e. given a row, a given number can appear in it in at most one column):

$$\forall r \forall n \forall c_1 \forall c_2 \ \{value(n, r, c_1) \wedge c_1 \neq c_2 \Rightarrow \neg value(n, r, c_2)\};$$

– **ST-cn**: in abstract cn-space, every cn-cell has at most one row as its value (i.e. given a column, a given number can appear in it in at most one row):

$$\forall c \forall n \forall r_1 \forall r_2 \ \{value(n, r_1, c) \wedge r_1 \neq r_2 \Rightarrow \neg value(n, r_2, c)\};$$

– **ST-bn**: in abstract bn-space, every bn-cell has at most one square as its value (i.e. given a block, a given number can appear in it in at most one square):

$$\forall b \forall n \forall s_1 \forall s_2 \ \{value'[n, b, s_1] \wedge s_1 \neq s_2 \Rightarrow \neg value'[n, b, s_2]\};$$

As in the general theory, the last axiom of ST says that the grid is complete:

– **ST-C**: the grid must be complete:

$$\forall r \forall c \exists n \ value(n, r, c).$$

At this point, it is important to notice that the first three of these axioms exhibit the symmetries and supersymmetries reviewed in chapter 2 (and they are block-free according to the definition in the next section), while the fourth exhibits analogy with the second and the third (and it is not block-free).

To better explicit the link with the general theory, let us introduce the following auxiliary predicate, with arity 7 and signature (Number, Row, Column, Number, Row, Column, Constraint-Type):
linked-by(n_1, r_1, c_1, n_2, r_2, c_2, lk) is defined as a shorthand for:
$[lk = rc \wedge r_1 = r_2 \wedge c_1 = c_2 \wedge n_1 \neq n_2] \vee$
$[lk = rn \wedge r_1 = r_2 \wedge n_1 = n_2 \wedge c_1 \neq c_2] \vee$
$[lk = cn \wedge c_1 = c_2 \wedge n_1 = n_2 \wedge r_1 \neq r_2] \vee$
$[lk = bn \wedge block(r_1, c_1) = block(r_2, c_2) \wedge n_1 = n_2 \wedge square(r_1, c_1) \neq square(r_2, c_2)].$

Then predicate "linked" of the general theory, with arity 6 and signature (Number, Row, Column, Number, Row, Column), can be shown to be equivalent to:
$[n_1 \neq n_2 \wedge r_1 = r_2 \wedge c_1 = c_2] \vee [n_1 = n_2 \wedge share\text{-}a\text{-}unit(r_1, c_1, r_2, c_2)]$
with auxiliary predicate share-a-unit(r_1, c_1, r_2, c_2) defined as:
$[r_1 = r_2 \vee c_1 = c_2 \vee block(r_1, c_1) = block(r_2, c_2)] \wedge [r_1 \neq r_2 \vee c_1 \neq c_2].$

3.5.2.3. *The axioms of LatinSquare Theory: LST*

One can define LatinSquare Theory (LST) as the Theory obtained from ST by forgetting any sort, function, predicate and axiom mentioning blocks and/or squares. In formal logic, we should normally have started with LST and specialised it to ST, but we are more interested in ST than in LST.

3.5.3 *Instance specific axioms (specifying the entries of a given puzzle)*

In order to be potentially consistent with any set of entries, ST includes no axioms on specific values. With any specific puzzle P we can associate the axiom E_P defined as the finite conjunction of the set of all the ground atomic formulæ **value(n_k, r_i, c_j)** such that there is an entry of P asserting that number n_k must occupy cell (r_i, c_j). Then, when added to the axioms of ST, axiom E_P defines the theory of the specific puzzle P.

3.6. Formalising the Sudoku symmetries

In this section, we introduce the concept of a block-free formula and we define three transformations on formulæ (in the language of ST) that will be used in chapter 4 to state and prove the formal versions of the intuitive meta-theorems 2.1, 2.2 and 2.3. We also prove a theorem that may be interesting in its own respect: it states that if a block-free formula (a formula that does not mention blocks or squares) can be proved in ST, then it can be proved without axiom ST-bn. As a result, a block-free formula is true for Sudoku (i.e. in ST) if and only if it is true for Latin Squares (i.e. in LST).

3.6.1. *Block-free predicates and formulæ*

The notion of a block-free formula is the formalisation of the natural language phrase ("mentioning only numbers, rows and columns") that we used in chapter 2 to express informally our Sudoku meta-theorems. Block-free formulæ play a major role in all that is related to Sudoku, because they are the formulæ to which these meta-theorems can be applied.

Definition: a function or predicate is called *block-free* if the sorts Block and Square do not appear in its sort or signature. "$=_n$", "$=_r$" and "$=_c$" are block-free predicates, and so are "label" and "value", whereas "$=_b$" and "$=_s$" are not.

Definition: a formula is called block-free if it is built only on block-free functions and predicates and it does not contain the bn constraint constant. For instance, "value" is block-free but "value' " is not.

3.6.2. The S_{rc}, S_{rn} and S_{cn} transformations of a block-free formula

In order to deal properly with the different kinds of symmetries reviewed in chapter 2, we need the following definitions. For any block-free formula F, we define inductively the three block-free formulæ $S_{rc}(F)$, $S_{rn}(F)$ and $S_{cn}(F)$. These formulæ have the same arity as F but they have different signatures.

Before giving the formal definitions, notice that they are just a pompous way of saying what was said informally in chapter 2, so that they can be skipped as technicalities of secondary interest:

– $S_{rc}(F)$ is the formula obtained from F by permuting systematically the words "row" and "column",

– $S_{rn}(F)$ is the formula obtained from F by permuting systematically the words "row" and "number",

– $S_{cn}(F)$ is the formula obtained from F by permuting systematically the words "column" and "number".

As is usual in logic, the formal definitions of $S_{rc}(F)$, $S_{rn}(F)$ and $S_{cn}(F)$ are given recursively, following the general construction of a formula:

– block-free terms (notice that the sorts cannot be permuted in functions, but the subscripts on the variables are permuted instead; this is technically important, especially when we deal with transformations of formulæ with different numbers of variables of different sorts):

F	$S_{rc}(F)$	$S_{rn}(F)$	$S_{cn}(F)$
$f(n_i, r_j, c_k)$	$f(n_i, r_k, c_j)$	$f(n_j, r_i, c_k)$	$f(n_k, r_j, c_i)$

– block-free atomic formulæ (as in functions, the sorts cannot be permuted in predicate "value", but the subscripts on the variables are permuted instead):

F	$S_{rc}(F)$	$S_{rn}(F)$	$S_{cn}(F)$
$n_i =_n n_j$	$n_i =_n n_j$	$r_i =_r r_j$	$c_i =_c c_j$
$r_i =_r r_j$	$c_i =_c c_j$	$n_i =_n n_j$	$r_i =_r r_j$
$c_i =_c c_j$	$r_i =_r r_j$	$c_i =_c c_j$	$n_i =_n n_j$
lk = rc	lk = rc	lk = cn	lk = rn
lk = rn	lk = cn	lk = rn	lk = rc
lk = cn	lk = rn	lk = rc	lk = cn
$value(n_i, r_j, c_k)$	$value(n_i, r_k, c_j)$	$value(n_j, r_i, c_k)$	$value(n_k, r_j, c_i)$

– logical connectives: each of the logical connectives merely commutes with each of S_{rc}, S_{rn}, S_{cn};

– quantifiers: they partly commute, with quantified variables exchanged:

F	$S_{rc}(F)$	$S_{rn}(F)$	$S_{cn}(F)$
$\forall n_i F, \exists n_i F$	$\forall n_i S_{rc}(F), \exists n_i S_{rc}(F)$	$\forall r_i S_{rn}(F), \exists r_i S_{rn}(F)$	$\forall c_i S_{cn}(F), \exists c_i S_{cn}(F)$
$\forall r_i F, \exists r_i F$	$\forall c_i S_{rc}(F), \exists c_i S_{rc}(F)$	$\forall n_i S_{rn}(F), \exists n_i S_{rn}(F)$	$\forall r_i S_{cn}(F), \exists r_i S_{cn}(F)$
$\forall c_i F, \exists c_i F$	$\forall r_i S_{rc}(F), \exists r_i S_{rc}(F)$	$\forall c_i S_{rn}(F), \exists c_i S_{rn}(F)$	$\forall n_i S_{cn}(F), \exists n_i S_{cn}(F)$

Notice that the three transformations are involutive, i.e. for any block-free formula F, one has $S_{rc} \bullet S_{rc}(F) = F$, $S_{rn} \bullet S_{rn}(F) = F$ and $S_{cn} \bullet S_{cn}(F) = F$.

3.6.3. S_{rcbs} transformation of a block-free formula

For a block-free formula F, its S_{rcbs} transform is also defined recursively by:

– block-free terms (notice again that the sorts cannot be permuted in the functions, but the subscripts on the variables are permuted instead; this is technically important, especially when we deal with transformations of formulæ with different numbers of variables of different sorts):

F	$S_{rcbs}(F)$
$f(n_i, r_j, c_k)$	$f(n_i, b_j, s_k)$

– block-free atomic formulæ:

F	$S_{rcbs}(F)$
$n_i =_n n_j$	$n_i =_n n_j$
$r_i =_r r_j$	$b_i =_b b_j$
$c_i =_c c_j$	$s_i =_s s_j$
$lk = rc$	$lk = rc$
$lk = rn$	$lk = bn$
$lk = cn$	$\perp$
$value(n_i, r_j, c_k)$	$value'[n_i, b_j, s_k]$

– logical connectives: all of them merely commute with S_{rcbs};

– quantifiers: they partly commute, (r, c) variables being changed to (b, s):

F	$S_{rcbs}(F)$
$\forall n_i F, \exists n_i F$	$\forall n_i S_{rcbs}(F), \exists n_i S_{rcbs}(F)$
$\forall r_i F, \exists r_i F$	$\forall b_i S_{rcbs}(F), \exists b_i S_{rcbs}(F)$
$\forall c_i F, \exists c_i F$	$\forall s_i S_{rcbs}(F), \exists s_i S_{rcbs}(F)$

3.6.4. Formal symmetries between the ST axioms

Using the above definitions, figure 3.1 shows all the symmetry, supersymmetry and analogy relationships between the four main axioms of ST.

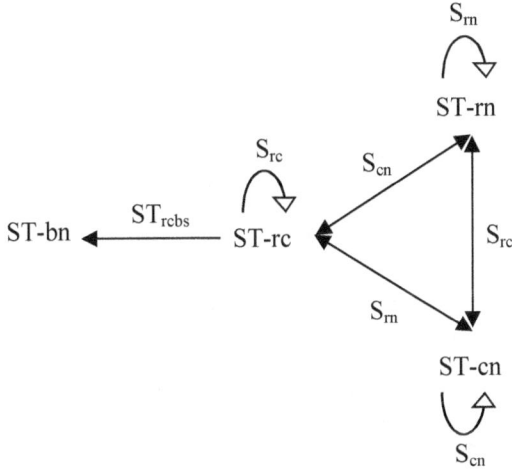

Figure 3.1. The symmetry relationships between the ST axioms

3.7. Formal relationship between Sudoku and Latin Squares

3.7.1. Block-free transform of a formula

With any formula G (not necessarily block-free) one can associate a well-defined block-free formula BF(G), called its block-free transform. It is defined recursively:

- if G is a block-free atomic formulæ, then BF(G) is F;
- if G is a non block-free atomic formulæ, then BF(G) is $\perp$;
- logical connectives $\neg$, $\wedge$, $\vee$, and $\Rightarrow$ merely commute with BF;

– if G is $\forall x G_1$, then BF(G) is $\forall x BF(G_1)$ if x is a block-free variable and it is merely G_1 if x is a non block-free variable;

– if G is $\exists x G_1$, then BF(F) is $\exists x BF(G_1)$ if x is a block-free variable and it is merely G_1 if x is a non block-free variable.

Remarks:

– the last two conditions are justified by the fact that non block-free variables are eliminated together with the non block-free atomic formulæ containing them;

– for any formula G (and not only the atomic ones), if G is block-free, then BF(G) is merely G.

3.7.2. Formal relationship between Sudoku and Latin Squares

Theorem 3.2: a block-free formula that is valid in ST has a block-free proof.

As an obvious corollary, we have:

Theorem 3.3: a block-free formula is valid for Sudoku (i.e. is a theorem of ST) if and only if it is valid for Latin Squares (i.e. it is a theorem of LST).

Proof of theorem 3.2: Remember the standard definition of a proof of F: it is a sequence of formulæ ending with F, where each formula in the sequence either is a logical axiom or is an axiom of ST or can be deduced from the previous ones by the rules of natural deduction.

Let F be a block-free formula and consider a proof of it in ST. It suffices to show that, if we apply BF to any step in this proof, we get a block-free proof of BF(F). This is an advantage of the Gentzen's formulation of logic adopted in this book: it is obvious (though tedious to check in detail) that all the rules of natural deduction in section 3.1 are stable under the BF transformation. (See *HLS1* for a slightly less obvious proof based on Hilbert's formalism instead of Gentzen's). The proof therefore reduces to the following obvious relationship between the sets of axioms of ST and LST: BF(ST) = LST.

4. CSP Resolution Theories

Before we try to capture CSP Resolution Theories in a logical formalism, we must establish a clear distinction between a logical theory of the CSP itself (as it has been formulated in chapter 3, with no reference to candidates) and theories related to the resolution methods (which we consider from now on as being based on the progressive elimination of candidates). These two kinds of theories correspond to two options: are we just interested in formulating a set of axioms describing the constraints a solution of a given CSP instance (if it has any) must satisfy or do we want a theory that somehow applies to intermediate states in the resolution process? To maintain this distinction as clearly as possible, we shall consistently use the expressions "*CSP Theory*" for the first type and "*CSP Resolution Theory*" for the second type. Section 4.1 elaborates on this distinction. Since it has been shown in chapter 3 that formulating the first theory is straightforward, theories of the second kind will remain as our main topic of interest in the present book. Nevertheless, it will be necessary to clarify the relationship between the two types of theories and between their respective basic notions ("value" and "candidate").

In section 4.2, we formalise the notion of a "resolution state". This provides the intuitive notion of a candidate with a clear logical status allowing to define precise relationships between the basic formal predicates "value" and "candidate".

As the first illustration of our logical formalism, section 4.3 shows that any CSP has a minimal CSP Resolution Theory (its Basic Resolution Theory or BRT(CSP)) and it expresses its axioms in this formalism. Here, "minimal" means that all the other resolution theories introduced in this book will be extensions of BRT(CSP) by additional axioms (logically speaking, they will thus be specialisations of BRT(CSP)). Section 4.4 then defines the general concepts of a CSP Resolution Theory. Section 4.5 defines a very important property a resolution theory can have (or not), the confluence property, and shows that BRT(CSP) has it in any CSP.

Finally, sections 4.6 and 4.7 deal with the Sudoku example. The latter proves the formal versions of meta-theorems 2.1, 2.2 and 2.3. It also proves an extension of the latter that will be very useful when we want to apply it in practice. Notice that, even without understanding the technicalities of these proofs, one can consider the meta-theorems as simple heuristics suggesting new potential rules and prove directly all the resolution rules deduced from them (this will generally be very easy).

4.1. CSP Theory vs CSP Resolution Theories; resolution rules

As our first approximation, we could say that CSP Theory is about *what* we want (a complete assignment of values to the CSP variables satisfying the general CSP constraints and the specific givens), with no consideration at all for the way it can be obtained, whereas a CSP Resolution Theory is about *how* we can reach this desired final state; but then we must correct the resulting erroneous suggestion that a theory of this second kind would be mainly concerned with resolution *processes*.

To state it formally, throughout this book, the status we grant a CSP Resolution Theory is *logical*, not *operational*, and we make a clear distinction between a Resolution Theory and possible *resolution methods* that may be built as operational counterparts or algorithmic computer implementations of it (e.g. by superimposing priorities on the pure logic of the resolution rules). Such resolution methods may themselves be considered from different points of view and different kinds of logic may be used to express these. For instance, one might be interested in the dynamics of the resolution processes associated with the method, in which case one could use temporal or dynamic logic for modelling them. This is not the point of view chosen in this book, where we consider a resolution method from the point of view of the "resolution states" underlying it and we adopt modal logic (logic of necessity and possibility) to model these. However, whereas the main part of this book deals with resolution theories themselves, these theories can have properties, such as confluence, that will be shown (in chapter 5) to be very important when one wants to define specific resolution methods based on them.

Then, from a logical standpoint, the only purpose of a Resolution Theory is to restrict the number of resolution states compatible with the axioms (i.e. the number of partial solutions, expressed in terms of values and candidates) and the relationships that exist between them. From an operational standpoint, it can be used as a reference for defining a resolution method that will dynamically modify the current information content; but, before a resolution theory can be used this way, there must be some operationalization process. This distinction is essential (and very classical in Artificial Intelligence) because a given set of logical axioms (a resolution theory) can often be operationalized in many different ways. (To be more specific: it can be expressed as very different sets of rules in an inference engine.)

Whereas CSP Theories, as developed in chapter 3 are very simple, CSP Resolution Theories require a more complex approach. All the CSP Resolution Theories should be restricted to satisfy two obvious general requirements: a) any of their rules should be a consequence of the CSP theory (under conditions, to be defined, on the relationship between values and candidates); b) they should apply to any set of givens. This is very far from being enough to constrain the possible theories of interest. But, as a consequence of these broad requirements, there are lots of aspects of the CSP solving that are excluded from our considerations. For instance, in the Sudoku case, we exclude any psychological bias, e.g. we do not take into account the physical proximity of rows or columns, although it is probably easier to see Hidden Triplets in three contiguous cells in a row than in three cells disseminated in this row.

4.2. The logical nature of CSP Resolution Theories

The analyses in this section constitute the central part of this chapter and they are the key to understanding the logical foundations of this book: given that the naive notion of a candidate is the basis for the various popular resolution rules in Sudoku and that it will also be the basis for the formulation of any resolution rule for any CSP, can one grant it a well defined logical status? Another point to be considered here is the relationship between the CSP Theory T(CSP), which does not use this notion, and related CSP Resolution Theories, which are based on it.

4.2.1. On the (non existent) problem on non-monotonicity

Let us first clarify the following point. One apparent problem in choosing the notion of a candidate as the basis for a logical formulation is that the set of candidates for any CSP variable is monotonically decreasing throughout the resolution process, whereas logic is usually associated with monotonically increasing sets: starting from what is initially assumed to be true (the axioms), each step in a proof adds new assertions to what has been proven to be true in the previous steps; there is no possibility in standard logic for removing anything.

Do we therefore need to use some sort of non-monotonic logic, as is often the case with AI problems? Not really: instead of considering candidates for a variable, we can consider the complementary set of "not-candidates" or excluded values, i.e. values that are effectively proven to be incompatible with all that is already known (in the Sudoku case, the crossed or erased candidates in the grid on the paper sheet) – and this is a monotonically increasing set. By "effectively proven", one should understand "proven by admissible reasoning techniques" (and the sequel will show that the informal word "admissible" must in turn be understood technically as "intuitionistically valid" or, equivalently, "constructively valid").

What is really important in logic is that the abstract information content is monotone increasing with the development of the proof. (In other terms, one should not confuse this information content with possibly varied representations of it.) In the sequel, when we write resolution rules, we shall conform to what we have done in *HLS* for the Sudoku example and we shall refer to candidates, but we must keep in mind that, when expressed with not-candidates, the underlying logic is always monotone increasing.

4.2.2. Resolution states and resolution models

Notwithstanding the above remarks on the informal notion of a candidate, can we grant it a precise logical status allowing us to use it consistently in the expression of the resolution rules? But, first of all, how is it related to the primary predicate "value"? Notice the vocabulary we used spontaneously: a value is asserted as being true, while a candidate is proven (or not proven) to be incompatible with all that is already proven. The most straightforward way of interpreting this is as an indication that the underlying logic of any CSP Resolution Theory based on candidates should be modal: it should be a logic of possibility/necessity as opposed to a logic of truth (such as standard logic or MS-FOL).

Before entering into the formal details, let us define the notions of a resolution state and of a resolution model. Defining the model theoretic aspects before the syntactic aspects is not the usual way to proceed in logic, but it is more intuitive.

4.2.2.1. Resolution states

Definitions (here $l°$ is a meta-variable designating a constant symbol for a label):

– a value datum is any ground atomic formula of the kind value($l°$);

– a candidate datum is any ground atomic formula of the kind candidate($l°$);

– a *resolution state* RS is any set of value data, of candidate data and of negated candidate data; it is not necessarily devoid of contradictions with respect to the CSP constraints, but it cannot contain both candidate($l°$) and ¬candidate($l°$) for the same $l°$; we write RS $\models$ value($l°$), RS $\models$ candidate($l°$) and RS $\models$ ¬candidate($l°$) to mean respectively that the value datum is present in RS, that the candidate datum or the negated candidate datum is present in RS;

– for a resolution state RS and a label $l°$, if RS $\models$ candidate($l°$) [respectively RS $\models$ ¬candidate($l°$), RS $\models$ value($l°$)], we say informally that $l°$ is a candidate [resp. is not a candidate, is a value] in RS.

Notice that:
a) we need not consider negated value data, because value data can only be asserted;
b) Notice that, instead of considering the absence of a candidate from RS (which may have an umbiguous interpretation), we consider in a the presence of its negation (the positive fact that the candidate has been "effectively eliminated" from RS).

Any resolution state is a finite set and the whole set **RS** of resolution states is therefore finite (and independent of any particular instance of the CSP) although very large.

As suggested in part by the name, a resolution state is intended to represent the totality of the ground atomic facts and their negations (in terms of value and candidate predicates) that are present in some possible state of reasoning for some instance of the CSP. This is what we called informally the information content of this state – in which all the "static" knowledge about the CSP, such as links between labels, is considered as background knowledge and is not explicitly listed, but is implicitly present. In the Sudoku CSP, a resolution state is a straightforward abstraction for something very concrete: the set of decided values, of candidates still present on the sheet of paper used to solve a puzzle and of candidates erased or crossed. (And the structure of the grid remains implicit.)

Vocabulary: if RS is a resolution state, "*a candidate l in RS*" is an informal way of saying "a label l such that RS $\models$ candidate(l)". Similarly, "a value in RS" is a way of saying "a label l such that RS $\models$ value(l)".

4.2.2.2. Resolution models

In order to be able to give the above interpretation of a resolution state in a way that respects our resolution paradigm, we must add some structure on the set **RS** of all the resolution states and on the way they are related. On **RS**, we define a natural partial order relation: $RS_1 \leq RS_2$ if and only if, for any constant symbol $l°$ for a label, one has:

– if $RS_1 \models$ value($l°$), then $RS_2 \models$ value($l°$), (assertion/addition of a value is not reversible),

– if $RS_1 \models$ ¬candidate($l°$), then $RS_2 \models$ ¬candidate($l°$) (negation/deletion of a candidate is not reversible),

– if $RS_2 \models$ candidate($l°$), then $RS_1 \models$ candidate($l°$) (new candidates cannot appear in a posterior resolution state).

Thus, the intended meaning of $RS_1 \leq RS_2$ is that when one passes from one resolution state to a "greater" or "posterior" one (according to this abstract order relation), the information content can only increase – the negation of a candidate being considered as an increase of this information content. The last condition says that no candidate absent from a resolution state can reappear in a posterior one. In practical terms, it also means that RS_2 is closer to a solution (or to the detection of a contradiction) than RS_1 is.

With any instance P of the CSP (considered as defined by a set of labels), one can associate a unique well-defined resolution state RS_P, called the initial resolution state of P, in which:

– for every given $l°$ in P, $RS_P \models$ value($l°$),

– for every label l^1 which has no direct contradiction with any of the givens $l°$ of P, i.e. such that linked($l°$, l^1) is not in the background axioms for any given $l°$ of P, $RS_P \models$ candidate(l^1),

– RS_P contains no other value or candidate data than those defined above (in particular, it contains no negated candidate data).

The resolution model of an instance P is then defined as the subset RS_P of RS (together with the order relation induced by RS) consisting of all the resolution states RS such that $RS_P \leq RS$. When trying to solve P, one can never escape RS_P, at least as long as one reasons consistently. Any solution of P must be in RS_P and it can only be a maximally consistent element of RS_P. But, conversely, a maximally consistent element of RS_P is not necessarily a solution (especially in case there is no solution). By exploring systematically all the states in RS_P, one is certain either to prove that P has no solution or to find all the solutions of P, if P has any. Of course, to find a solution, one does not have to explore all of RS_P. In some sense, the purpose of a resolution theory is to define a smart way of reducing RS_P to a relevant part as small as possible (without excluding any parts that may lead to a solution).

One can notice that our definition of RS_P already includes the deletion of candidates obviously contradictory with the givens of the problem instance. This amounts to restricting from the start the resolution model RS_P of P to a relevant part.

4.2.2.3. Remarks on the notions of a resolution state and a resolution model

Notice that the above notions of a resolution state and a resolution model are very narrow. For instance, a resolution state does not include any "mental" component such as having identified a pattern corresponding to the preconditions of a resolution rule. Similarly, the resolution model RS_P of an instance P defines only an abstract order relation on the set of resolution states reachable from the initial state RS_P, it does not indicate *how* to pass from one state to a posterior one. But this is the only way one can build a consistent semantics in case an instance has zero or several solutions.

Simplistic as they may seem, the above-defined notions allow us to state precisely what kind of resolution rules we are looking for. Given a resolution theory T, the application of any resolution rule R in T to an instance P should lead from one resolution state in RS_P to a posterior one, with the following interpretation: if, starting from a resolution state RS in RS_P, we notice a pattern (or configuration) of labels, links, values and candidates, satisfying the condition part of R, then R can be applied to this pattern; and, if we apply it, then, in the resulting resolution state RS_1 and in all the subsequent ones (still in RS_P), the value(s) and candidate(s) specified in the action part of R will respectively be asserted and negated (values can only be asserted, candidates can only be negated). Notice that the whole process of detecting a pattern, applying a rule and passing from RS to RS_1 is superimposed on RS_P but is not part of this abstract static model.

Now, still starting from the same resolution state RS, if we notice that the conditions of another resolution rule R' in T are also satisfied in RS and if we apply R' instead of R, we usually reach a resolution state RS_2 (still in RS_P) different from RS_1. For a real understanding of what a resolution theory is and is not, it is crucial to remark that the (relatively informal) definition we have just given does not *a priori* imply that the two states RS_1 and RS_2 are T-compatible, in the sense that there would be a resolution state RS_3 posterior to both RS_1 and RS_2 (i.e. such that $RS_1 \leq RS_3$, $RS_2 \leq RS_3$) and accessible from each of RS_1 and RS_2 *via rules in T* (see Figure 4.1). This is related to the fundamental question of the confluence property of a resolution theory T (see section 4.5 for a definition and an example of a theory with the confluence property).

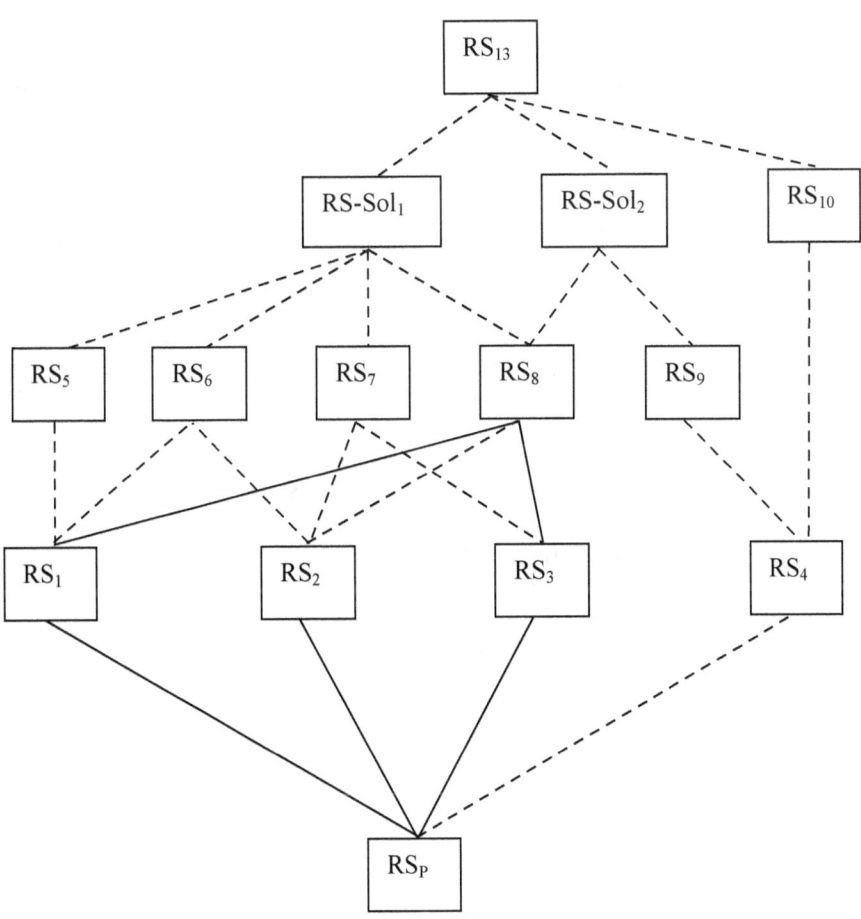

Figure 4.1. *The resolution model **RS_P** of an instance P with two solutions (RS-Sol₁ and RS-Sol₂) and the part of it accessible by some Resolution Theory T (full lines). Notice that the resolution states RS₁ and RS₂ (or RS₂ and RS₃) are not T-compatible, but RS₁ and RS₃ are.*

4.2.3. Logical interpretations of a resolution model

There are two possible logical interpretations of the above notions. The most straightforward one is in terms of modal logic. [In *HLS*, we used epistemic, instead of modal, logic; but the final interpretation of resolution theories (intuitionistic logic) is the same.]

4.2.3.1. The modal interpretation of a resolution model

Our notions of a resolution state and a resolution model appear to be a special case of the classical notions of a possible world and a Kripke model in modal logic.

In modal logic, there is a modal operator "□" of necessity (and a modal operator "◊" of possibility, which does not always appear explicitly, because it is equivalent to ¬□¬ in the most common modal theories); for any formula A, □A and ◊A are intended to mean respectively "A is necessary" and "A is possible".

Our notion of a resolution model coincides with that of a canonical Kripke model and the order relation we have defined on the set of resolution states corresponds to the accessibility relation between possible worlds in this model ([Kripke 1963]). We can apply Hintikka's interpretation of "□" ([Hintikka 1962]): RS |= □A if and only if RS' |= A for any possible world RS' accessible from RS (i.e., in our resolution model, such that RS ≤ RS').

Which (propositional) logical axioms for the modal operator □ should one adopt? This is the subject of much philosophical and scientific debate. It concerns the general relationship between truth, necessity and possibility and the axioms expressing this relationship. There are several modal theories in competition, the most classical of which are, in increasing order of strength: S4 < S4.2 < S4.3 < S4.4 < S5 (on this point and the following, see e.g. [Feys 1965] or [Fitting et al. 1999] or the Stanford Encyclopaedia of Philosophy: http://plato.stanford.edu/entries/logic-modal/).

Moreover, it is known that there is a correspondence between the axioms on □ and the properties of the accessibility relation between possible worlds (this is a form of the classical relationship between syntax and semantics). A very general expression of this correspondence was obtained by [Lemmon et al. 1977].

Here, we shall adopt the following rule of inference and set of axioms (in addition to the usual axioms of classical logic), which constitute the (most commonly used) propositional system S4 (we give them the names they are classically given in modal logic and, because of axioms M and 4, we write the accessibility relation "≤"):

– (Necessitation Rule) if A is a theorem, then so is □A;
– (Distribution Axiom) □(A ⇒ B) ⇒ (□A ⇒ □B);

– (axiom M) $\Box A \Rightarrow A$: "if a proposition is necessary then it is true" or "only true propositions can be necessary"; this axiom corresponds to the accessibility relation being reflexive (for all RS in **RS**, one has: $RS \le RS$);

– (axiom 4, reflection) $\Box A \Rightarrow \Box\Box A$: if a proposition is necessary then it is necessarily necessary; this axiom corresponds to the accessibility relation being transitive (for all RS_1, RS_2 and RS_3 in **RS**, one has: if $RS_1 \le RS_2$ and $RS_2 \le RS_3$, then $RS_1 \le RS_3$).

From our definition of a resolution model, it can easily be checked that it satisfies all the axioms of S4.

As for the predicate calculus part of our logic, quantifiers are generally a big problem in modal logic. But we must notice that in our CSPs we deal only with fixed domains; there is therefore no problem with quantifiers: we can merely adopt as axioms both Barcan Formula (BF) and its converse (CBF) ([Barcan 1946a and 1946b]), namely:

– (BF) $\forall x \Box A \Rightarrow \Box \forall x A$,

– (CBF) $\Box \forall x A \Rightarrow \forall x \Box A$.

One final thing should be noted: in modal logic, for any *ground atomic* formula A, "$A \vee \neg A$" is true in any resolution state and it is also necessarily true, i.e. one always has RS $\models \Box(A \vee \neg A)$, but this is not the case for "$\Box A \vee \Box \neg A$". For instance, given some definite place in space-time, it is always true that either it is raining (A) or it is not raining ($\neg A$) at this place, and this is necessarily true ($\Box(A \vee \neg A)$). But it is not true that either it is necessarily raining ($\Box A$) or it is necessarily not raining ($\Box \neg A$) at this place: the weather may change at this place. Said otherwise, "$\Box \neg A$" (A is necessarily false) and "$\neg \Box A$" (A is not necessarily true) are very different things and the first is much stronger than the second.

4.2.3.2. The intuitionistic interpretation of a resolution model

So far so good; but we are not very enthusiastic with the prospect of having to overload the formulation of our resolution rules with modal operators. Let us try to do one more step.

There is a well-known correspondence ([Fitting 1969]) between modal logic S4 and intuitionistic or constructive logic ([Bridges et al. 2006]). The language of a theory in intuitionistic logic is the same as in classical logic (there is no $\Box$ or $\Diamond$ logical operator). Given a formula A in intuitionistic logic, one defines a formula M(A) in S4 recursively by:

– for A atomic: $M(A) = \Box A$,

– $M(A \wedge B) = M(A) \wedge M(B)$,

– $M(A \vee B) = M(A) \vee M(B)$,

– M(¬A) = □¬M(A),

– M(A ⇒ B) = □(M(A) ⇒ M(B)),

– M(∀xA) = ∀xM(A).

Then, for every formula F with no modal operator, one has the well-known correspondence theorem: F is a theorem in intuitionistic logic if and only if M(F) is a theorem in modal logic S4.

In intuitionistic logic, although the formulæ are the same as in classical logic, their informal interpretation is different:

– A means that A is effectively proven;

– ¬A means that A is effectively proven to be contradictory;

– ¬¬A is not equivalent to A; it is weaker than A; it means that it is not effectively proven that A is contradictory, which does not imply that A is proven.

One main difference with classical logic is the "law of the excluded middle": A ∨ ¬A, which is not valid (it corresponds to formula □A ∨ □¬A in S4). A ∨ ¬A would mean that either A is proven or ¬A is proven. But there are propositions for which this is not true. Similarly, ∃xA is stronger than ¬∀x¬A; ∃xA means that a proof has effectively produced some x and shown it satisfies A; ¬∀x¬A only supposes that ∀x¬A leads to a contradiction.

The question for us is now: can we adopt intuitionistic instead of modal logic? It amounts to: can each of our resolution rules be written in the form M(A) for some formula A without modal operators? This raises the question of the intended meaning of the resolution rules.

4.2.4. Resolution theories are intuitionistic

Anticipating on our resolution rules (which will not refer explicitly to resolution states), in their naive formulations, their (non static) conditions will bear on the presence of some candidates and on the absence of others and their conclusions will always be the assertion of a value or the elimination of a candidate.

4.2.4.1. Analysing the intended meaning of resolution rules

Let us see how this can be used in the formulation of a CSP resolution theory:

– first, the entries of a CSP instance P, which are axioms, can be understood as necessarily true (in a formal way by the Necessitation rule, or in a semantic way because they will be present in all the resolution states): □value; this can be written as M(value), because "value" is atomic; intuitionistically, this is merely the tautology that axioms of T are effectively proven in T.

As for the resolution rules themselves:

– as links are part of the CSP structural background, they are also axioms of any Resolution Theory and a condition on the presence of a link between two labels can be understood as necessarily true (by the Necessitation rule): □linked-by(l_1, l_2, c); this can be written as M(linked(l_1, l_2, c)), because "linked-by" is atomic; the same can be said of "linked";

– a negative condition on a candidate [i.e. a condition ¬candidate(l)] in a resolution state RS implies that it is negated in any posterior resolution state; semantically, it must therefore be interpreted as: □¬candidate(l); this can be written as M(¬candidate(l)); intuitionistically, this means that this candidate has effectively been proven to be contradictory;

– a positive condition on a candidate in a resolution state RS could be intended to mean (in the modal sense) that "this label is still a possible value in RS": ◊value; but one should here anticipate on the final intended intuitionistic meaning: "this label has not yet been effectively proven to be an impossible value"; therefore, one should rather interpret such a condition in the sense of ¬¬value (in the intuitionistic meaning of it); in relation to the modal setting, this would appear to be the M transform of the stronger □¬□¬value or □◊value; (see section 4.2.4.3 below for comments);

– any ∧ and ∨ combination of such conditions remains of the form M(some formula with no □ symbol);

– a conclusion on the assertion of a value is intended to mean that the value becomes necessarily true: □value; this can be written as M(value), because "value" is atomic;

– a conclusion on the elimination of a candidate is intended to mean that this candidate becomes necessarily contradictory: □¬candidate; this can be written as M(¬candidate);

– any ∧ combination of such conclusions remains of the form M(some formula with no □ symbol);

– again by the Necessitation rule, the implication sign appearing in a resolution rule Cond ⇒ Act (which is an axiom in a Resolution Theory) can be understood as necessary: □(Cond ⇒ Act); this can be written as M(Cond ⇒ Act).

– finally, if the whole resolution rule ∀xR is surrounded by ∀ quantifiers, where R = M(A), it can be written as M(∀xA).

4.2.4.2. Resolution rules pertain to intuitionistic instead of classical logic

The above analysis shows that a resolution rule will always be of the form M(F) with no □ symbol in F. The general conclusion of all this is that a resolution rule is always the M transform of an MS-FOL formula and the MS-FOL formula can be used instead of the modal form, provided that we consider that **Resolution Theories pertain to intuitionistic (or constructive) logic**.

4.2.4.3. The meaning of positive conditions on candidates in resolution rules

Our interpretation of a positive condition on a candidate in the condition part of a resolution rule is worth some discussion. Our intuitionistic interpretation of "candidate" as "$\neg\neg$value", corresponding to the modal interpretation $\square\Diamond$value, rather than adopting the seemingly more natural (from the modal point of view) $\Diamond$value, is consistent with our definition of the order relation on **RS**: once a candidate has been eliminated, it can no longer re-appear in a posterior resolution state. So that, for any label l and resolution state RS, one can have RS $\models \Diamond$candidate(l) only if RS $\models$ candidate(l), i.e. if l is effectively present in RS as a candidate, which in turn implies that RS $\models \square\Diamond$candidate(l) .

Notice that the definition of **RS** and this interpretation together put a strong restriction on how resolution rules can be applied in a resolution state RS: a pattern mentioning non-negated candidates may only be instantiated if such candidates are effectively present in this resolution state. The condition part of the rule thus means: the pattern defined by this rule can be considered as present in RS only if the following candidates are still present in RS (i.e. have not yet been proven to be contradictory) and the other conditions of the rule are satisfied. From a computational point of view, the positive aspect is that, as candidates are progressively eliminated, it puts stronger and stronger conditions on patterns and it makes their potential number decrease while the resolution proceeds.

4.3. The Basic Resolution Theory of a CSP: BRT(CSP)

We can now define formally the Basic Resolution Theory of any CSP: BRT(CSP). Its logical language is an extension of the language defined in section 3.2 for CSP Theory T(CSP). In addition to it, it has only:

– two 0-ary predicates: solution-found and contradiction-found,

– a unary predicate: candidate, with signature (Label).

As for the axioms of BRT(CSP), they include all (the implicit sort axioms and) the background axioms of the CSP Theory defined in section(s) (3.2.2 and) 3.2.3. They cannot include the CSP constraint axioms of section 3.2.4 because they do not have the structure required of resolution rules: "meaning of links as constraints" is of the condition-action type, but it has the negation of a value in its conclusion (in a resolution rule, a value can never be negated); "completeness of solution" is not of the condition-action type. Instead, they contain the following:

– **ECP (Elementary Constraints Propagation)**: "if a value is asserted for a CSP variable (as is initially the case for the givens), then remove any candidate that is linked to this value by a direct contradiction":

ECP: $\forall l_1 \, \forall l_2 \, \{value(l_1) \wedge linked(l_1, l_2) \Rightarrow \neg candidate(l_2)\}$;
this is very close to "meaning of links as constraints", but the conclusion is about a candidate instead of a value;

– **S (Single):** "if a CSP variable V has only one candidate $<V, v>$ left, then assert it as the value of this variable":

S: $\forall l \, \forall V \, \forall v \, \{ \, [label(l, V, v) \wedge candidate(l)$
$\wedge \, \forall v' \neq v \, \forall l' \neq l \, (\neg label(l', V, v') \vee \neg candidate(l'))] \Rightarrow value(l) \, \}$;

this rule has no equivalent in the CSP Theory.

Axioms ECP and S together establish the correspondence between predicates "value" and "candidate". We define the set of value-candidate relationship axioms as **VCR = ECP $\cup$ S**.

BRT(CSP) also has a few technical axioms:

– **OOS (Only One Status):** "when a label is asserted as a value, it is no longer a candidate" (this rule has no equivalent in the CSP Theory):

OOS: $\forall l \, \{value(l) \Rightarrow \neg candidate(l)\}$;

– **SD (Solution Detection):** "if all the CSP variables have a unique decided value, then the problem is solved":

SD: $\forall V \, \exists! v \, \exists l \, \{[label(l, V, v) \wedge value(l)] \Rightarrow solution\text{-}found()\}$;

– **CD (Contradiction Detection):** "if there is a CSP variable with no decided value and no candidate left, then the problem has no solution":

CD: $\exists V \, \forall v \, \exists l \, \{[label(l, V, v) \wedge \neg value(l) \wedge \neg candidate(l)]$
$\Rightarrow contradiction\text{-}found()\}$.

Predicates "solution-found" and "contradiction-found" as well as rules SD and CD are not strictly necessary, but they illustrate how such situations can be written as resolution rules. They can be considered as hooks for external non-logical actions (such as displaying the solution). $\perp$ could be used instead of contradiction-found.

Finally, we define:

BRT(CSP) = {background axioms} $\cup$ ECP $\cup$ S $\cup$ {OOS, SD, CD}.

Two questions immediately come to mind. Can one solve all the instances of the CSP with only BRT(CSP)? No. How powerful is this Basic Resolution Theory? In Sudoku (with the strongest formulation including all the X_{rc}, X_{rn}, X_{cn}, X_{bn} variables), it allows to solve about 29% of the minimal puzzles. Notice that, if we considered only the X_{rc} variables, no (as of today known) minimal puzzle could be solved.

4.4. Formalising the general concept of a Resolution Theory of a CSP

Let us now state our final formal definitions. Given a CSP:

– a formula in the language of the CSP Basic Resolution Theory defined above, BRT(CSP), is said to be in the ***restricted condition-action form*** if it is written as $A \Rightarrow B$, possibly surrounded with universal quantifiers, where formula A does not contain the "$\Rightarrow$" sign and formula B is either value(z) or ¬candidate(z) for some variable z of sort Label, called the target of the rule, that already appears in the condition part (one can act only on what has been previously identified);

– a ***resolution rule*** is a formula written *in the restricted condition-action form*, with no constant symbols other than those already present in the constraint axioms of T(CSP), if any, and ***provable in the intuitionistic theory T(CSP)*** $\cup$ ***{ECP, S}***, i.e. the union of the CSP Theory (now considered as an intuitionistic theory) and the axioms on the value-candidate relationship;

– a resolution rule is ***instantiated*** in some resolution state RS when a value has been assigned to each of its variables in such a way that RS satisfies all the conditions of this rule; the rule can thus be applied; after its action part has been applied, another resolution state is reached in which its conclusion is valid;

– the condition part of a resolution rule is composed of two subparts: the pattern-conditions and the target-conditions;

– the ***pattern-conditions*** describe (in terms of labels, of well defined links between some of these labels and of value and candidate predicates for these labels) a factual situation that may occur in a resolution state (some of these conditions may depend on the target z);

– the ***target-conditions*** bear on label variable z; they always include the actual presence of this candidate in the resolution state (one cannot assert or eliminate something that is not present as a candidate; said otherwise, it is absurd to assert something that has already been proven to be impossible and it is useless to negate something that has already been negated); expressed in terms of its links with other labels mentioned in the pattern, they specify the conditions under which, in the action part of the rule, this candidate can be negated or asserted as a value;

– a ***Resolution Theory*** for a CSP is a *specialisation of its Basic Resolution Theory* in which all the additional axioms are *resolution rules*; it must be understood as a theory in intuitionistic logic.

In order to be concretely used to solve some instance of a CSP, a Resolution Theory must be completed with the same instance axioms as the corresponding

T(CSP) theory (see section 3.5). Nothing guarantees that a resolution theory can solve all the instances of the CSP, not even those that have a unique solution.

4.5. The confluence property of resolution theories

The confluence property is one of the most useful properties a resolution theory T can have. It justifies our principle according to which the instantiation of a rule in some resolution state RS depends on the effective presence of some candidates in RS (instead of depending only on relations between underlying labels); moreover, it allows to superimpose on T different resolution strategies.

4.5.1. Definition of the confluence property

Given a resolution theory T, consider all the strategies that can be built on it, e.g. by defining various implementations with different priorities on the rules in T. Given an instance P of the CSP and starting from the corresponding resolution state RS_P, the resolution process associated with a strategy S built on T consists of repeatedly applying resolution rules from T according to the additional conditions (e.g. the priorities) introduced by S. Considering that, at any point in the resolution process, different rules from T may be applicable (and different rules will be applied) depending on the chosen strategy S, we may obtain different resolution paths starting from RS_P when we vary S.

Definition: a CSP Resolution Theory T has the *confluence property* if, for any instance P of the CSP, any two resolution paths in T can be extended in T to meet in a common resolution state.

When a resolution theory has the confluence property, all the resolution paths starting from RS_P and associated with all the strategies built on T will lead to the same final state in **RS_P** (all explicitly inconsistent states are considered as identical; they mean contradictory constraints). If a resolution theory T does not have the confluence property, one must be careful about the order in which they apply the resolution rules (and they must try all the resolution paths if they want to find the "simplest"). But if T has this property, one may choose any resolution strategy, which makes finding a solution much easier, and one can define "simplest first" strategies if they want to find the simplest solution (see chapters 5 and 7).

Equivalent definitions:

– for any instance P of the CSP and any two resolution states RS_1 and RS_2 of P reachable from RS_P by resolution rules in T, there is a resolution state RS_3 such that RS_3 is reachable independently from both RS_1 and RS_2 by resolution rules in T;

– for any instance P of the CSP, the subset of **RS$_P$** consisting of the resolution states for P reachable by resolution rules in T, ordered by the reachability relation defined by T, is a DAG (Directed Acyclic Graph).

Consequence: if a resolution theory T has the confluence property, then for any instance P of the CSP, there is a single final state reachable by rules in T and all the resolution paths lead to this state. In particular, if T solves P, one cannot miss the solution by choosing to apply the "wrong" rule at any time.

The following property, a priori stronger than confluence, will often be useful to prove the confluence property of a resolution theory.

Definition: a CSP resolution theory T is *stable for confluence* if for any instance P of the CSP, for any resolution state RS_1 of P and for any resolution rule R in T applicable in state RS_1 for an elimination of a candidate Z, if any set Y of consistency preserving assertions and/or eliminations is done before R is applied, leading to a resolution state RS_2, and if it destroys the pattern of R (R can therefore no longer be applied to eliminate Z), then, there always exists a sequence of rules in T that will eliminate Z starting from RS_2 (if Z is still in RS_2). (Remark: in this definition, the assertions or eliminations in Y are not necessarily done by rules in T.)

It is obvious that: *if T is stable for confluence, then T has the confluence property*. A result that will be useful in Part III is the following (obvious):

Lemma 4.1: Let T_1 and T_2 be two resolution theories. If T_1 and T_2 are stable for confluence, then the union of T_1 and T_2 (considered as sets of rules) is stable for confluence (and therefore it has the confluence property).

4.5.2. The confluence property of BRT(CSP)

The following obvious case will be useful in many places, e.g. for defining T&E in section 5.5.

Theorem 4.1: The Basic Resolution Theory of any CSP is stable for confluence and it has the confluence property.

4.5.3. Resolution strategies and the strategic level

There are the resolution theories defined above and there are the many ways one can use them in practice to solve real instances of a CSP. From a strict logical standpoint, all the rules in a resolution theory are on an equal footing, which leaves no possibility for ordering them. But, when it comes to the practical exploitation of resolution theories and in particular to their implementation, e.g. in an inference engine (as in our SudoRules solver) or in any procedural algorithm, one question remains unanswered: can superimposing some ordering on the set of rules (using priorities or "saliences") prevent us from reaching a solution that the choice of another ordering might have made accessible? With resolution theories that have the confluence property such problems cannot appear and one can take advantage of this to define different resolution strategies.

Indeed, the confluence property allows to define a *strategic level* above the *logic level* (the level of the resolution rules) – which is itself above the implementation level in case the rules are implemented in a computer program.

Resolution strategies based on a resolution theory T can be defined in different ways and may correspond to different goals:

– implementation efficiency (in terms of speed, memory, …);

– giving a preference to some patterns over other ones: preference for bivalue-chains over whips, for whips over braids (see chapter 5 for the definitions);

– allowing the use of heuristics, such as focusing the attention on the elimination of some candidates (e.g. because they correspond to a bivalue variable or because they seem to be the key for further eliminations); but good heuristics are hard to define (in particular, the popular, intuitively natural heuristics consisting of focusing the attention on bivalue variables is blatantly unfit for hard Sudoku puzzles);

– finding the "simplest" resolution path and computing the rating of the instance according to some rating system; this will be the justification for the resolution strategies we shall introduce later; notice that this goal may be in strong opposition with a goal of pure efficiency.

4.6. Example: the Basic Sudoku Resolution Theory (BSRT)

After all the above general considerations, time has come to turn to the concrete Sudoku example and its Basic Resolution Theory, hereafter named BSRT. It will follow the general theory above, with the same adaptations as in ST for taking the basic sorts and their symmetries into better account.

4.6.1. Sorts, functions and predicates

As in the above general theory, the logical language of BSRT has the same sorts, functions and predicates as ST. In addition, it has predicates solution-found, contradiction-found and candidate. Indeed, as in the case of "value" in ST, we introduce a predicate **candidate** with signature (Number, Row, Column) and an auxiliary predicate **candidate'** with signature (Number, Block, Square) defined by the "change-of-coordinates axiom":

CC': $\forall n \forall b \forall s$ [candidate'[n, b, s] $\Leftrightarrow$ candidate(n, row(b,s), column(b,s))].

As can be seen from the signatures of predicates "value" and "candidate", they will be the basic support for the quasi-automatic expression of symmetry and super-symmetry in the Sudoku Theory and in all the Sudoku Resolution Theories.

4.6.2. The axioms of Basic Sudoku Resolution Theory (BSRT)

BSRT is defined *a priori* as being composed of the axioms of SGT plus CC, CC' and the following fourteen resolution rules.

The first group of four axioms expresses the mutual exclusion conditions on cells, rows, columns and blocks. They correspond to the ECP rule of the general theory (cut into four parts according to the type of constraint: rc, rn, cn or bn). These four rules, the elementary constraints propagation rules, can be considered as the direct operational transpositions of axioms ST-rc to ST-bn of ST. They can be used in practice to eliminate candidates as soon as a value is asserted. In this respect, they will be much more useful than rules such as ST-rc to ST-bn could be:

– **ECP(cell)**: unique value in a cell: if a number is effectively proven to be the value of a cell, then any other number is effectively proven to be excluded for this cell:

$$\forall r \forall c \forall n \forall n' \{value(n, r, c) \wedge n' \neq n \Rightarrow \neg candidate(n', r, c)\};$$

– **ECP(row)**: unique value in a row: if a number is effectively proven to be the value of a cell, then it is effectively proven to be excluded for other cells in this row:

$$\forall r \forall n \forall c \forall c' \{value(n, r, c) \wedge c' \neq c \Rightarrow \neg candidate(n, r, c')\};$$

– **ECP(col)**: unique value in a column: if a number is effectively proven to be the value of a cell, then it effectively proven to be excluded for other cells in this column:

$$\forall c \forall n \forall r \forall r' \{value(n, r, c) \wedge r' \neq r \Rightarrow \neg candidate(n, r', c)\};$$

– **ECP(blk)**: unique value in a block: if a number is effectively proven to be the value of a cell, then it is effectively proven to be excluded for other cells in this block:

$$\forall b \forall n \forall s \forall s' \{value'[n, b, s] \wedge s' \neq s \Rightarrow \neg candidate'[n, b, s']\}.$$

The second group of four axioms corresponds to the S rule of the general theory (again cut into four parts according to the type of constraint: rc, rn, cn or bn):

– **NS** or Naked-Single: assert a value whenever there is a unique possibility in an rc-cell:

$$\forall r \forall c \forall n \{[candidate(n, r, c) \wedge \forall n' \neq n \neg candidate(n', r, c)] \Rightarrow value(n, r, c)\};$$

– **HS(row)** or Naked-Single-in-a-row: assert a value whenever there is a unique possibility in an rn-cell:

$\forall r \forall n \forall c$ {[candidate(n, r, c) $\wedge$ $\forall c' \neq c$ ¬candidate(n, r, c')] $\Rightarrow$ value(n, r, c)};

– **HS(col)** or Naked-Single-in-a-column: assert a value whenever there is a unique possibility in a cn-cell:

$\forall c \forall n \forall r$ {[candidate(n, r, c) $\wedge$ $\forall r' \neq r$ ¬candidate(r', r, c)] $\Rightarrow$ value(n, r, c)};

– **HS(blk)** or Naked-Single-in-a-block: assert a value whenever there is a unique possibility in a bn-cell:

$\forall b \forall n \forall s$ {[candidate'[n, b, s] $\wedge$ $\forall s' \neq s$ ¬candidate'(n, b, s')] $\Rightarrow$ value'[n, b, s]};

The ninth axiom is the general axiom about uniqueness of status:

– **OOS (Only One Status)**: "when a label is asserted as a value, it is no longer a candidate":

$\forall n \forall r \forall c$ {value(n, r, c)] $\Rightarrow$ ¬candidate(n, r, c)};

The tenth axiom expresses solution detection (there could also be four axioms):

– **SD**: if every rc-cell has a value assigned, then the problem is solved:

$\forall r \forall c \exists n$ value(n, r, c) $\Rightarrow$ **solution-found**();

The last group of four axioms expresses contradiction detection (these axioms are redundant, but it is easier to have them all if we want to apply to Sudoku the general correspondence between braids and T&E in section 5.7):

– **CD-rc**: if there is an rc-cell such that all the numbers are proven to be excluded values for it, then the puzzle has no solution:

$\exists r \exists c \forall n$ [¬value(n, r, c) $\wedge$ ¬candidate(n, r, c)] $\Rightarrow$ **contradiction-found**();

– **CD-rn**: if there is an rn-cell such that all the columns are proven to be excluded values for it, then the puzzle has no solution:

$\exists r \exists n \forall c$ [¬value(n, r, c) $\wedge$ ¬candidate(n, r, c)] $\Rightarrow$ **contradiction-found**();

– **CD-cn**: if there is a cn-cell such that all the rows are proven to be excluded values for it, then the puzzle has no solution:

$\exists c \exists n \forall r$ [¬value(n, r, c) $\wedge$ ¬candidate(n, r, c)] $\Rightarrow$ **contradiction-found**();

– **CD-bn**: if there is a bn-cell such that all the squares are proven to be excluded values for it, then the puzzle has no solution:

$\exists b \exists n \forall s$ (¬value'[n, b, s] $\wedge$ ¬candidate'[n, b, s)]) $\Rightarrow$ **contradiction-found**();

Finally, we define the same sets of axioms as in the general theory (plus those associated with the existence of a double coordinate system):

ECP = {ECP(cell), ECP(row), ECP(col), ECP(blk)},

S = {NS, HS(row), HS(col), HS(blk)},

CD = {CD-rc, CD-rn, CD-cn, CD-bn},
VCR = ECP ∪ S (the value-candidate relationship axioms),
BSRT = SGT ∪ {CC, CC'} ∪ ECP ∪ S ∪ CD ∪ {OOS, SD}.

4.6.3. The axiom associated with the entries of a puzzle

As was the case for Sudoku Theory ST, with any specific puzzle P we can associate the axiom E_P defined as the finite conjunction of all the formulæ of type value(n_k, r_i, c_j) corresponding to each entry of P. Then, when added to the axioms of BSRT (or any extension of it), axiom E_P defines a Sudoku Resolution Theory for the specific puzzle P.

4.6.4. The Basic LatinSquare Resolution Theory: BLSRT

Lest us define the following sets of block-free axioms:
B(ECP) = {ECP(cell), ECP(row), ECP(col)},
B(S) = {NS, HS(row), HS(col)},
B(VCR) = B(ECP) ∪ B(S) (the block-free value-candidate relationship axioms),
BLSRT = LSGT ∪ B(ECP) ∪ B(S) ∪ {OOS, CD, SD}.

BLSRT is the Basic LatinSquare Resolution Theory: BRT(LatinSquare)

4.7. Sudoku symmetries and the three fundamental meta-theorems

Let us first extend the definition of the S_{rc}, S_{rn}, S_{cn} and S_{rcbs} transforms to predicate "candidate" and therefore to the whole language of BSRT:

F	$S_{rc}(F)$	$S_{rn}(F)$	$S_{cn}(F)$
candidate (n_i, r_j, c_k)	candidate (n_i, r_k, c_j)	candidate (n_j, r_i, c_k)	candidate (n_k, r_j, c_i)

F	$S_{rcbs}(F)$
candidate(n_i, r_j, c_k)	candidate[n_i, b_j, s_k]

We now have all the technical tools necessary for stating and proving our three fundamental meta-theorems.

4.7.1. Formal statement and proof of meta-theorem 2.1

Meta-theorem 4.1 (formal version of 2.1): if R is a resolution rule, then $S_{rc}(F)$ is a resolution rule (and it obviously has the same logical complexity as R). We shall express this as: the set of resolution rules is closed under symmetry.

Proof: If R is a resolution rule, then (by definition) R has a formal proof in ST ∪ VCR. From such a proof of R, a proof of $S_{rc}(R)$ in ST ∪ VCR can be obtained by replacing successively each step in the first proof (axioms included) by its transformation under S_{rc}. This is legitimate since:

– the set of axioms in ST ∪ VCR is invariant under S_{rc} symmetry;

– any application of a logical rule can be transposed.

The only technicality is that S_{rc} must be extended to non block-free formulæ. This is easily done by letting unchanged anything that is not of sort Row or Column.

4.7.2. Formal statement and proof of meta-theorem 2.2

Meta-theorem 4.2 (formal version of 2.2): if R is a block-free resolution rule, then $S_{rn}(R)$ and $S_{cn}(R)$ are resolution rules (and they obviously have the same logical complexity as R). We shall express this as: the set of resolution rules is closed under supersymmetry.

Proof: the proof (for S_{rn}) is similar to that of meta-theorem 4.1. By definition, R has a formal proof in ST ∪ VCR. Let T be the block-free theory consisting of the axioms in B(ST ∪ VCR) = B(ST) ∪ B(VCR) = LST ∪ B(VCR). Following the same lines as in the proof of theorem 3.2, there is a (second) proof of R, this time in LST ∪ B(VCR). From such a proof, a proof of $S_{rn}(R)$ in LST ∪ B(VCR) can be obtained by replacing successively each step in the second proof (axioms included) by its transformation under S_{rn}. This will also be a proof of $S_{rn}(R)$ in ST ∪ VCR.

4.7.3. Formal statement and proof of meta-theorem 2.3

Formally stating and proving meta-theorem 2.3 is done along the same lines as we did for meta-theorems 2.1 and 2.2.

Meta-theorem 4.3 (formal version of 2.3): if a block-free resolution rule R can be proved without using axiom ST-cn, then $S_{rcbs}(R)$ is a resolution rule (and it obviously has the same logical complexity as R). We shall express this as: the set of resolution rules is closed under analogy.

Proof: after the proof of theorem 4.2, there is a proof of R in LST ∪ B(VCR). This is not enough for our purpose, but the proof of theorem 4.2 can be transposed to show that there is a proof of R in LST ∪ B(VCR) that does not use axiom ST-cn (the transposition done in the proof of theorem 4.2 does not introduce axiom ST-cn if it was not used in the first proof); it is therefore a proof of R using only the axioms in the set {ST-rc, ST-rn, ST-C} ∪ B(VCR). From this proof of R, a proof of $S_{rcbs}(R)$ using only the axioms in the set {ST-rc, ST-bn, ST-C} ∪ B(VCR) is obtained by replacing each step in the first proof by its transformation under S_{rcbs}.

4.7.4. Symmetries, analogies and supersymmetries in BSRT

The above theorems are illustrated in Figure 4.2 with the various relationships existing between Singles. Similar figures could be drawn for ECP or CD rules.

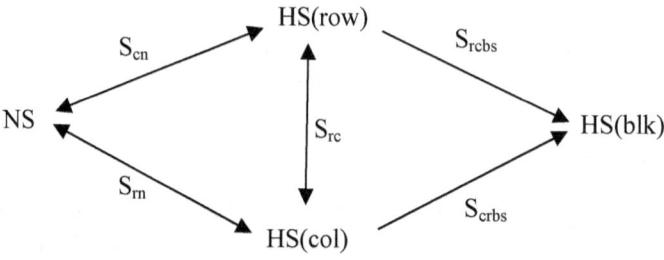

Figure 4.2. Symmetries, analogies and supersymmetries for Singles

4.7.5. Extension of meta-theorem 4.2

Finally, meta-theorem 4.2 can be modified and extended to a wider class of resolution rules by defining the notion of a block-positive formula. For an easier formulation, let us consider formulæ written without the logical symbol for implication ("⇒"), i.e. written with only the following logical symbols: ∧, ∨, ¬, ∀, ∃. Remember that the condition part of any resolution rule satisfies this restriction.

Definitions: A formula F is *block-positive* if it does not contain the logical symbol for implication ("⇒") and if any of its non block-free primary predicates is in the scope of an even number of negations (i.e. of "¬" symbols). A resolution rule A⇒B is said to be block-positive if B is block-free and A is block-positive.

Theorem 4.4: if F is a block-positive formula, then the validity of BF(F) entails the validity of F; in particular, if R is a block-positive resolution rule, then BF(R) is a resolution rule.

The proof of the first part is obvious. Notice that BF(R) is weaker than R, since it has stronger conditions; it might therefore be considered as totally uninteresting. But BF(R) is block-free and it can be submitted to meta-theorem 4.3. This is the way how, when we dealt with chains in *HLS1*, counterparts of all the chain rules in natural rc-space could be defined in rn- and cn-spaces, leading to entirely new types of chains (hidden xy-chains, hidden xyzt-chains, …).

Meta-theorem 4.5 (formal, extended version of 4.2): if R is a block-positive resolution rule, then $S_{rn} \bullet BF(R)$ and $S_{cn} \bullet BF(R)$ are resolution rules.

Part Two

GENERAL CHAIN RULES

5. Bivalue-chains, whips and braids

Now that our logical framework is completely set, this chapter – the central one of this book as for the types of resolution rules we shall meet – introduces very general types of chain patterns (of increasing complexity) giving rise to resolution rules for any CSP: bivalue chains and whips (together with a few intermediate cases). Braids, a pattern more general than chains, are also defined. We review a few properties of these patterns and of resolution theories based on them. In chapter 6, we shall show in detail that whips are very powerful, at least in the Sudoku case.

In this chapter, we give only examples related to the subsumption relationships between the whip and braid resolution theories. In the Sudoku case, many specialisations of the patterns introduced here (such as 2D chains and hidden chains) and many more examples can be found in *HLS*. In order not to overload the main text with long resolution paths, these are all grouped in the final section.

We now introduce the basic definitions needed for all the rules of this chapter.

Definition: a *chain* is a finite *sequence* of candidates (it is thus linearly ordered) such that any two consecutive candidates in the sequence are *linked* (we call this the "continuity condition" of chains), which implies that they are different.

Remarks:

– non consecutive candidates are not *a priori* forbidden to be identical, so that a chain may contain inner loops; for some specific types of chains, one can discard such loops as being "unproductive", an idea that will be explained in section 5.9;

– in case we need to specify the length of a chain, we shall speak of a chain[3], a chain[4], a chain[5]…, according to half the number of candidates it contains; if the number of candidates is odd, we round to the integer above (these conventions will be justified later);

– sequentiality (or linearity) and continuity are the two characteristic properties of all our types of chains; but chains must satisfy additional conditions in order to be usable for eliminations, such as given by the following definition.

Definition: *a regular sequence of length n associated with a sequence (V_1, ... V_n) of CSP variables* is a sequence of 2n or 2n-1 candidates (L_1, R_1, L_2, R_2, ... L_n, [R_n]) such that:

– any two consecutive candidates in the sequence are different;

– L_n [and R_n if present in the sequence] is a label for V_n: $<V_n, l_n>$ [and $<V_n, r_n>$];

– for any $1 \leq k < n$, both L_k and R_k are labels for V_k: $<V_k, l_k>$ and $<V_k, r_k>$; this condition, which we call "the strong L_k to R_k continuity condition", implies the "L_k to R_k continuity condition" of chains, i.e. that, for any $1 \leq k < n$, L_k and R_k are linked. The L_k's are called the *left-linking candidates* and the R_k's the *right-linking candidates*.

Definition: A *regular chain* is a regular sequence that satisfies all the R_{k-1} to L_k continuity conditions of chains (i.e. L_k is linked to R_{k-1} for all k).

5.1 Bivalue chains

Bivalue chains are the most basic chains that can be defined for any CSP.

Definition: a CSP variable V is said to be *bivalue* in a resolution state RS if it has exactly two candidates in RS. This could be formally defined by the auxiliary predicate "bivalue", with signature (CSP-Variable):

bivalue(V) $\equiv \exists \neq (v_1, v_2)\ \exists \neq (l_1, l_2)$ { label(l_1, V, v_1) $\wedge$ candidate(l_1)
$\wedge$ label(l_2, V, v_2) $\wedge$ candidate(l_2)
$\wedge\ \forall v \neq (v_1, v_2)\ \forall l\ \neg$[label(l, V, v) $\wedge$ candidate(l)]}.

Definition: in any CSP, a *bivalue-chain of length n* ($n \geq 1$) is a *regular chain* (L_1, R_1, L_2, R_2, ... L_n, R_n) associated with CSP variables (V_1, ... V_n) such that, for any $1 \leq k \leq n$, V_k is bivalue (L_k and R_k are thus the only two candidates for V_k).

Definition: a target of a bivalue chain is a candidate Z that does not belong to the chain and that is linked to both its endpoints (L_1 and R_n). Notice that these conditions imply that Z is a label for none of the CSP variables V_k.

Theorem 5.1 (bivalue-chain rule for a general CSP): in any resolution state of any CSP, if Z is a target of a bivalue-chain, then it can be eliminated (formally, this rule concludes ¬candidate(Z)).

Proof: the proof is short and obvious but it will be the basis for the proofs of all our forthcoming chain, whip and braid rules.

If Z was True, then L_1 would be eliminated by ECP; therefore R_1 would be asserted by S; but then L_2 would be eliminated by ECP and R_2 would be asserted by S... ; finally R_n would be asserted by S; which would contradict the hypothesis that Z was True. Therefore Z can only be False. qed.

Notation: a bivalue-chain of length n, together with a potential target elimination, is written symbolically as:

bivalue-chain[n]: {$L_1\ R_1$} – {$L_2\ R_2$} – ...... – {$L_n\ R_n$} $\Rightarrow$ ¬candidate(Z),

where the curly braces recall that the two candidates inside have representatives with the same CSP variable.

Re-writing the candidates as <variable, value> pairs and "factoring" the CSP variables out of the pairs, a bivalue chain will also be written symbolically in either of the more explicit forms:

bivalue-chain[n]: $V_1\{l_1 \; r_1\} - V_2\{l_2 \; r_2\} - ...... - V_n\{l_n \; r_n\} \Rightarrow \neg candidate(Z)$, or:

bivalue-chain[n]: $V_1\{l_1 \; r_1\} - V_2\{l_2 \; r_2\} - ...... - V_n\{l_n \; r_n\} \Rightarrow V_Z \neq v_Z$.

5.2 z-chains, t-whips and zt-whips (or whips)

5.2.1 Definitions

The definition of a bivalue-chain can be extended in different ways (z-extension, t-extension and zt-extension), as follows. We first introduced the following generalisations of bivalue-chains in *HLS*, in the Sudoku context. But everything works similarly for any CSP (see [Berthier 2008b]). It is convenient to start with the

Definition: a label C is *compatible* with a set S of labels if it does not belong to S and it is not linked to any element of S. This is a structural property, independent of any resolution state. We say that a candidate is compatible with a set S of candidates if its underlying label is compatible with all the underlying labels of elements of S.

Definition: given a candidate Z (which will be the target), a *z-chain* of length n ($n \geq 1$) built on Z is a *regular chain* (L_1, R_1, L_2, R_2, ... L_n, R_n) associated with a sequence (V_1, ... V_n) of CSP variables, such that:

– Z does not belong to $\{L_1, R_1, L_2, R_2, ... L_n, R_n\}$;

– L_1 is linked to Z;

– for any $1 \leq k < n$, R_k is the only candidate for V_k compatible with Z apart possibly for L_k;

– Z is not a label for V_n;

– L_n is the only candidate for V_n possibly compatible with Z (but V_n has more than one candidate – this is a non-degeneracy condition); in particular R_n is linked to Z.

Theorem 5.2 (z-chain rule for a general CSP [Berthier 2008b]): in any resolution state of any CSP, any target of a z-chain can be eliminated (formally, this rule concludes $\neg candidate(Z)$).

For the following "t-extension" of bivalue chains, it is natural to introduce *whips* instead of chains. Whips are also more general, because they are able to catch more contradictions than chains. A *target of a whip* is required to be linked to its first candidate, not necessarily to its last; the condition on the last variable is changed so that the final contradiction can occur with previous right-linking candidates.

Definition: given a candidate Z (which will be the target), a *t-whip* of length n ($n \geq 1$) built on Z is a *regular chain* (L_1, R_1, L_2, R_2, ... L_n) [notice there is no R_n] associated with a sequence (V_1, ... V_n) of CSP variables, such that:

– Z does not belong to {L_1, R_1, L_2, R_2, ... L_n};

– L_1 is linked to Z;

– for any $1 \leq k < n$, R_k is the only candidate for V_k compatible with all the previous right-linking candidates (i.e. with all the R_i for $i \leq k$), (in t-whips, compatibility with Z does not have to be checked for intermediate candidates; it allows to build t-whips before the target is fixed; this is a computational advantage over the zt-whips defined below, but they have a weaker resolution power);

– Z is not a label for V_n;

– V_n has no candidate compatible with Z and with all the previous right-linking candidates (but V_n has more than one candidate – this is a non-degeneracy condition).

Definition: given a candidate Z (which will be the target), a *zt-whip* (in short a *whip*) of length n ($n \geq 1$) built on Z is a *regular chain* (L_1, R_1, L_2, R_2, ... L_n) [notice there is no R_n] associated with a sequence (V_1, ... V_n) of CSP variables, such that:

– Z does not belong to {L_1, R_1, L_2, R_2, ... L_n};

– L_1 is linked to Z;

– for any $1 \leq k < n$, R_k is the only candidate for V_k compatible with Z and with all the previous right-linking candidates (i.e. with Z and with all the R_i, $1 \leq i < k$);

– Z is not a label for V_n;

– V_n has no candidate compatible with Z and with all the previous right-linking candidates (but V_n has more than one candidate – this is a non-degeneracy condition).

Theorem 5.3 (t- and zt-whip rules for a general CSP [Berthier 2008b]): in any resolution state of any CSP, if Z is a target of a t- or a zt- whip, then it can be eliminated (formally, this rule concludes ¬candidate(Z)).

Proof: the proof is a simple adaptation of that for bivalue-chains. If Z was True, then all the z- candidates would be eliminated by ECP and, progressively: all the left-linking candidates and all the t- candidates would be eliminated by ECP and all the right-linking ones would be asserted by S. The end is slightly different: the last condition on the whip entails that there would be no possible value for the last variable V_n (because it is not a CSP-Variable for Z), a contradiction.

Although these new chains or whips seem to be straightforward generalisations of bivalue-chains, their solving potential is much higher. In chapter 6, we shall give detailed statistics illustrating this in the Sudoku case.

Notation:

– a z-chain is written symbolically in the same two ways as a bivalue chain, but with prefix "z-chain" instead of "bivalue-chain";

– (a t-whip or) a whip of length n, together with a potential target elimination, is written symbolically as:

(t-)whip[n]: $\{L_1\ R_1\} - \{L_2\ R_2\} - \ldots\ldots - \{L_n\ .\} \Rightarrow \neg candidate(Z)$,

where the curly brackets recall that the two candidates inside them are relative to the same CSP variable; the dot inside the last curly brackets means the absence of a compatible candidate; as in the bivalue chains case, and for the same reasons, we shall also write whips in the form:

whip[n]: $V_1\{l_1\ r_1\} - V_2\{l_2\ r_2\} - \ldots\ldots - V_n\{l_n\ .\} \Rightarrow \neg candidate(Z)$, or:
whip[n]: $V_1\{l_1\ r_1\} - V_2\{l_2\ r_2\} - \ldots\ldots - V_n\{l_n\ .\} \Rightarrow V_Z \neq v_Z$.

Remarks:

– an alternative equivalent definition of a whip is available in section 11.1;

– as a consequence of the definition, Z is a label for none of the CSP variables in the whip;

– another consequence is that all the CSP variables of the whip are different;

– particular attention should be given to the whip[1] case. A given CSP may have whips of length 1 or not (without prejudice for longer ones – see section 5.11.7): Sudoku, n-Queens and n-SudoQueens have, LatinSquare does not. Having whips of length one has many consequences for the resolution theories of the CSP. Chapter 7 will be entirely dedicated to CSPs having such whips.

– very instructive whip[2] examples can be found in sections 8.7.1.1 and 8.8.1, where it is shown that they cannot be considered as S_2-subsets.

5.2.2. Formal definitions

As a mere exercise in logic, let us write the formal definitions of whips. We leave it as an exercise for the reader to write similar formulæ for all the other kinds of chains.

Recalling that CSP-Variable is a sub-sort of Constraint, let us first introduce two auxiliary predicates Linked-by and Linked (with capital "L"), with respective signatures (Label, Label, Constraint) and (Label, Label):

$Linked\text{-}by(l_1, l_2, c) \equiv linked\text{-}by(l_1, l_2, c) \wedge candidate(l_1) \wedge candidate(l_2)$
$Linked(l_1, l_2) \equiv linked(l_1, l_2) \wedge candidate(l_1) \wedge candidate(l_2)$
$\qquad\qquad \equiv \exists c\ Linked\text{-}by(l_1, l_2, c)$

We can now define a whip of length 1 based on target z by predicate whip[1] with signature (Label, Label, CSP-Variable):
$\text{whip[1]}(z, l_1, V_1) \equiv \text{Linked}(z, l_1) \wedge \forall l \neg [\text{Linked-by}(l_1, l, V_1) \wedge \neg \text{linked}(z, l)].$

The second condition above says that any candidate l linked to l_1 by CSP variable V_1 (i.e. any candidate value $l \neq l_1$ for CSP variable V_1) must be linked to z (by any constraint); notice that this condition could not be written as
$\forall l [\neg \text{Linked-by}(l_1, l, V_1) \vee \text{linked}(z, l)]$, because negating Linked-by(l_1, l, V_1) may negate a condition which is not the one we want, e.g. that l is a candidate.

For longer whips, we need auxiliary predicates whip[n] and partial-whip[n] with signatures (Label, [Label, Label, CSP-Variable]$^{n-1 \text{ times}}$, Label, CSP-Variable) and (Label, [Label, Label, CSP-Variable]$^{n \text{ times}}$) respectively; we define partial-whips and whips by simultaneous induction on n:
$\text{partial-whip[1]}(z, l_1, r_1, V_1) \equiv$
$\quad \text{Linked}(z, l_1) \wedge \text{Linked-by}(l_1, r_1, V_1) \wedge r_1 \neq z$
$\quad \wedge \neg \text{whip[1]}(z, l_1, V_1)$
$\quad \wedge \forall l \neq r_1 \neg [\text{Linked-by}(l_1, l, V_1) \wedge \neg \text{linked}(z, l)];$

$\text{whip[n+1]}(z, l_1, r_1, V_1, l_2, r_2, V_2, ..., l_n, r_n, V_n, l_{n+1}, V_{n+1}) \equiv$
$\quad \text{partial-whip[n]}(z, l_1, r_1, V_1, l_2, r_2, V_2, ..., l_n, r_n, V_n) \wedge$
$\quad \wedge \text{Linked}(r_n, l_{n+1}) \wedge l_{n+1} \neq z \wedge \text{Linked-by}(l_{n+1}, r_{n+1}, V_{n+1}) \wedge r_{n+1} \neq z$
$\quad \wedge \forall l \neg [\text{Linked-by}(l_{n+1}, l, V_{n+1})$
$\quad\quad\quad \wedge \neg \text{linked}(z, l) \wedge \neg \text{linked}(z, r_1) \wedge \neg ... \wedge \neg \text{linked}(z, r_n)];$

$\text{partial-whip[n+1]}(z, l_1, r_1, V_1, l_2, r_2, V_2, ..., l_n, r_n, V_n, l_{n+1}, r_{n+1}, V_{n+1}) \equiv$
$\quad \text{partial-whip[n]}(z, l_1, r_1, V_1, l_2, r_2, V_2, ..., l_n, r_n, V_n)$
$\quad \wedge \text{Linked}(r_n, l_{n+1}) \wedge l_{n+1} \neq z \wedge r_{n+1} \neq z$
$\quad \wedge \neg \text{whip[n+1]}(z, l_1, r_1, V_1, l_2, r_2, V_2, ..., l_n, r_n, V_n, l_{n+1}, V_{n+1})$
$\quad \wedge \forall l \neq r_{n+1} \neg [\text{Linked-by}(l_{n+1}, l, V_{n+1})$
$\quad\quad\quad \wedge \neg \text{linked}(z, l) \wedge \neg \text{linked}(z, r_1) \wedge \neg ... \wedge \neg \text{linked}(z, r_n)].$

Notice that, in these definitions, a whip is minimal, i.e. no initial segment is a shorter whip, due to the condition "$\neg \text{whip[n+1]}(z, ...$" in partial-whip[n+1].

5.3 Braids

We now introduce braids, a further generalisation of whips. Whereas whips have a sequential *and* continuous structure (a chain structure), braids still have a sequential structure but it is discontinuous (in restricted ways). In any CSP, braids are interesting for three reasons:

– they have an *a priori* greater solving potential than whips (at the cost of a more complex logical structure and *a priori* higher computational complexity);

– resolution theories based on them can be proven to have the very important confluence property, allowing to superimpose on them various resolution strategies (see section 5.5);

– their scope can be defined very precisely by a simple procedure: they can eliminate any candidate that can be eliminated by pure Trial-and-Error (T&E); they can therefore solve any puzzle that can be solved by T&E (and conversely – see section 5.6).

Definition: given a candidate Z (which will be the target), a *zt-braid* (in short a *braid*) of length n (n ≥ 1) built on Z is a *regular sequence* (L_1, R_1, L_2, R_2, ... L_n) [notice that there is no R_n] associated with a sequence (V_1, ... V_n) of CSP variables, such that:

– Z does not belong to {L_1, R_1, L_2, R_2, ... L_n};

– L_1 is linked to Z;

– for any 1 < k ≤ n, L_k is linked either to a previous right-linking candidate (some R_i, i < k) or to the target; this is the only (but major) structural difference with whips (for which the only linking possibility is R_{k-1}); the R_{k-1} to L_k continuity condition of chains is not satisfied by braids (a braid is defined as a regular *sequence*, a whip as a regular *chain*);

– for any 1 ≤ k < n, R_k is the only candidate for V_k compatible with Z and with all the previous right-linking candidates (i.e. with Z and with all the R_i, 1 ≤ i < k);

– Z is not a label for V_n;

– V_n has no candidate compatible with the target and with all the previous right-linking candidates (but V_n has more than one candidate – this is a non-degeneracy condition).

Remarks:

– an alternative equivalent definition is available in section 11.1;

– as in the case of whips, the t- and z- candidates are not considered as being part of the braid;

– in order to show the kind of restriction this definition implies on the nettish structure of a braid, the first of the following two structures can be part of a braid starting with{L_1 R_1} – {L_2 R_2} –... , whereas the second cannot:
{L_1 R_1} – {L_2 R_2 A_2} – ... where A_2 is linked to R_1 (or to Z);
{L_1 R_1 A_1} – {L_2 R_2 A_2} – ... where A_1 is linked to R_2 and A_2 is linked to R_1 but none of them is linked to Z. The only thing that could be concluded from this pattern if Z was True is (R_1 ∧ R_2) ∨ (A_1 ∧ A_2), whereas a braid should allow to conclude R_1 ∧ R_2.

The proof of the following theorem is almost the same as for whips, because the condition replacing R_{k-1} to L_k continuity still allows the elimination of L_k by ECP.

Theorem 5.4 (braid rule for a general CSP [Berthier 2008b]): in any resolution state of any CSP, if Z is a target of a braid, then it can be eliminated (formally, this rule concludes ¬candidate(Z)).

Notation: a braid is written symbolically in exactly the same ways as a whip, with prefix "braid" instead of "whip", but the "–" symbol must be interpreted differently:

braid[n]: $\{L_1\ R_1\} - \{L_2\ R_2\} - \ldots\ldots - \{L_n\ .\} \Rightarrow \neg\text{candidate}(Z)$, or
braid[n]: $V_1\{l_1\ r_1\} - V_2\{l_2\ r_2\} - \ldots\ldots - V_n\{l_n\ .\} \Rightarrow \neg\text{candidate}(Z)$, or:
braid[n]: $V_1\{l_1\ r_1\} - V_2\{l_2\ r_2\} - \ldots\ldots - V_n\{l_n\ .\} \Rightarrow V_Z \neq v_Z$.

Notice the double role played by the prefix in all of the above defined notations:

– it indicates how the curly brackets must be understood (pure bivalue or bivalue "modulo" the previous right-linking candidates and/or the target);

– it also indicates how the link symbol "–" must be understood.

The prefix of each resolution rule applied to solve any instance of the CSP should therefore always appear explicitly in any resolution path.

Definition: in any of the above defined chains, whips or braids, a candidate other than L_k for any of the CSP variables V_k is called a t- [respectively a z-] candidate if it is incompatible with a previous right-linking candidate [resp. with the target]. Notice that a candidate can be z- and t- at the same time and that the t- and z-candidates are not considered as being part of the pattern.

5.4. Whip and braid resolution theories; the W and B ratings

5.4.1. Whip resolution theories in a general CSP; the W rating

We are now in a position to define an increasing sequence of resolution theories based on whips. Recall that BRT(CSP) is the Basic Resolution Theory of the CSP, as defined in section 4.3.

Definition: for any $n \geq 0$, let W_n be the following resolution theory:

– $W_0 = \text{BRT(CSP)}$,

– $W_1 = W_0 \cup \{\text{rules for whips of length 1}\}$,

– ....

– $W_n = W_{n-1} \cup \{\text{rules for whips of length n}\}$,

– $W_\infty = \cup_{n \geq 0} W_n$.

Definition : the **W-rating** of an instance P of the CSP, noted W(P), is the smallest $n \leq \infty$ such that P can be solved within W_n. An instance P has B rating n if it can be solved using only whips of length no more than n but it cannot be solved using only whips of length strictly smaller than n. By convention, $W(P) = \infty$ means that P cannot be solved by whips.

The W rating has some good properties one can expect of a rating:

– it is defined in a purely logical way, independent of any implementation; the W rating of an instance is an intrinsic property;

– in the Sudoku case, it is invariant under symmetry and supersymmetry ; similar symmetry properties will be true for any CSP, if it has symmetries of any kind and they are properly formalised ;

– in the Sudoku case, it is well correlated with other measures of complexity.

5.4.2. Braid resolution theories in a general CSP; the B rating

One can define a similar increasing sequence of resolution theories, now based on braids.

Definition: for any $n \geq 0$, let B_n be the following resolution theory:

– $B_0 = BRT(CSP) = W_0$,

– $B_1 = B_0 \cup \{\text{rules for braids of length 1}\} = W_1$,

– $B_2 = B_1 \cup \{\text{rules for braids of length 2}\}$,

– ....

– $B_n = B_{n-1} \cup \{\text{rules for braids of length n}\}$,

– $B_\infty = \cup_{n \geq 0} B_n$.

Definition : the **B-rating** of an instance P of the CSP, noted B(P), is the smallest $n \leq \infty$ such that P can be solved within B_n. By convention, $B(P) = \infty$ means that P cannot be solved by braids.

The B rating has all the good properties one can expect of a rating:

– it is defined in a purely logical way, independent of any implementation; the B rating of an instance is an intrinsic property;

– as will be shown in the next section, it is based on an increasing sequence (B_n, $n \geq 0$) of resolution theories with the confluence property; this insures *a priori* better computational properties; in particular, one can define a "simplest first" resolution strategy able to find the B rating after following a single resolution path;

– in the Sudoku case, it is invariant under symmetry and supersymmetry ; similar symmetry properties will be true for any CSP;

– in the Sudoku case, it is well correlated with other measures of complexity.

5.4.3. Comparison of whip and braid resolution theories (and ratings)

Notice first that both the W and B ratings are measures of the hardest step in the simplest resolution paths, they do not take into account the whole path. An instance P with $W(P) = 12$ having a single step with such a long whip may be simpler (in some different, intuitive sense) than an instance Q with $W(Q) = 11$ but that has many steps with whips of length 11.

As a whip is a particular case of a braid, one has $W_n \subseteq B_n$ *and* $B(P) \leq W(P)$, *for any CSP, any instance P and any* $n \geq 1$. Moreover, as braids have a much more complex structure than whips, one may expect that the two ratings are very different in general. However, in the Sudoku case, it will be shown in chapter 6 that (although whip theories do not have the confluence property, they are not far from having it and) the W rating, when it is finite, is an excellent approximation of the B rating (fairly good approximations of W are easier to compute than the real value of B).

One always has $W_n \subseteq B_n$, but the converse is not true in general, except for $B_1 = W_1$ (obviously) and $B_2 = W_2$ (proof below): braids are a true generalisation of whips. Firstly, there are puzzles with $W(P) = 5$ and $B(P) = 4$ (example in section 5.10.1). Secondly, even in the Sudoku case (for which whips solve almost any puzzle), examples can be given (see one in section 5.10.2) of puzzles that can be solved with braids but not with whips, i.e. W_∞ is strictly included in B_∞. The case $n = 3$ remains open for the general CSP, but we have no example in Sudoku with $B(P) = 3$ and $W(P) > 3$. In section 7.4.2, we shall show that, for any CSP, one has $B_3 \subseteq gW_3$, where gW_n is the resolution theory for g-whips of length $\leq n$.

Theorem 5.5: In any CSP, any elimination done by a braid of length 2 can be done by a whip of same or shorter length; as a result, $B_2 = W_2$.

Proof: Let $B = V_1\{l_1\ r_1\} - V_2\{l_2\ .\} \Rightarrow V_z \neq v_z$ be a braid[2] with target $Z = <V_z, r_z>$ in some resolution state RS.

If variable V_2 has a candidate $<V_2, v'>$ (it may be $<V_2, l_2>$) such that $<V_2, v'>$ is linked to $<V_1, r_1>$, then $V_1\{l_1\ r_1\} - V_2\{v'\ .\} \Rightarrow V_z \neq v_z$ is a whip[2] with target Z. Otherwise, $<V_2, l_2>$ is linked to $<V_z, v_z>$ and $V_2\{l_2\ .\} \Rightarrow V_z \neq v_z$ is a shorter whip[1] with target Z.

5.5. Confluence of the B_n resolution theories; resolution strategies

We now consider the braid resolution theories B_n defined in section 5.4.2 and we prove that they have the confluence property. As a result, we can define a "simplest first strategy" allowing more efficient ways of computing the B rating of instances.

5.5.1. The confluence property of braid resolution theories

Theorem 5.6 [Berthier 2008b]: each of the B_n resolution theories, $0 \leq n \leq \infty$, is stable for confluence; therefore it has the confluence property.

Before proving this theorem, we must recall a convention about candidates. When one is asserted, its status changes: it becomes a value and it is "eliminated" (i.e. negated) as a candidate (axiom OOS). (This convention is very important for minimising the number of useless patterns, but the theorem does not really depend on it; the proof would only have to be slightly modified with other conventions.)

Let $n<\infty$ be fixed (the case $n=\infty$ is a corollary to all the cases $n<\infty$). We must show that, if an elimination of a candidate Z could have been done in a resolution state RS_1 by a braid B of length $m \leq n$ and with target Z, it will always still be possible, starting from any further state RS_2 obtained from RS_1 by consistency preserving assertions and eliminations, if we use a sequence of rules from B_n. Let B be: $\{L_1 R_1\} - \{L_2 R_2\} - .... - \{L_p R_p\} - \{L_{p+1} R_{p+1}\} - ... - \{L_m .\}$, with target Z.

Consider first the state RS_3 obtained from RS_2 by applying repeatedly the rules in BRT until quiescence. As BRT has the confluence property (theorem 4.1), this state is uniquely defined.

If target Z has been eliminated in RS_3, there remains nothing to prove. If target Z has been asserted, then the instance of the CSP is contradictory; if not yet detected in RS_3, this contradiction can be detected by CD in a state posterior to RS_3, reached by a series of applications of rules from BRT, following the braid structure of B.

Otherwise, we must consider all the elementary events related to B that can have happened between RS_1 and RS_3 (all the possibilities are marked by a letter for reference in further proofs). For this, we start from B' = what remains of B in RS_3. At this point, B' may not be a braid in RS_3. We repeat the following procedure, for $p = 1$ to $p = m$, producing in the end a new (possibly shorter) braid B' in RS_3 with target Z. All the references below are to the current B'.

a) If, in RS_3, the left-linking or any t- or z- candidate of CSP variable V_p has been asserted, then Z and/or the previous R_k('s) to which L_p is linked must have been eliminated by ECP in the passage from RS_2 to RS_3 (if it was not yet eliminated in RS_2); if Z is among these eliminations, there remains nothing to prove; otherwise, the procedure has already been successfully terminated by case f of the first such k.

b) If, in RS_3, left-linking candidate L_p has been eliminated (but not asserted) (it can therefore no longer be used as a left-linking candidate in a braid) and if CSP variable V_p still has a z- or a t- candidate C_p, then replace L_p by C_p; now, up to C_p, B' is a partial braid in RS_3 with target Z. Notice that, even if L_p was linked to R_{p-1} (as it would if B was a whip), this may not be the case for C_p; therefore trying to prove a similar theorem for whips would fail here (see section 5.10.3 for an example of non-confluence of the W_n theories). [As it missed this point, the proof given for zt-chains in *HLS1* was not correct.]

c) If, in RS_3, any t- or z- candidate of V_p has been eliminated (but not asserted), this has not changed the basic structure of B (at stage p). Continue with the same B'.

d) If, in RS_3, right-linking candidate R_p has been asserted (p can therefore not be the last index of B'), it can no longer be used as an element of a braid, because it is no longer a candidate. Notice that all the left-linking and t- candidates for CSP variables of B after p that were incompatible in B with R_p, i.e. linked to it, if still present in RS_2, must have been eliminated by ECP somewhere between RS_2 and RS_3. But, considering the braid structure of B upwards from p, more eliminations and assertions must have been done by rules from BRT between RS_2 and RS_3.

Let q be the smallest number strictly greater than p such that, in RS_3, CSP variable V_q still has a (left-linking, t- or z-) candidate C_q that is not linked to any of the R_i for $p \leq i < q$ (by definition of a braid, C_q is therefore linked to Z or to some R_i with $i < p$). Between RS_2 and RS_3, the following rules from BRT must have been applied for each of the CSP variables V_u of B with index u increasing from p+1 to q-1 included: eliminate its left-linking candidate (L_u) by ECP, assert its right-linking candidate (R_u) by S, eliminate by ECP all the left-linking and t-candidates for CSP variables after u that were incompatible in B with the newly asserted candidate (R_u).

In RS_3, excise from B' the part related to CSP variables p to q-1 (included) and (if L_q has been eliminated in the passage from RS_1 to RS_3) replace L_q by C_q; for each integer $s \geq p$, decrease by q-p the index of CSP variable V_s and of its candidates in B'; in RS_3, B' is now, up to p (the ex q), a partial braid in B_n with target Z.

e) If, in RS_3, left-linking candidate L_p has been eliminated (but not asserted) and if CSP variable V_p has no t- or z- candidate in RS_3 (complementary to case b), then V_p has only one possible value in RS_3, namely R_p; R_p must therefore have been asserted by S somewhere between RS_1 and RS_3; this case has therefore been dealt with by case d (because the assertion of R_p also entails the elimination of L_p).

f) If, in RS_3, right-linking candidate R_p of B has been eliminated (but not asserted), in which case p cannot be the last index of B', then replace B' by its initial part: $\{L_1 \, R_1\} - \{L_2 \, R_2\} - .... - \{L_p \, .\}$. At this stage, B' is in RS_3 a shorter braid with target Z. Return B' and stop.

Notice that this proof works only because the notion of being linked does not depend on the resolution state.

Notice also that what we have proven is indeed the following: given RS_1, B and RS_2 as above, if RS_3 is the resolution state obtained from RS_2 by the repeated application of rules from BRT until quiescence, then:

– either a contradiction has been detected by CD somewhere between RS_2 and RS_3 (and, due to consistency preservation between RS_1 and RS_2, it can only be because a contradiction inherent in the givens of P has been made manifest by CD);

– or Z has been eliminated by ECP somewhere between RS_2 and RS_3;

– or Z can be eliminated in RS_3 by a braid B' possibly shorter than B, with target Z, with CSP variables a sub-sequence W' of those of B, with right-linking candidates those of B belonging to the sub-sequence W', with left-linking candidates those of B belonging to the sub-sequence W', each of them possibly replaced by a t-candidate of B for the same CSP variable.

5.5.2. Braid resolution strategies consistent with the B rating

As explained in section 4.5.3, we can take advantage of the confluence property of braid resolution theories to define a "simplest first" strategy that will always find the simplest solution, in terms of the length of the braids it will use. As a result, it will also compute the B rating of an instance after following a single resolution path. The following order satisfies this requirement:
ECP < S <
bivalue-chain[1] < z-chain[1] < t-whip[1] < whip[1] < braid[1] < ... <
bivalue-chain[k] < z-chain[k] < t-whip[k] < whip[k] < braid[k] <
bivalue-chain[k+1] < z-chain[k+1] < t-whip[k+1] < whip[k+1] < braid[k+1] < ...

Notice that bivalue-chains, z-chains, t-whips and whips being special cases of braids of same length, their explicit presence in the set of rules does not change the final result. We put them here because when we look at a resolution path, it may be nicer to see simple patterns appear instead of more complex ones (braids). Also, it shows (in the Sudoku case) that braids that are not whips do not appear so often.

The above ordering defines a "simplest first" resolution strategy. It does not completely define a deterministic procedure: it does not set any precedence between different chains of same type and length. This could be done by using an ordering of the candidates instantiating them, based e.g. on their lexicographic order. But one can also decide that, for all practical purposes, which of these equally prioritised rule instantiations should be "fired" first will be chosen randomly.

5.6. The "T&E vs braids" theorem

For braids, the following "T&E vs braids" theorem is second in importance only to the confluence property. As it is easy to program very fast implementations of the T&E procedure, it allows to check quickly if a given instance P will be solvable by braids. This may be very useful: in case the answer is negative, we may not want to waste computation time on P. In case it is positive, it does not produce an explicit resolution path with braids and, even if we build one from the trace of this procedure, it will not be one with the shortest braids and it will not provide the B rating; but the computations with braids will then be guaranteed to give a solution.

5.6.1. Definition of the Trial-and-Error procedure T&E(T, P)

Definition: given a resolution theory T with the confluence property, a resolution state RS and a candidate Z in RS, *T&E(T, Z, RS) or Trial-and-Error based on T for Z in RS*, is the following procedure (notice: a procedure, not a resolution rule):
- make a copy RS_1 of RS; in RS_1, delete Z as a candidate and assert it as a value;
- in RS_1, apply repeatedly all the rules in T until quiescence;
- if RS_1 has become a contradictory state, then delete Z from RS (*sic*: RS, not RS_1); else do nothing (in particular if a solution is obtained in RS_1, merely forget it);
- return the (possibly) modified RS state.

Notice that this definition is meaningful only if T has the confluence property: otherwise, the result of "applying repeatedly in RS_1 all the rules in T until quiescence" may not be uniquely defined.

Definition: given a resolution theory T with the confluence property and a resolution state RS, we define the *T&E(T, RS)* procedure as follows:
a) in RS, apply the rules in T until quiescence; if the resulting RS is a solution or a contradictory state, then return it and stop;
b) mark all the candidates remaining in RS as "not-tried";
c) choose some "not-tried" candidate Z, un-mark it and apply T&E(T, Z, RS);
d) if Z has been eliminated from RS by step c,
    then goto a
    else if there remains at least one "not-tried" candidate in RS
        then goto c else return RS and stop.

Definition: given a resolution theory T with the confluence property and an instance P with initial resolution state RS_P, we define *T&E(T, P)* as T&E(T, RS_P).

Notice that this procedure always stays at depth 1 (i.e. only one candidate is tested at a time) but that a candidate Z may be tried several times for T&E(T, Z, RS_i) in different resolution states RS_i. This is normal, because the result may be different if other candidates have been eliminated in the meanwhile.

We say that P can be solved by *T&E(T)*, or that P is in T&E(T), if T&E(T, P) produces a solution for P.

When T is the Basic Resolution Theory of a CSP (which is known to always have the confluence property), we simply write *T&E* instead of T&E(BRT(CSP))).

5.6.2. The "T&E vs braids" theorem

It is obvious that any elimination that can be done by a braid B can be done by T&E (by applying rules from BRT following the structure of B). The converse is more interesting:

Theorem 5.7: for any instance of any CSP, any elimination that can be done by T&E can be done by a braid. Any instance of a CSP that can be solved by T&E can be solved by braids.

Proof: Let RS be a resolution state and let Z be a candidate eliminated by T&E(BRT, Z, RS) using some auxiliary resolution state RS'. Following the steps of BRT in RS', we progressively build a braid in RS with target Z. First, remember that BRT contains three types of rules: ECP (which eliminates candidates), S (which asserts a value for a CSP variable) and CD (which detects a contradiction on a CSP variable).

Consider the first step of BRT in RS', which is an application of rule S, asserting some label R_1 as a value. As R_1 was not a value in RS, there must have been in RS' some elimination of a candidate, say L_1, for a CSP variable V_1 of which R_1 is a candidate, and the elimination of L_1 (which made the assertion of R_1 by S possible in RS') can only have been made possible in RS' by the assertion of Z. But if L_1 has been eliminated in RS', it can only be by ECP and because it is linked to Z. Then $\{L_1 \ R_1\}$ is the first pair of candidates of our braid in RS and V_1 is its first CSP variable. (Notice that there may be other z-candidates for V_1, but this is pointless, we can choose any of them as L_1 and consider the remaining ones as z-candidates).

The sequel is done by recursion. Suppose we have built a braid in RS corresponding to the part of the BRT resolution in RS' up to its k-th assertion step. Let R_{k+1} be the next candidate asserted in RS'. As R_{k+1} was not a value in RS, there must have been in RS' some elimination of a candidate, say L_{k+1}, for a CSP variable V_{k+1} of which R_{k+1} is a candidate, and the elimination of L_{k+1} (which made the assertion of R_{k+1} possible in RS') can only have been made possible in RS' by the assertion of Z and/or of some of the previous R_i. But if L_{k+1} has been eliminated in RS', it can only be by ECP and because it is linked to Z or to some of the previous R_i, say C. Then our partial braid in RS can be extended to a longer one, with $\{L_{k+1} \ R_{k+1}\}$ added to its candidates, L_{k+1} linked to C, and V_{k+1} added to its sequence of CSP variables.

End of the procedure: as Z is supposed to be eliminated by T&E(Z, RS), a contradiction must have been obtained by BRT in RS'. As, in BRT, only ECP can eliminate a candidate, a contradiction is obtained if a value asserted in RS', i.e. Z or one of the R_i, $i < n$, eliminates in RS' (via ECP) a candidate, say L_n, that was the last one for a corresponding variable V_n and that is linked to Z or one of the R_i, $i < n$. L_n

and V_n are thus the last left-linking candidate and CSP variable of the braid we were looking for in RS.

Here again (as in the proof of confluence), this proof works only because the existence of a link between two candidates does not depend on the resolution state. Finally, notice that it is very unlikely that the T&E procedure followed by the construction in this proof would produce the shortest available braid in resolution state RS (and this intuition is confirmed by experience).

5.6.3. Comments on T&E and on the "T&E vs braids" theorem

As using T&E(T) leads to examining arbitrary hypotheses for the creation of auxiliary resolution states, it could be considered as blind search. Nevertheless, as T has the confluence property, the final result of T&E(T) applied to any instance does not depend in any way on the sequence of tested candidates.

5.6.3.1. T&E versus structured search: no-guessing

Moreover, it is important for our purposes to notice that, contrary to the usual structured search algorithms [depth-first or breadth-first search, with search paths pruned by the rules in T – DFS(T) or BFS(T)], T&E(T) includes no "guessing": if a solution is obtained in an auxiliary state RS', then it is not taken into account. This notion of "guessing" is inherent to the DFS or BFS procedures. Closely related to it is the idea of a "backdoor" of an instance (see section 11.5): a minimal set of labels such that adding them as values to the instance would give a solution within T, i.e. with no search at all. But this idea is totally alien to T&E(T).

As a result of the "no guessing" and no recursion, there is a major difference between T&E(T) and general DFS(T) and BFS(T): whereas, given any instance P, the latter algorithms can always find a solution (if there is any) or prove that it has none, T&E(T) cannot: if P has multiple solutions, T&E(T) can only find what is common to all its solutions. Given the "T&E(T) vs T-braids" theorem (this theorem will be proved for many resolution theories T) and the correspondence between a solution of P in a resolution theory T' and a model of T' $\cup$ E_P, this is just the basic fact that what can be proved in a FOL theory (here T' = T-braids) is (and is only) what is true in all the models of this theory. Notice that another consequence of this basic property of FOL is that, given T, there cannot exist any resolution theory TT such that one would get a "DFS(T) vs TT" or a "BFS(T) vs TT" theorem.

5.6.3.2. Comments on the "acceptability" of braids

In the Sudoku community, T&E (which had always been the topic of heated debates, although it had never been precisely defined before *HLS*) is generally not accepted by advocates of "pattern-based" solutions. But the "T&E vs braids" theorem shows that a solution based on T&E can always be replaced by a pattern-based solution, more precisely by a solution based on braids. The question naturally arises, for any CSP: can one reject T&E and nevertheless accept solutions based on braids? There are three main reasons for a positive answer, both related to the goals one pursues.

Firstly, as shown in section 5.5, resolution theories based on braids have the confluence property and many different resolution strategies can be super-imposed on them. One may prefer a solution with the shortest braids and adopt the "simplest first" strategy defined in section 5.5. The T&E procedure cannot provide this (unless it is drastically modified, in ways that would make it computationally very inefficient).

Secondly, in each of the B_n resolution theories based on braids, one can add rules corresponding to special cases, such as whips or bivalue-chains of same length, and one can decide to give a natural preference to such special cases. This is still a "simplest first" principle. In Sudoku, this entails that non-whip braids appear very rarely in the solution of randomly generated puzzles; in a sense, this is a measure of how powerful whips are: although they are structurally much more "beautiful" and simpler (they are continuous chains with no "branching") and computationally much better than braids, they can solve almost all the puzzles that can be solved by T&E (i.e. almost all the randomly generated ones). One could say that the "T&E vs braids" theorem (together with the statistical results of chapter 6 and the subsumption results of chapter 8) is the best advertisement for whips.

Thirdly, in the Sudoku world, in spite of what some Sudoku addicts would like to believe, the reality is that most of the Sudoku players use T&E as their main and most natural resolution strategy. Trying to find a braid or a whip justifying an elimination in a simpler way than what they have first found by T&E may thus be an entertaining idea. The same remarks may be applied to any CSP.

5.7. The objective properties of chains and braids

Chains should not be confused with chain rules. A chain rule can only be valid or not valid, which depends neither on the way it has been proven nor on any of the properties defined below for the underlying chain. A non valid chain rule is merely useless. But a valid chain rule can be more a less general (giving rise to subsumption relationships) or more a less useful, easy to apply, acceptable. As (apart from the first) these are purely subjective criteria, they can only lead to confusion if they cannot be grounded in objective ones.

We have therefore devised a few, purely objective (or descriptive, or factual) properties of chains that may be relevant to estimate their usefulness, desirability or understandability. Even these objective properties can give rise to much debate when it comes to subjectively evaluating their impact on usefulness or acceptability; it all depends on which criteria of acceptability are adopted.

5.7.1. Linearity (sequentiality)

We use the words *linearity* or *sequentiality* as synonyms to mean that the candidates composing the chain are sequentially ordered; it is supposed that this order is essential in the definition of the chain (i.e. not arbitrarily super-imposed on it) and in the proof of the associated chain rule. Linearity is what makes the difference with a net: in a net, only a partial ordering of the candidates is required, while there may be branching and merging of different paths in the net. Both whips and braids are linear.

5.7.2. Continuity

Continuity supposes linearity and means that consecutive candidates are linked. In this definition, possible additional t- or z- candidates of whips and braids, which are not considered as part of the pattern, do not alleviate in any way this requirement (they are considered as inessential). Continuity is what distinguishes whips from braids: braids satisfy linearity but not continuity.

5.7.3. Homogeneity

Homogeneous means that the pattern is a sequence of similar bricks. This property is obvious from their definitions for all the chains and braids introduced in this book.

5.7.4. Reversibility

Although it had never been defined before *HLS2*, the word "reversibility" has been the pretext for the most poisonous debates on Sudoku Web forums. There is nevertheless an obvious definition, valid for any CSP:

− given a chain, the reversed chain is the chain obtained by reversing the order of the candidates; in the process, when used in the definition of some types of chains, left- [respectively right-] linking candidates become right- [resp. left-] linking candidates;

− a chain type is called *reversible* if for any chain of this type, the reversed chain is of this type.

These definitions will be extended in chapters 9 and 10 to chains with more general right-linking objects.

Theorem 5.8: bivalue-chains and z-chains are reversible.

Proof: obvious (left and right-linking candidates are interchanged).

The advantage of reversibility is that, in general, one can find other chains by "circulating along the chain" (i.e. making circular permutations of the candidates and changing accordingly the end points); these often allow other eliminations.

Notice that chains using the t-extension (and braids) are not reversible. This is a weak point for them. But the sequel will show that they satisfy properties (left-extendibility and composability) that partially palliate this weakness.

5.7.5. Non anticipativeness (or no look-ahead)

Definition: a given type of chain is called *non-anticipative* or *no look-ahead* if, when a chain of this type is built from left to right, all that needs be checked when the next candidate is added depends only on the previous candidates (and not on the potential future ones) and possibly on the target (for chains that have to be built around a target, such as whips or braids). Notice that this does not imply that adding a candidate will always allow to finally get a full chain of this type, but it guarantees that, up to the new candidate added, the chain satisfies the conditions on chains of this type whatever will be added to it later.

Comment: this seems to be a strong criterion for acceptability of chains, from both points of view of human solvers and programmers, because it is the practical condition necessary for being able to build the chain progressively from left to right, instead of having to spot it globally at once. It is a major computational property, the opposite of which is look-ahead.

Theorem 5.9: a reversible chain is non-anticipative.

Theorem 5.10: all the chains defined in this chapter, from bivalue-chains to whips and braids, are non-anticipative.

Proofs: obvious. Indeed, we had implicitly this condition in mind when we introduced the first types of chains in *HLS1*.

5.7.6. Left-extendibility and composability

Definition: a given type of chain is called *left-extendable* if, when given a partial chain of this type, candidates can be added not only to its right but also to its left (of course, respecting the linking conditions on left- and right- linking candidates for chains of this type at the junction and having the same target in case they are built around a target).

Theorem 5.11: a reversible chain is left-extendable.

Theorem 5.12: a non-anticipative chain is left-extendable.

Theorem 5.13: all the chains defined in this chapter, from bivalue-chains to whips and braids, are left-extendable.

Proof: obvious. The idea is that, when the presence of a t-candidate can be justified by previous right-linking candidates in a partial chain, it will remain justified by them if we add candidates to the left of this partial chain (and justifications of z-candidates will not be changed). This notion and the last theorem were first suggested by Mike Barker.

Definition: a given type of chain is called *composable* if, when two partial chains of this type are given, they can be combined into a single chain of this type (of course, respecting the linking conditions on left- and right- linking candidates for chains of this type at the junction and having the same target in case they are built around a target).

Theorem 5.14: all the chains defined in this chapter, from bivalue-chains to whips and braids, are composable.

The practical impact of this theorem is mainly for chains with the t-extension: when t-candidates are justified by previous right-linking candidates of a partial chain, they will still be justified by the same candidates if another partial chain of the same type is added to its left. Of course, not all the chains with the t-extension can be obtained by combining shorter chains of the same type, but looking first for combinations of such shorter sub-chains before chains with longer distance t-interactions may be a valuable strategy, different from the one described in section 5.5.2 (and it can also be combined with it in order to keep taking advantage of the confluence property).

5.7.7. Complexity

As whips or braids are much more general than bivalue-chains, the search for whips or braids in a real resolution state of a real instance of a CSP is likely to be more difficult than the search for the simplest bivalue-chains of same length. The counterpart is, the former can solve many more instances (see chapter 6).

Unfortunately, defining an objective complexity measure for the instances of a CSP is a very difficult task. Whereas worst case analysis is not very meaningful, mean case analysis is more meaningful but is very difficult in practice, as will be illustrated by the Sudoku case in chapter 6, where the W_n and B_n ratings will be shown to be reasonable measures of complexity.

5.8. About loops in bivalue- and z- chains, in whips and in braids

We say that there is a loop in a chain or braid if it has two identical candidates.

In this section, we review the usefulness of accepting various kinds of loops in the different chains or braids introduced in this chapter.

5.8.1 Global loops are useless in bivalue-chains and z-chains

Define a global loop as a chain with same first and last candidates; this is the broadest definition of a global loop one can give.

Let $\{L_1 R_1\} - \{L_2 R_2\} - \dots - \{L_n R_n\}$ be a bivalue-chain [respectively a z-chain] and suppose it has a global loop, i.e. $R_n = L_1$. And let Z be a target.

Consider the shorter bivalue-chain [resp. z-chain] obtained by excising the last pair of candidates: $\{L_1 R_1\} - \{L_2 R_2\} - \dots - \{L_{n-1} R_{n-1}\}$. As L_n is linked to $L_1 = R_n$ and to its other endpoint, R_{n-1}, this chain admits L_n as a target. L_n can therefore be eliminated by a chain of the same type as our original chain, of shorter length and with no global loop.

If our chain was a bivalue one, then the CSP variable V_n of $\{L_n R_n\}$ was bivalue and R_n can be asserted by rule S, present in BRT(CSP) and in any resolution theory; then Z will be deleted by rule ECP, also present in any resolution theory.

In case our chain was a z-chain, then either the CSP variable V_n of $\{L_n R_n\}$ was bivalue and we can apply the previous reasoning, or V_n was bivalue modulo Z. In the latter case, we can still consider the same subchain $\{L_1 R_1\} - \{L_2 R_2\} - \dots - \{L_{n-1} R_{n-1}\}$, which is of the same type as the original one, but shorter and with no global loop, and we can eliminate L_n. Let C be a z-candidate for the last variable of the original chain. Then $\{C, R_n\}$ is a z-chain (or a bivalue-chain, which is a particular case of a z-chain) of length 1 with target Z. Z can therefore be eliminated by rules of same type, of shorter length and with no global loop.

As a result, we have:

Theorem 5.15: Any elimination that could be done by a bivalue-chain or a z-chain with a global loop can be done by BRT(CSP) and by shorter chains of the same type with no global loop. Practical statement: global loops are useless in bivalue-chains and z-chains.

5.8.2. Inner loops are useless in bivalue-chains and z-chains

We say that a chain as an inner loop if it has two equal candidates, but at most one of them is an endpoint.

Let $\{L_1 R_1\} - \{L_k R_k\} - \dots - \{L_p R_p\} - \dots - \{L_n R_n\}$ be a bivalue-chain [respectively a z-chain] and suppose it has an inner loop. Let Z be a target. There are two possibilities for an inner loop.

The first possibility is the equality of two left-linking candidates or of two right-linking candidates: $L_k = L_p$ or $R_k = R_p$. Then, by excision of the inner loop, we get a shorter chain:
$$\{L_1\ R_1\} - \{L_{k-1}\ R_{k-1}\} - \{L_p\ R_p\} - ... - \{L_n\ R_n\}\ \text{or}$$
$$\{L_1\ R_1\} - \{L_k\ R_k\} - \{L_{p+1}\ R_{p+1}\} - ... - \{L_n\ R_n\}$$
of same (or simpler) type and with Z as a target.

The second possibility is the equality of a right-linking candidate with a subsequent or a previous left-linking candidate: $R_k = L_p$ or $L_k = R_p$.

In the first case, by excision of the extremities, we get a shorter chain of same (or simpler) type:
$\{L_{k+1}\ R_{k+1}\} - ... - \{L_{p-1}\ R_{p-1}\}$ with $R_k = L_p$ as a target. Once it has eliminated L_p, rules S and ECP from BRT(CSP) will assert all the right-linking candidates and eliminate all the left-linking candidates after p until a variable V_q which was not bivalue in the original chain (i.e. which had a z-candidate) is encountered. If q = n which is always the case if the original chain was bivalue, R_q is asserted by S and Z is eliminated by ECP. Otherwise, L_q is eliminated by ECP and what remains is a shorter z-chain $\{C_q\ R_q\} - ... - \{L_n\ R_n\}$, q > p, where C_q is a z-candidate of the initial chain. This chain has Z as a target and can eliminate it.

The second case can be dealt with in exactly the same way, after reversing the original chain (which reverses the role of candidates: left-linking become right-linking and conversely).

In case the original chain had several inner loops, all these reductions can be applied iteratively to as many subparts of the chain as necessary; every iteration eliminates one loop, until there remains none. Finally, we get:

Theorem 5.16: Any elimination that could be done by a bivalue-chain or a z-chain with inner loops can be done by BRT(CSP) and by shorter chains of same type with no inner loop. Practical statement: inner loops are useless in bivalue-chains and z-chains.

5.8.3. Bivalue-chains and z-chains should have no loops

As a general conclusion of all the preceding cases, we have:

Theorem 5.17: resolution rules that might be obtained from bivalue-chains or z-chains with (global or inner) loops are subsumed by BRT(CSP) together with rules for shorter chains of same type with no loops. Practical statement: bivalue-chains and z-chains should have no loops.

5.8.4. Should one allow loops in whips and braids?

In whips or braids, equality of a left-linking and a right-linking candidate is the closest notion we can have of a global loop; but such equality would produce the final whip contradiction.

As for other kinds of inner loops in whips (equality of two left-linking or of two right-linking candidates), nothing allows to eliminate them. *A priori*, one can consider that, as we go forward along a whip, we accumulate knowledge about the consequences of assuming its target, which allows more possibilities of extending it; allowing such inner loops could therefore lead to accumulate more knowledge and to find more whips.

However, although the general definition of a whip does not exclude loops, experience with the Sudoku example shows that they do not bring much more generality but they bring more computational complexity. Moreover, in any CSP, whips with loops are subsumed by braids and it may be more interesting to use braids than whips with inner loops. Therefore, we do not add any *a priori* no-loop condition in the general definition of a whip, but, unless otherwise stated, all the whips we shall consider will be loopless. In particular, the statistical results for Sudoku in chapter 6 are about loopless whips.

As for braids, the notion of an inner loop is pointless: the same left-linking or right-linking candidate can be used several times for sprouting new branches, which has the same "accumulation" result as loops, but without the useless parts that may be needed to join the endpoints of a loop; i.e. for any possible inner loop, there is always, obviously, a shorter braid without this loop.

5.9. Forcing whips and braids, a bad idea?

Consider a bivalue variable (in any resolution state), with its two possible values x_1 and x_2 corresponding to candidates Z_1 and Z_2. Suppose there are two partial whips/braids, say W_1 and W_2, one with target Z_1, the other with target Z_2. In case W_1 and W_2 share a left-linking candidate L [respectively a right-linking candidate R], L can be deleted [resp. R can be asserted]: this is reasoning by cases, which is perfectly valid in intuitionistic logic. Do we get interesting new patterns this way (to be called forcing whips / forcing braids because they force a conclusion that could not be obtained by a single whip/braid pattern)? Obviously, the answer would be negative for reversible chains: it would suffice to reverse one of the chains and to link the two by the bivalue variable in order to obtain a single chain of same type as the two given ones. Notice that in this process the chain thus obtained should be assigned length n_1+n_2+1, where the n_i are the lengths of the two chains.

But as whips and braids are not reversible, it seems one could get new patterns, more general than whips and braids. What is the resolution power of such patterns? We have no general answer. But, in the Sudoku case, if these patterns are assigned length n_1+n_2+1 and given smaller priority than whips/braids of same length, we have found no occurrence in a random sample of 1,300 puzzles. As the memory requirements for such combinations are very high, we did not try on larger samples.

There is still the possibility of starting from trivalue variables and considering three whips/braids instead of two, but the complexity increases accordingly.

5.10. Exceptional examples

As mentioned in the Introduction, all the resolution rules defined in this book have been implemented in our SudoRules solver in such a way that they could be applied to any CSP (except the S_p-whips, S_p-braids, W_p-whips and B_p-braids in Part III). Each of them can be activated or de-activated independently. Different strategies can be chosen. In the examples below, we systematically apply the "simplest first" resolution strategy defined in section 5.5.2, with whips [respectively whips and braids] activated, in order to get the W [respectively the B] rating.

The longest whip or braid of each resolution path appears in bold characters. As for the notation (the "nrc notation"), it is self explaining and consistent with the general representation for whips and braids introduced in sections 5.2 and 5.3; the only adaptations are: 1) outside curly brackets, CSP Variables X_{rc}, X_{rn}, ... are merely written as rc, rn, ...; 2) within curly brackets, the s value of an X_{bn} variable is replaced by its equivalent in rc-coordinates; the reason is better readability on the standard rc-grid. Apart from some hand editing, the following is the raw output from SudoRules. Hand made changes (in addition to those mentioned in the Introduction) have the only purpose of using less paper; they consist mainly of writing several whips in the same line (even if, because they have different targets, the justifications of their z-candidates may be different).

A general warning is in order about our Sudoku examples: because they are intended to illustrate exceptional properties, most of them are far more difficult than the vast majority of puzzles (see the classification results in chapter 6); as a result, they have exceptionally long resolution paths with exceptionally long whips/braids and they may give a very wrong idea of the much simpler typical resolution paths.

5.10.1. Proof of $B_4 \neq W_4$: an instance with $W(P) = 5 > B(P) = 4$

The example in Figure 5.1 is one of the rare (in percentage) puzzles with a B rating smaller than its W rating.

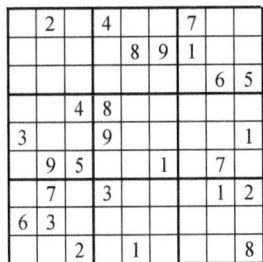

Figure 5.1. *A puzzle P with B(P) = 4 and W(P) = 5*

1) The resolution path with whips shows that W(P) = 5:

***** SudoRules version 15b.1.12-W *****
26 givens, 196 candidates and 1151 nrc-links
singles ==> r8c9 = 7, r8c3 = 1, r1c1 = 1, r4c2 = 1, r3c4 = 1, r8c6 = 8
whip[1]: c9n6{r6 .} ==> r4c7 ≠ 6, r5c7 ≠ 6, r6c7 ≠ 6
whip[1]: r1n5{c5 .} ==> r2c4 ≠ 5
whip[1]: c4n5{r9 .} ==> r8c5 ≠ 5, r7c6 ≠ 5, r7c5 ≠ 5, r9c6 ≠ 5
whip[1]: b4n6{r5c2 .} ==> r5c6 ≠ 6, r5c5 ≠ 6
whip[2]: c8n8{r1 r5} - c2n8{r5 .} ==> r1c3 ≠ 8
hidden-single-in-a-row ==> r1c8 = 8
whip[3]: r9c2{n4 n5} - r7n5{c1 c7} - b9n6{r7c7 .} ==> r9c7 ≠ 4
whip[4]: r6c1{n2 n8} - r5n8{c3 c7} - r5n2{c7 c8} - r2n2{c8 .} ==> r6c4 ≠ 2
singles ==> r6c4 = 6, r4c9 = 6, r1c9 = 9
whip[1]: r2n6{c2 .} ==> r1c3 ≠ 6
naked-single ==> r1c3 = 3
whip[1]: r2n3{c9 .} ==> r3c7 ≠ 3
whip[4]: r6n8{c7 c1} - r6n2{c1 c5} - c6n2{r4 r3} - r3c7{n2 .} ==> r6c7 ≠ 4
whip[4]: b9n3{r9c8 r9c7} - r9n6{c7 c6} - r7c6{n6 n4} - r8n4{c5 .} ==> r9c8 ≠ 4
whip[4]: r6n4{c5 c9} - c8n4{r5 r2} - r2n2{c8 c4} - b8n2{r8c4 .} ==> r8c5 ≠ 4
whip[1]: r8n4{c8 .} ==> r7c7 ≠ 4
whip[4]: b9n4{r8c8 r8c7} - r3c7{n4 n2} - r2n2{c8 c4} - r8c4{n2 .} ==> r8c8 ≠ 5
whip[4]: b8n9{r7c5 r8c5} - r8c8{n9 n4} - r8c7{n4 n5} - r7n5{c7 .} ==> r7c1 ≠ 9
whip[4]: b7n9{r9c1 r7c3} - b7n8{r7c3 r7c1} - r7n5{c1 c7} - b9n6{r7c7 .} ==> r9c7 ≠ 9
whip[4]: r9n9{c1 c8} - b9n3{r9c8 r9c7} - b9n6{r9c7 r7c7} - r7n5{c7 .} ==> r9c1 ≠ 5
;;; Resolution state RS₁
whip[5]: c2n5{r2 r9} - r9c4{n5 n7} - r2c4{n7 n2} - b3n2{r2c8 r3c7} - b3n4{r3c7 .} ==> r2c2 ≠ 4
whip[4]: c2n8{r5 r3} - c1n8{r3 r7} - b7n5{r7c1 r9c2} - c2n4{r9 .} ==> r5c3 ≠ 8
whip[2]: c3n9{r3 r7} - c3n8{r7 .} ==> r3c3 ≠ 7
whip[3]: r3c7{n2 n4} - r3c2{n4 n8} - r5n8{c2 .} ==> r5c7 ≠ 2
whip[3]: c6n2{r4 r3} - r2n2{c4 c8} - r5n2{c8 .} ==> r4c5 ≠ 2
whip[3]: c6n2{r5 r3} - r2n2{c4 c8} - r5n2{c8 .} ==> r6c5 ≠ 2

whip[2]: r6c9{n3 n4} - r6c5{n4 .} ==> r6c7 ≠ 3
whip[4]: b6n9{r4c7 r4c8} - c8n2{r4 r2} - c8n3{r2 r9} - c7n3{r9 .} ==> r4c7 ≠ 2
whip[4]: b9n5{r9c7 r9c8} - r9c2{n5 n4} - r3c2{n4 n8} - r5n8{c2 .} ==> r5c7 ≠ 5
whip[3]: r3c7{n4 n2} - r6c7{n2 n8} - r5c7{n8 .} ==> r8c7 ≠ 4
hidden-single-in-a-block ==> r8c8 = 4
whip[3]: r3c2{n4 n8} - r5n8{c2 c7} - c7n4{r5 .} ==> r3c1 ≠ 4
whip[3]: b1n5{r2c1 r2c2} - r9c2{n5 n4} - c1n4{r9 .} ==> r2c1 ≠ 7
whip[2]: c5n7{r4 r3} - c1n7{r3 .} ==> r4c6 ≠ 7
whip[4]: b7n9{r9c1 r7c3} - b7n8{r7c3 r7c1} - r7n5{c1 c7} - r8c7{n5 .} ==> r9c8 ≠ 9
singles ==> r4c8 = 9, r9c1 = 9, r7c3 = 8, r3c3 = 9
whip[3]: r4c7{n3 n5} - r5c8{n5 n2} - b5n2{r5c5 .} ==> r4c6 ≠ 3
hidden-single-in-a-column ==> r3c6 = 3
whip[1]: c6n2{r5 .} ==> r5c5 ≠ 2
whip[2]: c3n7{r5 r2} - b2n7{r2c4 .} ==> r5c5 ≠ 7
whip[3]: c7n4{r5 r3} - c7n2{r3 r6} - r5n2{c8 .} ==> r5c6 ≠ 4
whip[1]: c6n4{r9 .} ==> r7c5 ≠ 4
whip[3]: c8n5{r5 r9} - r9c4{n5 n7} - c6n7{r9 .} ==> r5c6 ≠ 5
whip[3]: r4c6{n5 n2} - r5n2{c6 c8} - b6n5{r5c8 .} ==> r4c5 ≠ 5
whip[3]: r9n6{c7 c6} - r1c6{n6 n5} - r4n5{c6 .} ==> r9c7 ≠ 5
whip[4]: r9c2{n4 n5} - r9c8{n5 n3} - b3n3{r2c8 2c9} - b3n4{r2c9 .} ==> r3c2 ≠ 4
singles to the end

2) The resolution path with braids shows that B(P) = 4:

***** SudoRules version 15b.1.12-B *****
;;; same path up to RS$_1$ (no braid appears before); after, the two paths diverge:
braid[4]: r2c4{n7 n2} - r4c1{n7 n2} - c8n2{r2 r5} - c6n2{r5 .} ==> r2c1 ≠ 7
whip[2]: c5n7{r4 r3} - c1n7{r3 .} ==> r4c6 ≠ 7
whip[3]: c2n4{r2 r9} - c2n5{r9 r2} - r2c1{n5 .} ==> r3c1 ≠ 4
whip[4]: b1n6{r2c2 r2c3} - r2n7{c3 c4} - r9c4{n7 n5} - c2n5{r9 .} ==> r2c2 ≠ 4
whip[4]: c2n8{r5 r3} - c1n8{r3 r7} - b7n5{r7c1 r9c2} - c2n4{r9 .} ==> r5c3 ≠ 8
whip[2]: c3n9{r3 r7} - c3n8{r7 .} ==> r3c3 ≠ 7
whip[3]: r3c7{n2 n4} - r3c2{n4 n8} - r5n8{c2 .} ==> r5c7 ≠ 2
whip[3]: c6n2{r4 r3} - r2n2{c4 c8} - r5n2{c8 .} ==> r4c5 ≠ 2
whip[3]: c6n2{r5 r3} - r2n2{c4 c8} - r5n2{c8 .} ==> r6c5 ≠ 2
whip[2]: r6c9{n3 n4} - r6c5{n4 .} ==> r6c7 ≠ 3
whip[4]: b6n9{r4c7 r4c8} - c8n2{r4 r2} - c8n3{r2 r9} - c7n3{r9 .} ==> r4c7 ≠ 2
whip[4]: b9n5{r9c7 r9c8} - r9c2{n5 n4} - r3c2{n4 n8} - r5n8{c2 .} ==> r5c7 ≠ 5
whip[3]: r3n4{c7 c2} - c2n8{r3 r5} - r5c7{n8 .} ==> r8c7 ≠ 4
hidden-single-in-a-block ==> r8c8 = 4
whip[4]: b7n9{r9c1 r7c3} - b7n8{r7c3 r7c1} - r7n5{c1 c7} - r8c7{n5 .} ==> r9c8 ≠ 9
singles ==> r4c8 = 9, r9c1 = 9, r7c3 = 8, r3c3 = 9
whip[3]: r4c7{n3 n5} - r5c8{n5 n2} - b5n2{r5c5 .} ==> r4c6 ≠ 3
hidden-single-in-a-column ==> r3c6 = 3
whip[1]: c6n2{r5 .} ==> r5c5 ≠ 2
whip[2]: c3n7{r5 r2} - b2n7{r2c4 .} ==> r5c5 ≠ 7
whip[3]: c7n4{r5 r3} - c7n2{r3 r6} - r5n2{c8 .} ==> r5c6 ≠ 4

whip[1]: c6n4{r9 .} ==> r7c5 ≠ 4
whip[3]: c8n5{r5 r9} - r9c4{n5 n7} - c6n7{r9 .} ==> r5c6 ≠ 5
whip[3]: r4c6{n5 n2} - r5n2{c6 c8} - b6n5{r5c8 .} ==> r4c5 ≠ 5
whip[3]: r9n6{c7 c6} - r1c6{n6 n5} - r4n5{c6 .} ==> r9c7 ≠ 5
whip[4]: r2c9{n3 n4} - r2c1{n4 n5} - r7n5{c1 c7} - r4c7{n5 .} ==> r6c9 ≠ 3
singles to the end

5.10.2. Proof of $B_\infty \neq W_\infty$: an instance with $W(P) = \infty$ and $B(P) = 12$

After the previous example, one may still wonder: if a puzzle can be solved by braids, cannot one always find whips, though longer than the braids, such that they will also solve it? Said otherwise, is not B_∞ equal to W_∞? The answer is negative; there are puzzles that can be solved by braids but not by whips of any length. The example in Figure 5.2 is one of the exceptional (in percentage) puzzles in this case (see statistics in chapter 6); it is the only one in the whole "Magictour top 1465" collection; its B rating is 12 but its W rating is ∞.

The following exceptionally long resolution paths are typical of what one gets with the "simplest first" strategy: braids that are not whips appear much less often than whips. For puzzles P solvable by whips, if both whips and braids are activated, braids appear even more rarely – and they very rarely change the rating, i.e. $W(P) = B(P)$ most of the time. In both resolution paths below, one can also notice the long streaks of eliminations necessary before a new value can be asserted.

		3		5		
	5		1		3	
	7		4			1
2				4		
	6		9			
	1		6			2
8		7		2		
	9		8		5	
	5		9			7

9	1	4	3	7	8	5	2	6
6	5	8	9	1	2	7	3	4
3	2	7	5	6	4	9	8	1
2	8	9	1	3	7	4	6	5
4	6	3	2	9	5	1	7	8
5	7	1	8	4	6	3	9	2
8	4	6	7	5	3	2	1	9
7	9	2	4	8	1	6	5	3
1	3	5	6	2	9	8	4	7

Figure 5.2. *Puzzle Magictour top 1465 #89 and its solution; W = ∞ and B = 12*

1) The resolution path with whips shows that $W(P) = \infty$; it also gives an example of a very long whip[18] (but there are much longer ones in other puzzles):

***** SudoRules version 15.1.8W *****
24 givens, 218 candidates and 1379 nrc-links
hidden-single-in-a-row ==> r8c1 = 7
whip[2]: r5n7{c8 c6} - r2n7{c6 .} ==> r6c7 <> 7
whip[3]: c2n2{r1 r9} - c5n2{r9 r3} - b3n2{r3c8 .} ==> r1c3 <> 2

whip[3]: c5n2{r1 r9} - c2n2{r9 r3} - b3n2{r3c8 .} ==> r1c6 <> 2
whip[3]: c1n9{r2 r6} - c7n9{r6 r3} - r1n9{c8 .} ==> r2c3 <> 9
whip[3]: r2n2{c6 c3} - c2n2{r1 r9} - c5n2{r9 .} ==> r3c4 <> 2
whip[3]: b6n5{r4c9 r5c9} - b4n5{r5c1 r6c1} - b4n9{r6c1 .} ==> r4c9 <> 9
whip[3]: b7n2{r9c2 r8c3} - r2n2{c3 c6} - b5n2{r5c6 .} ==> r9c4 <> 2
whip[4]: b6n6{r4c8 r4c9} - b6n5{r4c9 r5c9} - b4n5{r5c1 r6c1} - b4n9{r6c1 .} ==> r4c8 <> 9
hidden-single-in-a-row ==> r4c3 = 9
whip[7]: b9n9{r7c9 r7c8} - r1n9{c8 c1} - r3n9{c1 c4} - b2n5{r3c4 r3c5} - b8n5{r7c5 r7c6} - r7n1{c6 c2} - c1n1{r9 .} ==> r2c9 <> 9
;;; Resolution state RS$_1$
whip[9]: c7n7{r2 r5} - c8n7{r4 r1} - r1n9{c8 c1} - c1n1{r1 r9} - c7n1{r9 r8} - b8n1{r8c6 r7c6} - b8n5{r7c6 r7c5} - b2n5{r3c5 r3c4} - b2n9{r3c4 .} ==> r2c7 <> 9
whip[11]: b3n7{r2c7 r1c8} - r1c6{n7 n8} - c3n8{r1 r5} - c9n8{r5 r4} - c8n8{r4 r9} - c8n4{r9 r7} - b9n9{r7c8 r7c9} - r1n9{c9 c1} - c1n1{r1 r9} - r7c2{n1 n3} - c3n3{r8 .} ==> r2c7 <> 8
whip[18]: r1c6{n8 n7} - r2n7{c6 c7} - r5n7{c7 c8} - r6c8{n7 n9} - c7n9{r6 r3} - r1n9{c9 c1} - c1n1{r1 r9} - c2n1{r7 r1} - r1n2{c2 c5} - r2n2{c4 c3} - c3n8{r2 r5} - c9n8{r5 r4} - c7n8{r5 r9} - c7n6{r9 r8} - b7n6{r8c3 r7c3} - c3n3{r7 r8} - b9n3{r8c7 r7c9} - b9n9{r7c9 .} ==> r1c8 <> 8

After this very long whip, there is no more elimination. (Whips are programmed up to length 36 in SudoRules and there is a mechanism for detecting the need for longer ones – it never fired! The same programmed maximum length is true of the braids and of the g-whips and g-braids introduced in chapter 7.)

2) The resolution path with braids shows that B(P) = 12:

***** SudoRules version 15.1.8-B *****
;;; same path up to state RS$_1$
;;; the next two eliminations were done by slightly longer whips (length +1) in the previous path
braid[8]: r1n9{c9 c1} - r3n9{c1 c4} - b2n5{r3c4 r3c5} - b8n5{r7c5 r7c6} - c1n1{r1 r9} - c7n7{r2 r5} - c7n1{r5 r8} - b8n1{r9c4 .} ==> r2c7 ≠ 9
braid[10]: b3n7{r2c7 r1c8} - r1c6{n7 n8} - c3n8{r1 r5} - b9n8{r9c7 r9c8} - c8n4{r1 r7} - b9n9{r7c8 r7c9} - r1n9{c8 c1} - c1n1{r1 r9} - r7c2{n1 n3} - c3n3{r8 .} ==> r2c7 ≠ 8
;;; now the two paths diverge completely
braid[11]: c9n9{r7 r1} - c7n9{r3 r6} - b6n3{r6c7 r5c7} - c3n3{r5 r8} - c3n2{r8 r2} - c7n7{r5 r2} - r2c6{n2 n8} - r1c6{n8 n7} - r5n7{c6 c8} - r6c8{n7 n8} - c9n8{r5 .} ==> r7c9 ≠ 3
braid[10]: b6n6{r4c8 r4c9} - b6n5{r4c9 r5c9} - c9n3{r5 r8} - r5n7{c8 c6} - r1c6{n7 n8} - c9n8{r5 r2} - r3n8{c8 c2} - r4c2{n8 n3} - c6n3{r8 r7} - c3n3{r8 .} ==> r4c8 ≠ 7
braid[12]: c7n9{r3 r6} - b9n8{r9c7 r9c8} - r6c8{n9 n7} - r5c8{n8 n1} - r5c7{n8 n3} - b9n3{r9c7 r8c9} - r5n7{c8 c6} - r1c6{n7 n8} - r2c6{n8 n2} - c3n3{r8 r7} - c5n2{r1 r9} - r9n3{c7 .} ==> r3c7 ≠ 8
whip[8]: r2c7{n7 n6} - r3c7{n6 n9} - r1n9{c8 c1} - c1n1{r1 r9} - c1n6{r9 r3} - r2c1{n6 n4} - r1c3{n4 n8} - r1c6{n8 .} ==> r2c6 ≠ 7
hidden-single-in-a-row ==> r2c7 = 7
whip[4]: r5n7{c8 c6} - r1c6{n7 n8} - r3n8{c4 c2} - b4n8{r6c2 .} ==> r5c8 ≠ 8
braid[9]: b2n9{r2c4 r3c4} - r3c7{n9 n6} - r2n9{c4 c1} - r2n6{c9 c3} - c3n2{r2 r8} - r2n4{c3 c9} - r8n4{c9 c4} - c4n6{r8 r9} - b7n6{r9c1 .} ==> r2c4 ≠ 2

braid[9]: b2n9{r2c4 r3c4} - r3c7{n9 n6} - r2n9{c4 c1} - r2n6{c9 c3} - c3n2{r2 r8} - r2c9{n8 n4} - r8n4{c9 c4} - c4n6{r8 r9} - b7n6{r9c1 .} ==> r2c4 ≠ 8

braid[11]: c4n2{r5 r8} - b6n1{r5c8 r4c8} - b6n6{r4c8 r4c9} - b6n5{r4c9 r5c9} - c9n3{r5 r8} - r8n4{c9 c3} - r8n6{c9 c7} - b9n1{r9c8 r9c7} - c1n1{r9 r1} - r3c7{n6 n9} - r1n9{c9 .} ==> r5c4 ≠ 1

whip[8]: r8n1{c4 c7} - r5n1{c7 c8} - r5n7{c8 c6} - c6n5{r5 r4} - r4c5{n5 n3} - b8n3{r9c5 r8c6} - c9n3{r8 r5} - b6n5{r5c9 .} ==> r7c6 ≠ 1

whip[8]: c1n1{r1 r9} - r7n1{c2 c8} - b9n9{r7c8 r7c9} - r1n9{c9 c8} - r3c7{n9 n6} - c1n6{r3 r2} - r1c3{n6 n8} - r1c9{n8 .} ==> r1c1 ≠ 4

whip[8]: c1n1{r1 r9} - r7n1{c2 c8} - b9n9{r7c8 r7c9} - r1n9{c9 c8} - r3c7{n9 n6} - r2n6{c9 c4} - r9c4{n6 n4} - c8n4{r9 .} ==> r1c1 ≠ 6

whip[10]: b6n6{r4c8 r4c9} - b6n5{r4c9 r5c9} - c9n3{r5 r8} - r8c7{n3 n1} - r9c7{n1 n8} - r5c7{n8 n3} - c3n3{r5 r7} - b7n6{r7c3 r8c3} - b8n6{r8c4 r7c5} - r1n6{c5 .} ==> r9c8 ≠ 6

braid[10]: r3c7{n6 n9} - r1n9{c8 c1} - c1n1{r1 r9} - r9c4{n1 n4} - r9c8{n4 n8} - r3n8{c8 c2} - c1n6{r9 r2} - r1c3{n8 n4} - c8n4{r9 r7} - b7n4{r9c2 .} ==> r3c4 ≠ 6

whip[4]: c4n6{r9 r2} - c1n6{r2 r3} - r3c7{n6 n9} - b2n9{r3c4 .} ==> r9c5 ≠ 6

whip[10]: b6n6{r4c8 r4c9} - b6n5{r4c9 r5c9} - c9n3{r5 r8} - r8c7{n3 n1} - b8n1{r8c6 r9c4} - r9n6{c4 c1} - r8n6{c3 c4} - r8n4{c4 c3} - c3n2{r8 r2} - r2n6{c3 .} ==> r7c8 ≠ 6

whip[11]: r3c7{n6 n9} - r1n9{c8 c1} - c1n1{r1 r9} - c2n1{r9 r1} - r1n2{c2 c5} - r2c6{n2 n8} - r2c9{n8 n4} - r1n4{c8 c3} - r8n4{c3 c4} - b8n2{r8c4 r8c6} - b8n1{r8c6 .} ==> r1c8 ≠ 6

whip[5]: c8n6{r4 r3} - r3c7{n6 n9} - r1n9{c8 c1} - c1n1{r1 r9} - r7n1{c2 .} ==> r4c8 ≠ 1

whip[1]: r4n1{c4 .} ==> r5c6 ≠ 1

whip[5]: r5n1{c7 c8} - r5n7{c8 c6} - r1c6{n7 n8} - c9n8{r1 r2} - c3n8{r2 .} ==> r5c7 ≠ 8

whip[4]: r5n1{c7 c8} - b6n7{r5c8 r6c8} - b6n9{r6c8 r6c7} - c7n8{r6 .} ==> r9c7 ≠ 1

whip[5]: r5c7{n1 n3} - b9n3{r9c7 r8c9} - c3n3{r8 r7} - c6n3{r7 r4} - c6n1{r4 .} ==> r8c7 ≠ 1

singles ==> r5c7 = 1, r5c8 = 7

whip[1]: r8n1{c4 .} ==> r9c4 ≠ 1

braid[7]: r8c7{n6 n3} - c3n2{r8 r2} - r2c6{n2 n8} - r8c9{n6 n4} - r2c9{n8 n6} - b9n6{r8c9 r9c7} - c4n6{r9 .} ==> r8c3 ≠ 6

whip[5]: b7n6{r7c3 r9c1} - r9c4{n6 n4} - r5n4{c4 c1} - c2n4{r6 r1} - c8n4{r1 .} ==> r7c3 ≠ 4

whip[5]: b9n9{r7c9 r7c8} - r7n1{c8 c2} - r7n4{c2 c5} - r9c4{n4 n6} - r8n6{c4 .} ==> r7c9 ≠ 6

whip[2]: r7n6{c3 c5} - c4n6{r9 .} ==> r2c3 ≠ 6

braid[6]: b7n6{r9c1 r7c3} - r8c7{n6 n3} - c3n3{r8 r5} - r6n3{c7 c5} - r9c4{n6 n4} - c5n4{r9 .} ==> r9c7 ≠ 6

whip[1]: b9n6{r8c7 .} ==> r8c4 ≠ 6

whip[8]: b4n7{r6c2 r4c2} - b4n8{r4c2 r5c3} - r5n4{c3 c4} - r9c4{n4 n6} - r7n6{c5 c3} - r1c3{n6 n4} - r2n4{c3 c9} - r8n4{c9 .} ==> r6c2 ≠ 4

braid[6]: b4n4{r5c1 r5c3} - b7n6{r9c1 r7c3} - c3n3{r7 r8} - b7n2{r8c3 r9c2} - r9c5{n4 n3} - b9n3{r9c7 .} ==> r9c1 ≠ 4

whip[8]: r7c3{n3 n6} - r9c1{n6 n1} - b9n1{r9c8 r7c8} - c2n1{r7 r1} - c2n4{r1 r9} - c2n2{r9 r3} - b3n2{r3c8 r1c8} - c8n4{r1 .} ==> r7c2 ≠ 3

whip[3]: r7c9{n4 n9} - r7c8{n9 n1} - r7c2{n1 .} ==> r7c5 ≠ 4

whip[6]: r8c7{n3 n6} - r3c7{n6 n9} - r1n9{c8 c1} - c1n1{r1 r9} - b7n6{r9c1 r7c3} - r7n3{c3 .} ==> r8c6 ≠ 3

whip[7]: r2c6{n8 n2} - c5n2{r3 r9} - c5n4{r9 r6} - b5n7{r6c5 r4c5} - r4c2{n7 n3} - r6c1{n3 n5} - r6c4{n5 .} ==> r4c6 ≠ 8

whip[6]: b4n8{r6c2 r5c3} - c6n8{r5 r2} - r2n2{c6 c3} - r3c2{n2 n3} - r6c2{n3 n7} - r4c2{n7 .} ==> r1c2 ≠ 8 whip[7]: b5n2{r5c4 r5c6} - r8c6{n2 n1} - b5n1{r4c6 r4c4} - b5n8{r4c4 r6c4} - r3c4{n8 n9} - c7n9{r3 r6} - r6c8{n9 .} ==> r5c4 ≠ 5

whip[8]: c4n1{r4 r8} - r8c6{n1 n2} - r2c6{n2 n8} - r3c4{n8 n9} - c7n9{r3 r6} - r6c8{n9 n8} - b5n8{r6c4 r5c4} - b5n2{r5c4 .} ==> r4c4 ≠ 5

whip[7]: c4n6{r9 r2} - b2n9{r2c4 r3c4} - c4n5{r3 r6} - b4n5{r6c1 r5c1} - r5n4{c1 c3} - c1n4{r6 r2} - r2n9{c1 .} ==> r9c4 ≠ 4

singles ==> r9c4 = 6, r2c4 = 9, r7c3 = 6

whip[1]: r7n3{c6 .} ==> r9c5 ≠ 3

whip[5]: c3n3{r5 r8} - r8c7{n3 n6} - r8c9{n6 n4} - c8n4{r9 r1} - r1c3{n4 .} ==> r5c3 ≠ 8

whip[1]: c3n8{r1 .} ==> r3c2 ≠ 8

whip[2]: b4n7{r4c2 r6c2} - c2n8{r6 .} ==> r4c2 ≠ 3

whip[2]: b4n7{r6c2 r4c2} - c2n8{r4 .} ==> r6c2 ≠ 3

whip[3]: b3n2{r1c8 r3c8} - r3n8{c8 c4} - r2c6{n8 .} ==> r1c5 ≠ 2

whip[4]: b8n4{r8c4 r9c5} - c8n4{r9 r1} - b3n2{r1c8 r3c8} - c5n2{r3 .} ==> r8c9 ≠ 4

whip[2]: r8c7{n3 n6} - r8c9{n6 .} ==> r9c7 ≠ 3

naked-single ==> r9c7 = 8

r9n3{c1 .} ==> r8c3 ≠ 3

hidden-single-in-a-column ==> r5c3 = 3

b4n4{r6c1 .} ==> r2c1 ≠ 4

naked-single ==> r2c1 = 6

whip[4]: c4n8{r6 r3} - c4n5{r3 r6} - r4n5{c5 c9} - r5c9{n5 .} ==> r5c6 ≠ 8

whip[1]: c6n8{r1 .} ==> r3c4 ≠ 8 ; singles to the end

5.10.3. An example of non-confluence for the W_4 whip resolution theory

As mentioned in the proof of the confluence property for the B_n resolution theories (section 5.5), there is one step in this proof (step b) that would not work for the W_n theories. But this did not prove that the W_n do not have the confluence property. The puzzle in Figure 5.3 (Sudogen0_1M #279845) provides the missing proof, for the Sudoku CSP. n = 4 is the smallest n we could find with a counter-example to confluence.

9	8	1	7		3	2	5	
7	5	2			1	9		
3	6	4		9				8
	1	7	3				9	2
	4	3			9		6	
	9				7			
4			1				2	9
	2	9			8			
			9			5		

9	8	1	7	6	3	2	5	4
7	5	2	4	8	1	9	3	6
3	6	4	5	9	2	1	7	8
8	1	7	3	5	6	4	9	2
2	4	3	8	1	9	7	6	5
6	9	5	2	4	7	8	1	3
4	7	8	1	3	5	6	2	9
5	2	9	6	7	8	3	4	1
1	3	6	9	2	4	5	8	7

Figure 5.3. An example of non confluence of W_4: puzzle Sudogen0_1M #279845

***** SudoRules version 15b.1.12-W *****
37 givens, 146 candidates and 792 nrc-links
whip[1]: c7n6{r7 .} ==> r9c9 ≠ 6, r8c9 ≠ 6
whip[2]: c1n8{r6 r9} - c8n8{r9 .} ==> r6c3 ≠ 8
whip[1]: c3n8{r9 .} ==> r9c1 ≠ 8
whip[3]: c7n4{r6 r8} - c4n4{r8 r2} - b3n4{r2c9 .} ==> r6c9 ≠ 4
whip[3]: b7n5{r8c1 r7c3} - r7n8{c3 c7} - b9n6{r7c7 .} ==> r8c1 ≠ 6
whip[3]: b8n2{r9c5 r9c6} - r3c6{n2 n5} - r7c6{n5 .} ==> r9c5 ≠ 6
whip[3]: c6n4{r9 r4} - r4c7{n4 n8} - b9n8{r7c7 .} ==> r9c8 ≠ 4
whip[2]: r1n4{c5 c9} - b9n4{r9c9 .} ==> r8c5 ≠ 4
;;; Resolution state RS$_1$

	c1	c2	c3	c4	c5	c6	c7	c8	c9	
r1	9	8	1	7	n4 n6	3	2	5	n4 n6	r1
r2	7	5	2	n4 n6 n8	n4 n6 n8	1	9	n4	n3 n4 n3 n6	r2
r3	3	6	4	n2 n5	9	n2 n5	n1 n7	n1 n7	8	r3
r4	n5 n6 n8	1	7	3	n4 n5 n6 n8	n4 n5 n6	n4 n8	9	2	r4
r5	n2 n5 n8	4	3	n2 n5 n8	n1 n2 n5 n8	9	n1 n7 n8	6	n1 n5 n7	r5
r6	n2 n5 n6 n8	9	n5 n6	n2 n4 n5 n6 n8	n1 n2 n4 n5 n6 n8	7	n1 n3 n4 n8	n1 n3 n4 n8	n1 n3 n5	r6
r7	4	n3 n7	n5 n6 n8	1	n3 n5 n6 n7	n5 n6	n3 n6 n7 n8	2	9	r7
r8	n1 n5	2	9	n4 n5 n6	n3 n5 n6 n7	8	n1 n3 n4 n7	n1 n3 n4 n6 n7	n1 n3 n4 n7	r8
r9	n1 n6	n3 n7	n6 n8	9	n2 n3 n4 n7	n2 n4 n6	5	n1 n3 n4 n7 n8	n1 n3 n4 n7	r9
	c1	c2	c3	c4	c5	c6	c7	c8	c9	

Figure 5.4. *Resolution state RS$_1$ of puzzle Sudogen0_1M #279845*

The resolution state RS$_1$ at this point is shown in Figure 5.4. After RS$_1$ has been reached, there are (at least) the following two resolution paths.

1) The first path starts with a general whip:

whip[4]: c6n4{r4 r9} - c6n6{r9 r7} - r8c4{n6 n5} - c5n5{r8 .} ==> r4c6 ≠ 5

It is worth analysing this whip by adding it a few details:

whip[4]: c6n4{r4 r9} - c6n6{r9 r7 r4*} - r8c4{n6 n5 n4#2} - c5n5{r8 . r4* r5* r6* r7#3} ==> r4c6≠5

The * sign corresponds to z-candidates, the # sign corresponds to t-candidates and the number following this # sign is the number of the right-linking candidate linked to this t-candidate (remember however that, by definition, these z- and t-candidates do not belong to the whip; we display them here for the only sake of showing how a whip deals with these additional candidates).

Notice that there is an alternative whip, for the same target, with the same first two cells and the last cell replaced by the slightly simpler: r3n5{c4 . c6*}. Using it instead would not change the sequel.

The end of this first resolution path has nothing noticeable:

whip[2]: b7n5{r7c3 r8c1} - r4n5{c1 .} ==> r7c5 ≠ 5
whip[4]: r7c6{n5 n6} - r4c6{n6 n4} - r4c7{n4 n8} - r7n8{c7 .} ==> r7c3 ≠ 5
singles ==> r8c1 = 5, r6c3 = 5, r5c9 = 5, r4c5 = 5, r3c4 = 5, r3c6 = 2, r9c5 = 2, r7c6 = 5, r5c7 = 7, r3c7 = 1, r3c8 = 7, r5c5 = 1, r9c1 = 1
whip[2]: b8n3{r7c5 r8c5} - b8n7{r8c5 .} ==> r7c5 ≠ 6
whip[2]: r7c2{n3 n7} - r7c5{n7 .} ==> r7c7 ≠ 3
whip[2]: b8n3{r8c5 r7c5} - b8n7{r7c5 .} ==> r8c5 ≠ 6
whip[2]: r9n4{c9 c6} - r4n4{c6 .} ==> r8c7 ≠ 4
whip[1]: c7n4{r4 .} ==> r6c8 ≠ 4
whip[3]: r9n4{c9 c6} - b8n6{r9c6 r8c4} - r8c7{n6 .} ==> r9c9 ≠ 3
whip[3]: r8c7{n3 n6} - r7c7{n6 n8} - r9c8{n8 .} ==> r8c9 ≠ 3, r8c8 ≠ 3
whip[3]: r6n2{c4 c1} - r6n6{c1 c5} - r4c6{n6 .} ==> r6c4 ≠ 4
whip[2]: c8n4{r2 r8} - c4n4{r8 .} ==> r2c9 ≠ 4, r2c5 ≠ 4
whip[2]: r1n4{c9 c5} - c4n4{r2 .} ==> r8c9 ≠ 4
whip[4]: b9n6{r7c7 r8c7} - r8c4{n6 n4} - c6n4{r9 r4} - r4c7{n4 .} ==> r7c7 ≠ 8
singles to the end

Now, if we activate braids and we re-start with our usual "simplest first" strategy, we get exactly the same path (there appears no non-whip braid). Thanks to the confluence property of B_4, we do not have to consider any other resolution path to claim that the correct B rating is B = 4. As $W(P) \leq B(P)$ for any P and we have found a resolution path for P with whips of lengths no more than 4, we can claim that $W(P) = 4$.

2) Let us now consider what would have happened if we had followed an alternative resolution path. In state RS_0, before using the first whip[4] above, we could have chosen a whole sequence of simpler whips – "simpler" in the sense that they are special subtypes of whips, not in the sense of shorter (these subtypes were introduced in *HLS*, but it is not necessary here to know their precise definitions, they are whips anyway, with the lengths indicated in square brackets):

***** SudoRules version 13.7wter2 *****
;;; same path up to RS$_1$
xyzt-chain[4] r7c6{n6 n5} - r3c6{n5 n2} - r9c6{n2 n4} - r8c4{n4 n6} ==> r8c5 ≠ 6, r7c5 ≠ 6
nrc-chain[4] b6n7{r5c7 r5c9} - b6n5{r5c9 r6c9} - c3n5{r6 r7} - r7n8{c3 c7} ==> r7c7 ≠ 7, r5c7 ≠ 8
naked-pairs-in-a-column c7{r3 r5}{n1 n7} ==> r8c7 ≠ 7, r8c7 ≠ 1, r6c7 ≠ 1
;;; Resolution state RS$_2$
nrc-chain[4] r9c3{n6 n8} - b9n8{r9c8 r7c7} - r4c7{n8 n4} - c6n4{r4 r9} ==> r9c6 ≠ 6
;;; Resolution state RS$_3$
interaction row r9 with block b7 ==> r7c3 ≠ 6
nrct-chain[5] c6n4{r4 r9} - c6n2{r9 r3} - r3n5{c6 c4} - r8c4{n5 n6} - r7c6{n6 n5} ==> r4c6≠5
nrc-chain[2] r4n5{c5 c1} - b7n5{r8c1 r7c3} ==> r7c5 ≠ 5
naked-pairs-in-a-row r7{c2 c5}{n3 n7} ==> r7c7 ≠ 3
xy-chain[3] r7c7{n6 n8} - r4c7{n8 n4} - r4c6{n4 n6} ==> r7c6 ≠ 6
singles to the end

Until we reach resolution state RS$_2$, the whip[4] of the first path is still available; but if we apply the nrc-chain[4] rule before this whip[4], it deletes the left-linking candidate n6r9c6 for its second CSP variable. Then, in the resulting state RS$_3$, there remains no whip[4]; the simpler whip available is a slightly longer nrct-chain[5]; it makes the same r4c6 ≠ 5 elimination.

Conclusion: if we considered only this second resolution path, we would find, erroneously, a rating W = 5 for this puzzle. This example is thus not only a clear case of non confluence for whip theories, it is also a case in which this non confluence leads to a bad evaluation of the W rating if we do not try all the paths. This is a very rare case.

Final remark: if we allow braids, even after the nrc-chain[4] is applied, there is a replacement braid for the missing whip[4] (and it is as provided in section 5.5.1 by the general proof of confluence for braid resolution theories):

braid[4] c6n4{r4 r9} - c6n6{r4 r7} - r8c4{n6 n5 n4#1} - c5n5{r8 . r4* r5* r6* r7#3} ==> r4c6 ≠ 5

The z-candidate n6r4c6 in cell 2 of the whip[4] is now used as a left-linking candidate in the braid, in which it is linked to the target.

5.10.4. A puzzle P with W(P) = 24

What is the largest finite W rating one can find? This is a very difficult question.

The largest W rating we could obtain with random generators is 16 (but only one puzzle with W=16 was found in more than 10,000,000). However, one can find an example with much larger W rating. The puzzle P in Figure 5.5 has W(P) = 24.

1								
	4		6		2			
		8				5		
	9		3				2	
				7		9		
			2		9		6	
5								1
	6				4		9	
		7						8

1	7	6	4	3	5	2	8	9
9	4	5	6	8	2	1	7	3
2	3	8	1	9	7	5	4	6
6	9	4	3	5	1	8	2	7
3	5	2	8	7	6	9	1	4
7	8	1	2	4	9	3	6	5
5	2	9	7	6	8	4	3	1
8	6	3	5	1	4	7	9	2
4	1	7	9	2	3	6	5	8

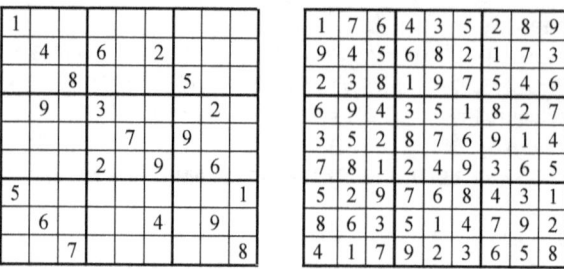

Figure 5.5. *A puzzle P with W(P) = 24*

***** SudoRules version 15b.1.12-W *****
21 givens, 248 candidates and 1711 nrc-links
whip[1]: c9n6{r3 .} ==> r1c7 ≠ 6
whip[2]: b7n4{r7c3 r9c1} - b7n9{r9c1 .} ==> r7c3 ≠ 3, r7c3 ≠ 2
whip[2]: b7n4{r9c1 r7c3} - b7n9{r7c3 .} ==> r9c1 ≠ 3, r9c1 ≠ 2
whip[5]: b1n6{r1c3 r3c1} - r3n2{c1 c9} - b3n6{r3c9 r1c9} - b3n9{r1c9 r2c9} - b1n9{r2c1 .} ==> r1c3 ≠ 2
whip[5]: b3n6{r3c9 r1c9} - c9n2{r1 r8} - c3n2{r8 r5} - c1n2{r5 r3} - b1n6{r3c1 .} ==> r3c9 ≠ 4, r3c9 ≠ 3, r3c9 ≠ 7, r3c9 ≠ 9
whip[4]: r1n6{c3 c9} - b3n9{r1c9 r2c9} - b1n9{r2c1 r3c1} - b1n6{r3c1 .} ==> r1c3 ≠ 5, r1c3 ≠ 3
whip[6]: r3n9{c4 c1} - r1n9{c3 c9} - b3n6{r1c9 r3c9} - c9n2{r3 r8} - c3n2{r8 r5} - c1n2{r5 .} ==> r2c5 ≠ 9
whip[6]: c3n2{r8 r5} - c1n2{r5 r3} - c9n2{r3 r1} - b3n9{r1c9 r2c9} - b1n9{r2c3 r1c3} - b1n6{r1c3 .} ==> r8c5 ≠ 2
whip[6]: c3n2{r8 r5} - c1n2{r5 r3} - b3n2{r3c9 r1c9} - b3n9{r1c9 r2c9} - b1n9{r2c3 r1c3} - b1n6{r1c3 .} ==> r8c7 ≠ 2
whip[6]: c3n2{r5 r8} - c1n2{r8 r3} - c9n2{r3 r1} - b3n9{r1c9 r2c9} - b1n9{r2c3 r1c3} - b1n6{r1c3 .} ==> r5c2 ≠ 2
whip[11]: b9n6{r9c7 r7c7} - c7n2{r7 r1} - b9n2{r9c7 r8c9} - b9n5{r8c9 r9c8} - r9n4{c8 c1} - r7n4{c3 c8} - b3n4{r3c8 r1c9} - b3n9{r1c9 r2c9} - c1n9{r2 r3} - c1n2{r3 r5} - c3n2{r5 .} ==> r9c7 ≠ 3
whip[11]: b9n6{r7c7 r9c7} - c7n2{r9 r1} - b9n2{r7c7 r8c9} - b9n5{r8c9 r9c8} - r9n4{c8 c1} - r7n4{c3 c8} - b3n4{r3c8 r1c9} - b3n9{r1c9 r2c9} - c1n9{r2 r3} - c1n2{r3 r5} - c3n2{r5 .} ==> r7c7 ≠ 3
whip[11]: c9n4{r6 r1} - r1n6{c9 c3} - r3n6{c1 c9} - c9n2{r3 r8} - r9c7{n2 n6} - r7c7{n6 n7} - r8n7{c9 c4} - r8n5{c4 c5} - r8n1{c5 c3} - r4c3{n1 n5} - r2n5{c3 .} ==> r4c7 ≠ 4
whip[10]: b9n6{r7c7 r9c7} - b9n2{r9c7 r8c9} - c3n2{r8 r5} - c1n2{r5 r3} - b3n2{r3c9 r1c7} - c7n4{r1 r6} - c9n4{r6 r1} - b3n9{r1c9 r2c9} - b1n9{r2c1 r1c3} - b1n6{r1c3 .} ==> r7c7 ≠ 7
whip[5]: c9n4{r6 r1} - b3n6{r1c9 r3c9} - b3n2{r3c9 r1c7} - r9c7{n2 n6} - r7c7{n6 .} ==> r6c7 ≠ 4
whip[22]: b4n2{r5c3 r5c1} - c1n6{r5 r3} - r3c9{n6 n2} - r8n2{c9 c3} - b7n1{r8c3 r9c2} - c2n2{r9 r1} - b1n5{r1c2 r2c3} - c3n3{r2 r6} - b4n1{r6c3 r4c3} - c3n4{r4 r7} - b7n9{r7c3 r9c1} - r9c4{n9 n5} - c8n5{r9 r5} - r5c2{n5 n8} - r7c2{n8 n3} - r7c8{n3 n7} - b8n7{r7c6 r8c4} - b8n1{r8c4 r8c5} - r6n1{c5 c7} - r6n8{c7 c5} - r2c5{n8 n3} - b1n3{r2c1 .} ==> r5c3 ≠ 6

whip[18]: b8n2{r7c5 r9c5} - c5n6{r9 r4} - c3n6{r4 r1} - c3n9{r1 r2} - b3n9{r2c9 r1c9} - b3n6{r1c9 r3c9} - c9n2{r3 r8} - b9n5{r8c9 r9c8} - r9c4{n5 n1} - r9c2{n1 n3} - b7n2{r9c2 r7c2} - r3c2{n2 n7} - r2c1{n7 n3} - r2c9{n3 n7} - c8n7{r3 r7} - r8n7{c9 c4} - b8n5{r8c4 r8c5} - r2n5{c5 .} ==> r7c5 ≠ 9

whip[18]: r7n9{c3 c4} - r3n9{c4 c5} - r1n9{c4 c9} - b3n6{r1c9 r3c9} - c9n2{r3 r8} - b9n5{r8c9 r9c8} - r9c4{n5 n1} - r8n1{c4 c3} - c3n2{r8 r5} - c3n3{r5 r6} - c3n5{r6 r4} - r2n5{c3 c5} - b8n5{r9c5 r8c4} - r8n7{c4 c7} - b9n3{r8c7 r7c8} - b6n3{r5c8 r5c9} - r2c9{n3 n7} - b6n7{r6c9 .} ==> r2c3 ≠ 9

whip[6]: r9c1{n4 n9} - c3n9{r7 r1} - b1n6{r1c3 r3c1} - r3c9{n6 n2} - b9n2{r8c9 r7c7} - b9n6{r7c7 .} ==> r9c7 ≠ 4

whip[3]: c7n4{r1 r7} - b9n6{r7c7 r9c7} - c7n2{r9 .} ==> r1c7 ≠ 3, r1c7 ≠ 7, r1c7 ≠ 8

whip[7]: r3n6{c1 c9} - r1n6{c9 c3} - b1n9{r1c3 r2c1} - b3n9{r2c9 r1c9} - c9n2{r1 r8} - r3n2{c9 c2} - b7n2{r9c2 .} ==> r3c1 ≠ 3

whip[7]: r3n6{c1 c9} - r1n6{c9 c3} - b1n9{r1c3 r2c1} - b3n9{r2c9 r1c9} - c9n2{r1 r8} - r3n2{c9 c2} - b7n2{r9c2 .} ==> r3c1 ≠ 7

whip[7]: r1c7{n4 n2} - b9n2{r9c7 r8c9} - c3n2{r8 r5} - c1n2{r5 r3} - b1n6{r3c1 r1c3} - b1n9{r1c3 r2c1} - b3n9{r2c9 .} ==> r1c9 ≠ 4

whip[1]: c9n4{r6 .} ==> r5c8 ≠ 4

whip[7]: r1n6{c9 c3} - r3n6{c1 c9} - c9n2{r3 r8} - c3n2{r8 r5} - c1n2{r5 r3} - b1n6{r3c1 r2c1} - b3n9{r2c9 .} ==> r1c9 ≠ 7

whip[7]: r1n6{c9 c3} - r3n6{c1 c9} - c9n2{r3 r8} - c3n2{r8 r5} - c1n2{r5 r3} - b1n6{r3c1 r2c1} - b3n9{r2c9 .} ==> r1c9 ≠ 3

whip[15]: c3n4{r5 r7} - c3n9{r7 r1} - r1n6{c3 c9} - r3c9{n6 n2} - r1n2{c7 c2} - r3c1{n2 n6} - c1n9{r3 r9} - r9n4{c1 c8} - c8n5{r9 r5} - c2n5{r5 r6} - b4n7{r6c2 r4c1} - r4c9{n7 n4} - r5n4{c9 c4} - r3n4{c4 c5} - c5n9{r3 .} ==> r6c1 ≠ 4

whip[16]: b8n2{r9c5 r7c5} - c5n6{r7 r4} - r5n6{c6 c1} - r3n6{c1 c9} - r1n6{c9 c3} - c3n9{r1 r7} - r9c1{n9 n4} - r9c8{n4 n3} - b9n5{r9c8 r8c9} - c9n2{r8 r1} - b3n9{r1c9 r2c9} - b3n3{r2c9 r2c7} - r2c1{n3 n7} - b4n7{r6c1 r6c2} - r6n5{c2 c3} - r2n5{c3 .} ==> r9c5 ≠ 5

whip[21]: b4n2{r5c1 r5c3} - c3n4{r5 r7} - c3n9{r7 r1} - r1n6{c3 c9} - r3c9{n6 n2} - r1c7{n2 n4} - b9n4{r7c7 r9c8} - c8n5{r9 r5} - c9n5{r6 r8} - r8n2{c9 c1} - b7n8{r8c1 r7c2} - c2n2{r7 r1} - r3c1{n2 n6} - r5n6{c1 c6} - r5n8{c6 c4} - r5n1{c4 c2} - b7n1{r9c2 r8c3} - r8c4{n1 n7} - b9n7{r8c9 r7c8} - r1n7{c8 c6} - c6n8{r1 .} ==> r5c1 ≠ 4

whip[24]: b3n2{r1c7 r3c9} - r3n6{c9 c1} - r1c3{n6 n9} - r7c3{n9 n4} - b9n4{r7c7 r9c8} - c8n5{r9 r5} - c2n5{r5 r6} - c2n7{r6 r3} - r2c1{n7 n3} - c1n9{r2 r9} - c5n9{r9 r3} - r3n4{c5 c4} - r5n4{c4 c9} - r4c9{n4 n7} - r6c9{n7 n3} - r6c3{n3 n1} - r6c7{n1 n8} - c7n3{r6 r8} - r8n7{c7 c4} - r8n1{c4 c5} - r9c4{n1 n5} - r1c4{n5 n8} - r2c5{n8 n5} - r1n5{c5 .} ==> r1c2 ≠ 2

whip[1]: r1n2{c9 .} ==> r3c9 ≠ 2

singles ==> r3c9 = 6, r1c3 = 6, r7c3 = 9, r9c1 = 4

whip[9]: c8n5{r9 r5} - c9n5{r4 r8} - c9n2{r8 r1} - b3n9{r1c9 r2c9} - b3n3{r2c9 r2c7} - r2c1{n3 n7} - b4n7{r6c1 r6c2} - c2n5{r6 r1} - r2c3{n5 .} ==> r9c8 ≠ 3

naked-single ==> r9c8 = 5

whip[4]: b9n3{r8c7 r7c8} - b7n3{r7c2 r9c2} - r1n3{c2 c6} - r3n3{c6 .} ==> r8c5 ≠ 3

whip[8]: c8n7{r1 r7} - r8c7{n7 n3} - r8c9{n3 n2} - r1c9{n2 n9} - r2c9{n9 n3} - b6n3{r6c9 r5c8} - c1n3{r5 r6} - c3n3{r6 .} ==> r2c7 ≠ 7

whip[11]: b3n7{r3c8 r2c9} - r2n9{c9 c1} - r3c1{n9 n2} - b4n2{r5c1 r5c3} - r8n2{c3 c9} - b9n3{r8c9 r8c7} - r8n7{c7 c4} - b8n5{r8c4 r8c5} - r2n5{c5 c3} - c3n3{r2 r6} - c1n3{r5 .} ==> r7c8 ≠ 7

whip[1]: c8n7{r1 .} ==> r2c9 ≠ 7

whip[1]: r7n7{c4 .} ==> r8c4 ≠ 7

whip[7]: r2c9{n3 n9} - b1n9{r2c1 r3c1} - b1n2{r3c1 r3c2} - r3n3{c2 c8} - r1n3{c8 c2} - r9n3{c2 c6} - r7n3{c5 .} ==> r2c5 ≠ 3

whip[7]: r8c7{n3 n7} - r8c9{n7 n2} - r1c9{n2 n9} - r2c9{n9 n3} - c7n3{r2 r6} - c3n3{r6 r5} - c3n2{r5 .} ==> r8c1 ≠ 3

whip[3]: c1n3{r5 r2} - c7n3{r2 r8} - b7n3{r8c3 .} ==> r6c2 ≠ 3

whip[4]: b7n1{r9c2 r8c3} - r8n2{c3 c9} - r8n7{c9 c7} - r8n3{c7 .} ==> r9c2 ≠ 2

whip[4]: b8n9{r9c4 r9c5} - r1n9{c5 c9} - c9n2{r1 r8} - r9n2{c7 .} ==> r3c4 ≠ 9

whip[4]: b7n3{r9c2 r8c3} - b1n3{r2c3 r2c1} - c9n3{r2 r6} - c7n3{r6 .} ==> r5c2 ≠ 3

whip[6]: r1c7{n4 n2} - r9n2{c7 c5} - c5n9{r9 r3} - r3n4{c5 c8} - r7c8{n4 n3} - c5n3{r7 .} ==> r1c5 ≠ 4

whip[6]: r2c9{n3 n9} - b1n9{r2c1 r3c1} - b1n2{r3c1 r3c2} - b1n3{r3c2 r1c2} - b7n3{r9c2 r8c3} - b9n3{r8c9 .} ==> r2c8 ≠ 3

whip[8]: r2c9{n3 n9} - r1c9{n9 n2} - r1c7{n2 n4} - b9n4{r7c7 r7c8} - c8n3{r7 r5} - c1n3{r5 r6} - c9n3{r6 r8} - c3n3{r8 .} ==> r2c7 ≠ 3

whip[4]: c7n3{r6 r8} - c9n3{r8 r2} - c1n3{r2 r6} - c3n3{r6 .} ==> r5c8 ≠ 3

whip[6]: b4n7{r6c2 r4c1} - b4n6{r4c1 r5c1} - b4n2{r5c1 r5c3} - r5n3{c3 c9} - c7n3{r6 r8} - b9n7{r8c7 .} ==> r6c9 ≠ 7

whip[7]: r9c2{n3 n1} - r9c6{n1 n6} - r5n6{c6 c1} - b4n2{r5c1 r5c3} - r8c3{n2 n3} - b4n3{r5c3 r6c1} - c7n3{r6 .} ==> r9c5 ≠ 3

whip[5]: r2n3{c1 c9} - b3n9{r2c9 r1c9} - b2n9{r1c4 r3c5} - b2n3{r3c5 r3c6} - r9n3{c6 .} ==> r1c2 ≠ 3

whip[4]: r3n2{c2 c1} - b1n9{r3c1 r2c1} - r2c9{n9 n3} - b1n3{r2c3 .} ==> r3c2 ≠ 7

whip[9]: b8n2{r9c5 r7c5} - c5n6{r7 r4} - r5n6{c6 c1} - b4n2{r5c1 r5c3} - r5n3{c3 c9} - r5n4{c9 c4} - c5n4{r6 r3} - c5n9{r3 r1} - c5n3{r1 .} ==> r9c5 ≠ 1

whip[9]: b8n2{r7c5 r9c5} - c5n6{r9 r4} - r5n6{c6 c1} - b4n2{r5c1 r5c3} - r5n3{c3 c9} - c7n3{r6 r8} - r8c3{n3 n1} - r8c5{n1 n5} - r8c4{n5 .} ==> r7c5 ≠ 8

whip[10]: r9c4{n9 n1} - r9c2{n1 n3} - r9c6{n3 n6} - r5n6{c6 c1} - b4n2{r5c1 r5c3} - r5n3{c3 c9} - r5n4{c9 c4} - c4n5{r5 r8} - r8c5{n5 n8} - c4n8{r7 .} ==> r1c4 ≠ 9

hidden-single-in-a-column ==> r9c4 = 9

whip[2]: r9c7{n6 n2} - r9c5{n2 .} ==> r9c6 ≠ 6

whip[5]: r1n9{c5 c9} - r2c9{n9 n3} - b1n3{r2c1 r3c2} - r9n3{c2 c6} - b2n3{r1c6 .} ==> r1c5 ≠ 8

whip[5]: r1n9{c5 c9} - r2c9{n9 n3} - b1n3{r2c1 r3c2} - r9n3{c2 c6} - b2n3{r1c6 .} ==> r1c5 ≠ 5

whip[6]: r5n6{c6 c1} - b4n2{r5c1 r5c3} - r5n3{c3 c9} - c7n3{r6 r8} - r8c3{n3 n1} - b8n1{r8c4 .} ==> r5c6 ≠ 1

whip[8]: c5n9{r3 r1} - c5n3{r1 r7} - r9c6{n3 n1} - b5n1{r4c6 r5c4} - c8n1{r5 r2} - r2n7{c8 c1} - b4n7{r6c1 r6c2} - c2n1{r6 .} ==> r3c5 ≠ 1

whip[8]: r5c8{n1 n8} - r5c2{n8 n5} - r5c4{n5 n4} - c5n4{r6 r3} - c5n9{r3 r1} - c5n3{r1 r7} - r9n3{c6 c2} - b7n1{r9c2 .} ==> r5c3 ≠ 1

whip[9]: c3n2{r5 r8} - b7n1{r8c3 r9c2} - r5c2{n1 n8} - r5c8{n8 n1} - r5c4{n1 n4} - c5n4{r6 r3} - c5n9{r3 r1} - c5n3{r1 r7} - r9n3{c6 .} ==> r5c3 ≠ 5

whip[9]: b4n2{r5c1 r5c3} - r5n3{c3 c9} - r5n4{c9 c4} - c5n4{r6 r3} - c5n9{r3 r1} - c5n3{r1 r7} - b8n2{r7c5 r9c5} - c5n6{r9 r4} - r5n6{c6 .} ==> r5c1 ≠ 8

whip[9]: r5n6{c6 c1} - b4n2{r5c1 r5c3} - r5n3{c3 c9} - r5n4{c9 c4} - c5n4{r6 r3} - c5n9{r3 r1} - c5n3{r1 r7} - b8n6{r7c5 r9c5} - b8n2{r9c5 .} ==> r4c6 ≠ 6 whip[7]: b5n6{r5c6 r4c5} - r9c5{n6 n2} - r7c5{n2 n3} - r9c6{n3 n1} - r8n1{c4 c3} - r4n1{c3 c7} - r5c8{n1 .} ==> r5c6 ≠ 8

whip[5]: b4n6{r4c1 r5c1} - r5c6{n6 n5} - r5c2{n5 n1} - r9n1{c2 c6} - r4c6{n1 .} ==> r4c1 ≠ 8

whip[5]: b1n5{r2c3 r1c2} - c2n7{r1 r6} - r4c1{n7 n6} - r5n6{c1 c6} - c6n5{r5 .} ==> r4c3 ≠ 5

whip[6]: c2n7{r6 r1} - c2n5{r1 r5} - b4n8{r5c2 r6c1} - b4n7{r6c1 r4c1} - b4n6{r4c1 r5c1} - r5c6{n6 .} ==> r6c2 ≠ 1

whip[5]: r4n8{c6 c7} - r2c7{n8 n1} - r2c5{n1 n5} - c3n5{r2 r6} - r6n1{c3 .} ==> r6c5 ≠ 8

whip[7]: r5c8{n1 n8} - r5c2{n8 n5} - r1c2{n5 n7} - r2n7{c1 c8} - c8n1{r2 r3} - c6n1{r3 r9} - c2n1{r9 .} ==> r5c4 ≠ 1

whip[3]: c4n1{r3 r8} - r9n1{c6 c2} - r5n1{c2 .} ==> r3c8 ≠ 1

whip[1]: r3n1{c4 .} ==> r2c5 ≠ 1

whip[5]: r7c8{n4 n3} - b3n3{r3c8 r2c9} - r2n9{c9 c1} - r2n7{c1 c8} - r3c8{n7 .} ==> r1c8 ≠ 4

whip[5]: r2c5{n8 n5} - b8n5{r8c5 r8c4} - r5c4{n5 n4} - r6c5{n4 n1} - r8c5{n1 .} ==> r4c5 ≠ 8

whip[4]: b7n8{r8c1 r7c2} - r5n8{c2 c8} - r1n8{c8 c6} - b5n8{r4c6 .} ==> r8c4 ≠ 8

whip[4]: r7c4{n7 n8} - c5n8{r8 r2} - b2n5{r2c5 r1c6} - r1c2{n5 .} ==> r1c4 ≠ 7

whip[5]: c5n8{r2 r8} - b8n5{r8c5 r8c4} - r8n1{c4 c3} - c2n1{r9 r5} - c8n1{r5 .} ==> r2c8 ≠ 8

whip[2]: b5n8{r4c6 r5c4} - c8n8{r5 .} ==> r1c6 ≠ 8

whip[4]: r1c2{n5 n7} - r1c6{n7 n3} - r9c6{n3 n1} - r8c4{n1 .} ==> r1c4 ≠ 5

whip[3]: c6n8{r4 r7} - c5n8{r8 r2} - b2n5{r2c5 .} ==> r4c6 ≠ 5

whip[4]: c2n7{r1 r6} - r4c1{n7 n6} - r5n6{c1 c6} - c6n5{r5 .} ==> r1c2 ≠ 5

singles to the end

5.11. Whips in n-Queens and Latin Squares; definition of SudoQueens

In this final section, mainly about the n-Queens problem, we show that the rules introduced in this chapter work concretely for other CSPs than Sudoku or Latin Squares. We also show that n-Queens has whips of length 1 and how they look like.

Using the LatinSquare CSP, we also show that a CSP with no whips of length 1 can nevertheless have longer ones. Finally, we introduce the n-SudoQueens CSP.

5.11.1. The n-Queens CSP

Given an n×n chessboard, the n-Queens CSP consists of placing n queens on it in such a way that no two queens appear in the same row, column or diagonal.

Here again, as in the Sudoku case, we introduce redundant sets of CSP variables:
- for each $r°$ in {r1, r2, ..., rn}, CSP variable $Xr°$ with values in {c1, c2, ..., cn};
- for each $c°$ in {c1, c2, ..., cn}, CSP variable $Xc°$ with values in {r1, r2, ..., rn}.

We define CSP-Variable-Type as the sort with domain {r, c} and Constraint-Type as the super-sort of CSP-Variable-Type with domain {r, c, f, s} corresponding to the four types of constraints: along row, column, first diagonal and second diagonal. Notice that there are now other constraints (f and s) than those taken care of by the CSP variables (corresponding to the r and c in Constraint-Type). And there is no possibility of adding CSP variables for the constraints along the diagonals: although no two queens may appear in the same diagonal, there are diagonals with no queen (there are 2n-1 diagonals of each kind); if we tried to define them as CSP variables, some of them would have no value.

For each $r°$ in $\{r1, r2, ..., rn\}$ and each $c°$ in $\{c1, c2, ..., cn\}$, we define label $(r°, c°)$ or $r°c°$ as corresponding to the two <variable, value> pairs <$Xr°, c°$> and <$Xc°, r°$> (which is equivalent to the implicit axiom: $Xr° = c° \Leftrightarrow Xc° = r°$). A label can be assimilated with a cell in the grid.

Easy details of the model (in particular the writing of the constraints along rows, columns and diagonals) are left as an exercise for the reader. Similarly, the explicit writing of the Basic Resolution Theory BRT(n-Queens) is considered as obvious. As for whips, they need no specific definition; they are part of the general theory.

In all the forthcoming figures for n-Queens, the * signs represent the given queens; the small ° signs represent the candidates eliminated by ECP at the start of the resolution process; the A, B, C, ... letters represent the candidates eliminated by resolution rules after the first ECP, in this order; the + signs represent the queens placed by the Single rule (at any time in the resolution process).

Notice that all our solutions for n-Queens were obtained manually; therefore, the resolution path for some of them may not be the shortest possible and the resolution theory in which the solution is obtained may not be the weakest possible. For lack of a generator of minimal instances, all our examples were built manually and they remain elementary. Our only ambition with respect to the n-Queens CSP is to illustrate how our general concepts can be applied and how our patterns look in them; it is not to produce any classification results.

5.11.2. Simple whips of length 1 and 2 in 8-Queens

For the 8-Queens CSP, consider the instance described in Figure 5.6, with 3 queens already given (in positions r1c2, r2c7 and r3c5). After the first obvious ECP eliminations, the Single rule cannot be applied. But we have the following resolution path with whips of lengths 1 and 2.

***** Manual solution *****
whip[1]: r6{c4 .} $\Rightarrow \neg$r8c4 (A eliminated)
whip[2]: r6{c4 c6} – r8{c6 .} $\Rightarrow \neg$r7c3, $\neg$r7c4 (B and C eliminated)
single in c4: r6c4; single in c6: r7c6; single in r5: r5c1; single in r4: r4c8; single in r8: r8c3
Solution found in W_2.

Notice the first whip[1], in the grey cells, with an interaction of a column and a diagonal occurring in a row at a relatively small distance from the target; it proves that there are whips of length 1 in n-Queens and it shows how they can look.

	c1	c2	c3	c4	c5	c6	c7	c8
r1	o	*	o	o	o	o	o	o
r2	o	o	o	o	o	o	*	o
r3	o	o	o	o	*	o	o	o
r4		o		o	o	o	o	+
r5	+	o	o	o	o	o	o	
r6	o	o	o	+	o		o	o
r7		o	B	C	o	+	o	o
r8	o	o	+	A	o		o	

Figure 5.6. An 8-Queens instance solved by whips

5.11.3. Whips[1] in 10-Queens with long distance interactions

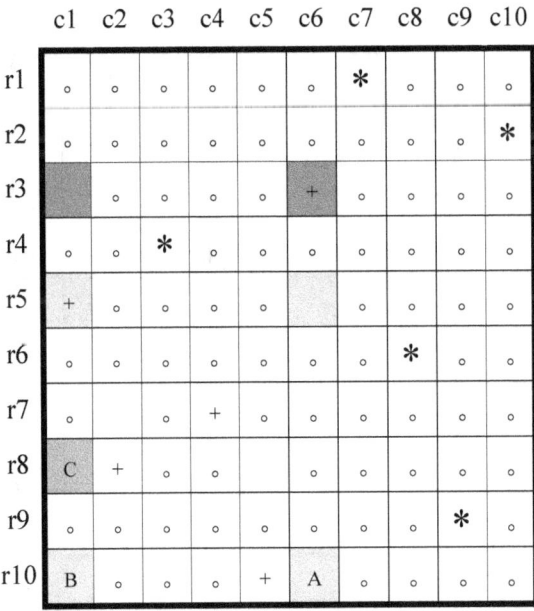

Figure 5.7. A 10-Queens instance, with 3 whips[1] based on long distance interactions

The instance of 10-Queens in Figure 5.7 shows that whip[1] interactions can happen on much longer distances than in the previous example. (They can also happen at distance 0, i.e. in the row or column adjacent to the target, or at still longer distances in n-Queens for large n.)

This puzzle has five queens already given (in r1c7, r2c10, r4c3, r6c8 and r9c9). Its first three whips[1] have interactions of a column and a diagonal in rows at long distances from their targets. After them, it can be solved by singles.

***** Manual solution *****
whip[1]: r5{c6 .} $\Rightarrow$ ¬r10c6 (A eliminated); whip in light grey cells with target A
whip[1]: r5{c6 .} $\Rightarrow$ ¬r10c1 (B eliminated); "same" whip in light grey cells, but with target B
whip[1]: r3{c1 .} $\Rightarrow$ ¬r8c1 (C eliminated); whip in dark grey cells with target C
single in r10: r10c5; single in r8: r8c2; single in r7: r7c4; single in r5: r5c1; single in r3: r3c6
Solution found in W_1.

5.11.4. Another kind of whip[1] in n-Queens

The instance of 9-Queens in Figure 5.8, with three queens already given (in r3c3, r6c2 and r9c7) has three whips[1] of another kind, relying on the interaction of three different constraints in a row or a column at a medium distance from the target. It can be solved in W_4.

	c1	c2	c3	c4	c5	c6	c7	c8	c9
r1	○	○	○	+	○		○		F
r2	+	○	○	○			○	○	H
r3	○	○	∗	○	○	○	○	○	○
r4		○	○	○	B	+	○		D
r5	○	○	○		○	A	○	C	+
r6	○	∗	○	○	○	○	○	○	○
r7	○	○	○	G	○		○	+	○
r8		○	○	○	+	○	○	○	E
r9	○	○	○	○	○	○	∗	○	○

Figure 5.8. *A 9-Queens instance, with another kind of whip[1]*

***** Manual solution *****
whip[1]: r7{c6 .} ⇒ ¬r5c6 (A eliminated, whip on light grey cells)
whip[1]: r8{c5 .} ⇒ ¬r4c5 (B eliminated, whip on medium grey cells and r8c5)
whip[1]: c5{r2 .} ⇒ ¬r5c8 (C eliminated, whip on dark grey cells)
whip[1]: c6{r4 .} ⇒ ¬r4c9 (D eliminated)
whip[2]: r5{c9 c4} - r7{c4 .} ⇒ ¬r8c9 (E eliminated)
whip[3]: r5{c9 c4} - r2{c1 c5} - r8{c5 .} ⇒ ¬r1c9 (F eliminated)
whip[4]: r5{c9 c4} - r4{c8 c6} - r1{c6 c8} - r8{c1 .} ⇒ ¬r2c9 (G eliminated)
whip[4]: r5{c9 c4} - r1{c4 c6} - r4{c6 c8} - r7{c8 .} ⇒ ¬r2c9 (H eliminated)
single in c9: r5c9; single in c4: r1c4; single in r4: r4c6; single in r2: r2c1; single in r8: r8c5; single in r7: r7c8.
Solution found in W_4 or gW_3.

5.11.5. An instance of 8-Queens with two solutions

Whips can also be used to produce a readable proof that an instance has two (or more) solutions. For the 8-Queens CSP, consider the instance displayed in Figure 5.9, with 3 queens already given (in positions r2c7, r3c5 and r4c8). Although it has the same solution as the example in section 5.11.1, we shall prove that it has two solutions.

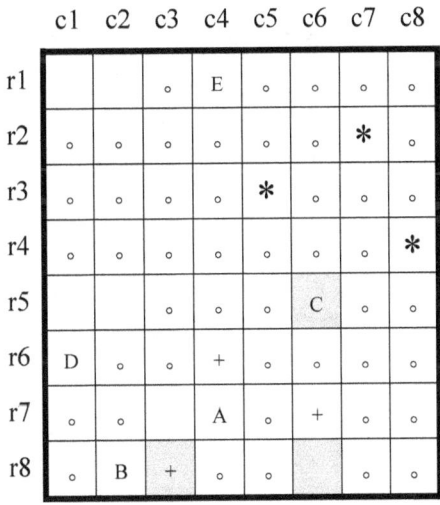

Figure 5.9. *An instance of 8-Queens with two solutions, partially solved by whips*

***** Manual solution *****
;;; The first two whips[1] display an interaction of a row and a diagonal in a column at the shortest possible distance from the target:
whip[1]: c3{r7 .} ⇒ ¬r7c4 (A eliminated)

whip[1]: c3{r8 .} $\Rightarrow$ ¬r8c2 (B eliminated)

;;; The third whip[1], in the grey cells, appearing after B has been eliminated, has an interaction of a column and a diagonal in a row at a longer distance from the target:
whip[1]: r8{c6 .} $\Rightarrow$ ¬r5c6 (C eliminated)

;;; The fourth whip[1], appearing after C has been eliminated, has an interaction of a column and a diagonal in a row, again at the shortest possible distance from the target:
whip[1]: r5{c1 .} $\Rightarrow$ ¬r6c1 (D eliminated)
whip[2]: c1{r1 r5} - c2{r5 .} $\Rightarrow$ ¬r1c4 (E eliminated)
single in r6 $\Rightarrow$ r6c4 ; single in c3 $\Rightarrow$ r8c3 ; single in r7 $\Rightarrow$ r7c6

At this point, the resolution path cannot go further because there appears to be two obvious solutions: r1c2+r5c1 (as in section 5.11.1) and r1c1+r5c2; but we have shown that whips can be used to lead from a situation where this was not obvious to one where it is.

5.11.6. An instance of 6-Queens with no solution

Whips can also provide a readable proof that an instance has no solution. Of course, this is not specific to n-Queens but true for any CSP. And the proof that an instance has no solution can be as hard as finding a solution when there is one.

Consider Figure 5.10, an instance of 6-Queens, with only two queens given in cells r4c5 and r5c2. Although these data show no direct contradiction with the constraints, a unique elimination by a whip[3] and two Singles are enough to make it obvious that there can be no solution.

	c1	c2	c3	c4	c5	c6
r1		o		+	o	o
r2		o	o		o	
r3	A	o	+	o	o	o
r4	o	o	o	o	*	o
r5	o	*	o	o	o	o
r6	o	o	o		o	

Figure 5.10. An instance of 6-Queens with no solution; proven by a whip[3]

***** Manual solution *****

whip[3]: r6{c4 c6} – r2{c6 c4} – r1{c4 .} ⇒ ¬r3c1 (A eliminated)

single in r3 ⇒ r3c3

single in r1 ⇒ r1c4

This puzzle has no solution: no value for Xr6

5.11.7. The absence of whip[1] does not preclude the existence of longer whips

The non-existence of whips of length 1 in a CSP does not preclude the existence of longer whips. Figure 5.11 gives an example of a partial whip[3] in LatinSquare.

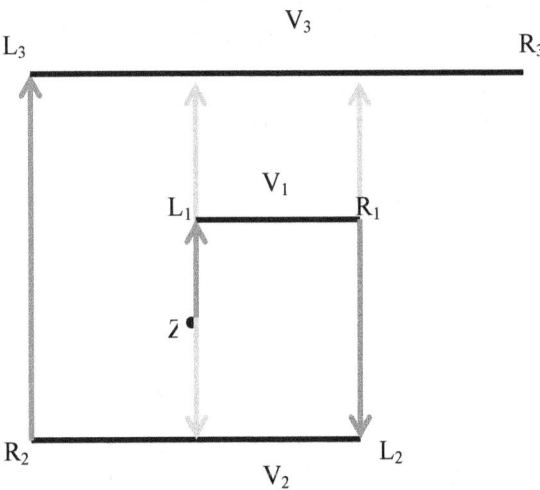

Figure 5.11. *A symbolic representation of a partial whip[3] in LatinSquare.*

In this Figure, black horizontal lines represent CSP variables (V_1, V_2, V_3); they are supposed to have candidates only at their extremities (L_k and R_k candidates) or at their meeting points with arrows (z- and t- candidates). Dark grey vertical arrows represent links from Z to L_1 or from R_k to L_{k+1}. Light grey arrows represent links to z- or t- candidates. Here, arrows represent only the flow of reasoning in the proof of the whip rule (by themselves, links are not orientated).

A particular interpretation of Figure 5.11 can be obtained by considering only labels (n, r, c) with a fixed Number n and by interpreting horizontal lines as rows and vertical lines as columns. Similarly, one can fix Row r or Column c. But these restricted visions of the symbolic representation, limited to rc-space (or cn-space, or rn-space), do not take into account the 3D symmetries of this CSP.

Similar symbolic representations, for whips in a general CSP (Figure 11.1) and for generalised whips (Figures 9.1 and 11.2) can be seen in chapters 9 and 11.

5.11.8. Defining SudoQueens

Given an integer n that is a square ($n = m^2$) and starting from the n-Queens CSP, one can define the n-SudoQueens CSP by the additional constraint that there should not be two queens in the same m×m block, where blocks are defined as in Sudoku.

In this new CSP, we can use the same two coordinate systems as in Sudoku, with the same relations between them. Because it implies that there must be one queen in each square, the new constraint can be taken care of by n new CSP-Variables b1, ..., bn, all with domain {s1, ..., sn} and/or by a new CSP-Variable-Type: b.

It is easy to check that n-SudoQueens has no instances for n=2 or n=4 (i.e. m=1 or m=2). But, as shown by the example in Figure 5.12, it has for $n \geq 9$ (m≥3).

In n-SudoQueens, one can find two types of whips[1]: the same as in n-Queens and the same as in Sudoku[n].

	c1	c2	c3	c4	c5	c6	c7	c8	c9
r1	*	o	o	o	o	o	o	o	o
r2	o	o	o	o	o	o	o	*	o
r3	o	o	o	o	*	o	o	o	o
r4	o	o	*	o	o	o	o	o	o
r5	o	o	o	o	o	*	o	o	o
r6	o	o	o	o	o	o	o	o	*
r7	o	*	o	o	o	o	o	o	o
r8	o	o	o	*	o	o	o	o	o
r9	o	o	o	o	o	o	*	o	o

Figure 5.12. *A complete grid for 9-SudoQueens*

6. Unbiased statistics and whip classification results

In the previous chapter, we gave a pure logic definition of the W and B ratings of an instance P, as the smallest n ($0 \leq n \leq \infty$) such that P can be solved by resolution theory W_n [respectively B_n]. Because these theories involve longer whips [resp. braids] as n increases, it is *a priori* meaningful for any CSP to chose W(P) [resp. B(P)] as a measure of complexity for P. In the Sudoku case, there are additional justifications, based on results[1] obtained with our SudoRules solver:

– W [resp. B] is strongly correlated with the logarithm of the number of partial whips [resp. braids] one must check before finding the solution when the "simplest first" strategy is adopted[2,3];

– for W ≤ 9,[4] W [resp. B] is strongly correlated with SER[5], the Sudoku Explainer rating, a rating widely used in the Sudoku community; it often gives some rough idea of the difficulty of a puzzle for a human player (at least for SER ≤ 9.3);

– W is also strongly correlated with less popular ratings (see our website).

It should however be noted that a rating based on the hardest step (instead of e.g. the whole resolution path) can only be meaningful statistically. (This applies also to SER.) In particular, there remains much variance in the number of partial chains needed to solve Sudoku puzzles with W(P) = n, n fixed. Based on the thousands of resolution paths we observed in detail, one explanation is that a puzzle P with W(P) = n can be hard to solve with whips [or any other type of pattern: braids, g-whips, …] for two opposed reasons: either because it does not have enough smaller whips [patterns of this type] or because it has too many useless ones.

---

[1] Details and additional correlation results can be found on our website.

[2] Although this number is not completely independent of implementation (it depends in part on the resolution path chosen), it is statistically meaningful.

[3] In this situation, W is also strongly correlated with the logarithm of the resolution time, but this is mainly a consequence of the previous correlation (and computation times are too dependent on implementation to be good indicators).

[4] For larger values of W, the number of available instances in our unbiased samples is too small to compute meaningful correlations.

[5] This rating is the most widely referred to on Sudoku Web forums, though it is defined only by non-documented Java code, it is not invariant under symmetry and it is based neither on any general theory nor (for the most part of it) on popular resolution techniques.

The results[6] reported in this chapter required several months of (2.66 GHz) CPU time (for the generation of unbiased samples and for the computation of ratings). They will show that:

– building unbiased uncorrelated samples of minimal instances of a CSP and obtaining unbiased statistics can be very hard;

– (loopless) whips have a very strong resolution power, at least for Sudoku: the ten million puzzles we have produced using different kinds of random generators could all be solved by whips (of relatively short length: 93.9% by whips of length no more than 4, 99.9% by whips of length no more than 7 and 99.99% by whips of length no more than 9 – see Table 6.4).

Only the main results of direct relevance to the topic of this book are provided here; many additional statistical results for Sudoku can be found on our website.

Although we can only present such results in the specific context of the Sudoku CSP, the sample generation methods described here (bottom-up, top-down and controlled-bias) could be extended to many CSPs. The specific $P(n+1)/P(n)$ formula proven in section 6.2.2 for the controlled-bias generator will not hold in any CSP, but the same approach can in many cases help understand the existence of a very strong bias in the samples with respect to the number of clues. Probably, it can also help explain the well-known fact that, for many CSPs, it is very difficult to generate hard instances.

The number of clues may not be a criterion of much interest in itself, but the existence of such a strong bias in it suggests the possibility of a bias with respect to many other different classification criteria, even if they are weakly correlated with the number of clues: in the Sudoku case, preliminary analyses showed that the correlation coefficient between the W rating and the number of clues is only 0.12, but tables 6.3 and 6.4 below show that the bias in the generators has nevertheless a very noticeable impact on the classification of instances according to the W rating.

Even in the very structured and apparently simple Sudoku domain, none of this was clear before the present analysis. In particular, as the results in *HLS* were based on a top-down generator, they were biased.

Acknowledgements: Thanks are due to "Eleven" for implementing the first modification (suexg-cb) of a well-known top-down generator (suexg, written in C) compliant with the specification of controlled-bias defined below, and then several faster versions of it; this allowed to turn the whole idea into reality. Thanks to Paul Isaacson for adapting Brian Turner's fast solver so that it could be used instead of that of suexg. Thanks to Glenn Fowler (gsf) for providing an *a priori* unbiased

---

[6] first published on the Sudoku Player's Forum (July to October 2009) and in [Berthier 2009]

source of complete grids: the full (compressed) collection of their equivalence classes together with a fast decompressor. Thanks also, for discussions and/or various contributions, to Allan Barker, Coloin, David P. Bird, Mike Metcalf, Red Ed (who was first to suggest the existence of a bias in the current generators). The informal collaboration that the controlled-bias idea sprouted on the Sudoku Player's Forum was very productive: due to several independent optimisations, the last version of suexg-cb (which does not retain much of the original suexg code) is 200 times faster than the first.

All the generators mentioned below are available on our website.

6.1 Classical top-down and bottom-up generators

There is a very simple procedure for generating an unbiased sample of n uncorrelated minimal Sudoku puzzles:

```
1) set p = 0 and list = ();
2) if p = n then return list;
3) randomly choose a complete grid P;
4) for each cell in P, delete its value with probability 0.5, thus
obtaining a puzzle Q;
5) if Q is minimal then add Q to list and set p = p+1 else goto 3.
```

Unfortunately, the probability of getting a valid puzzle this way is infinitesimal for each complete grid tried as a starting point (see last column of Table 6.2, which should be combined for each n with the probability of obtaining 81-n deletions). One has no choice but rely on more efficient generators. Before going further, let us introduce the two classical algorithms that have been widely used in the Sudoku community for generating minimal puzzles: bottom-up and top-down.

A standard bottom-up generator works as follows to produce n minimal puzzles:

```
1) set p = 0 and list = ();
2) if p = n then return list;
3a) set p = p+1 and start from an empty grid P;
3b) in P, randomly choose an undecided cell and a value for it, thus
getting a puzzle Q with one more clue than P;
3c) if Q is minimal, then add it to list and goto 2;
3d) if Q has several solutions, then set P = Q and goto 3b;
3e) if Q has no solution, then goto 3b (i.e. backtrack: forget Q and
try another cell from P).
```

A standard top-down generator works as follows to produce n minimal puzzles:

```
1) set p = 0 and list = ();
2) if p = n then return list;
3a) set p = p+1 and randomly choose a complete grid P;
```

```
3b) randomly choose one clue from P and delete it, thus obtaining a
puzzle Q;
3c) if Q still has only one solution but is not minimal, set P=Q and
goto 3b (for trying to delete one more clue);
3d) if Q is minimal, then add it to list and goto 2;
3e) otherwise, i.e. if Q has several solutions, then goto 3b (i.e.
reinsert the clue just deleted and try deleting another clue from P).
```

Notice that, in both cases, a minimal puzzle is produced from each complete random grid. Backtracking (i.e. clause 3e in both cases) makes any formal analysis of these algorithms very difficult. However, at first sight, it seems that it causes the generator to look for puzzles with fewer clues (this intuition will be confirmed in section 6.3). It may thus be suspected of introducing a strong, uncontrolled bias with respect to the number of clues, which, in turn, may induce a bias with respect to other properties of the collection of puzzles generated.

6.2 A controlled-bias generator

No unbiased generator of uncorrelated minimal puzzles is currently known and building such a generator with reasonable computation times seems out of reach. We therefore decided to proceed differently: taking the generators (more or less) as they are and applying corrections for the bias, if we can estimate it.

This idea was inspired by an article we read in a newspaper about what is done in digital cameras: instead of complex optimisations of the lenses to reduce typical anomalies (such as chromatic aberration, purple fringing, barrel or pincushion distortion…) – optimisations that lead to large and expensive lenses –, some camera makers now accept a small amount of these in the lenses and they take advantage of the huge computational power available in the processors to correct the result in real time with dedicated software before recording the photo.

The main question was then: can we determine the bias of the classical top-down or bottom-up generators? The answer was negative. But there appeared to be a medium way between "improving the lens to make it perfect" and "correcting its small defects by software": we devised a modification of the top-down generator that allows a precise mathematical computation of the bias.

6.2.1. Definition of the controlled-bias generator

Consider the following, modified top-down generator, the *controlled-bias generator* for producing n minimal uncorrelated puzzles:

```
1) set p = 0 and list = ();
2) if p = n then return list;
3a) randomly choose a complete grid P;
```

3b) randomly choose one clue from P and delete it, thus obtaining a puzzle Q;
3c) if Q still has only one solution but is not minimal, set P=Q and goto 3b (for trying to delete one more clue);
3d) if Q is minimal, then add it to list, set p = p+1 and goto 2;
3e) otherwise, i.e. if Q has several solutions, then goto 3a (i.e. forget everything about P and restart with another complete grid).

The only difference with the top-down algorithm is in clause 3e: if a multi-solution puzzle is encountered, instead of backtracking to the previous state, the current complete grid is merely discarded and the search for a minimal puzzle is restarted with another complete grid.

Notice that, contrary to the standard bottom-up or top-down generators, which produce one minimal puzzle per complete grid, the controlled-bias generator will generally use several complete grids before it outputs a minimal puzzle. The efficiency question is: how many? Experimentations show that many complete grids (approximately 257,514 in the mean) are necessary before a minimal puzzle is reached. But this question is about the efficiency of the generator, it is not a conceptual problem.

The controlled-bias generator has the same output and will therefore produce minimal puzzles according to the same probability distribution as its following "virtual" counterpart:

1) set p = 0 and list = ();
2) if p = n then return list;
3a) randomly choose a complete grid P;
3b) if P has no more clue, then goto 2 else randomly choose one clue from P and delete it, thus obtaining a puzzle Q;
3c) if Q is minimal, add Q to list, set P=Q, set p=p+1 and goto 3b;
3d) otherwise, set P=Q and goto 3b.

The only difference with the controlled-bias generator is that, once it has found a minimal or a multi-solution puzzle, instead of exiting, this virtual generator continues along a useless path until it reaches the empty grid.

But this virtual generator is interesting theoretically because it works similarly to the random uniform search defined in the next section and according to the same transition probabilities; and it outputs minimal puzzles according to the probability Pr on the set B of minimal puzzles defined below.

6.2.2. Analysis of the controlled-bias generator

We now build our formal model of the controlled-bias generator.

Let us first introduce the notion of a *doubly indexed puzzle*. We consider only (single or multi solution) consistent puzzles P. The double index of a doubly indexed puzzle P has a clear intuitive meaning: the first index is one of its solution grids and the second index is a sequence (notice: not a set, but a sequence, i.e. an ordered set) of clue deletions leading from this complete grid to P. In a sense, the double index keeps track of the full generation process.

Given a doubly indexed puzzle Q, there is an underlying singly-indexed puzzle: the ordinary puzzle obtained by forgetting the second index of Q, i.e. by remembering the solution grid from which it came and by forgetting the order of the deletions leading from this solution to Q. Given a doubly indexed puzzle Q, there is also a non indexed puzzle, obtained by forgetting the two indices.

For a single solution doubly indexed puzzle, the first index is useless as it can be computed from the puzzle; in this case singly indexed and non-indexed are equivalent. This is true in particular for minimal puzzles. In terms of the generator, it could equivalently output minimal puzzles or couples (minimal-puzzle, solution).

Consider now the following layered structure (a forest, in the graph-theoretic sense, i.e. a set of disjoint trees, with branches pointing downwards), the nodes being (single or multi solution) doubly indexed puzzles:

– floor 81 : the N different complete solution grids (considered as puzzles), each indexed by itself and by the empty sequence; notice that all the puzzles at floor 81 have 81 clues;

– recursive step: given floor $n+1$, where each doubly indexed puzzle has $n+1$ clues and is indexed by a complete grid that solves it and by a sequence of length $81-(n+1)$, build floor n as follows:
each doubly indexed puzzle Q at floor $n+1$ sprouts $n+1$ branches; for each clue C in Q, there is a branch leading to a doubly indexed puzzle R at floor n: R is obtained from Q by removing clue C; its first index is identical to that of Q and its second index is the $(81-n)$-element sequence obtained by appending C to the end of the second index of Q; notice that all the doubly indexed puzzles at floor n have n clues and the length of their second index is equal to $1 + (81-(n+1)) = 81-n$.

It is easy to see that, at floor n, each doubly indexed puzzle has an underlying singly indexed puzzle identical to that of $(81 - n)!$ doubly indexed puzzles with the same first index at the same floor (including itself).

This is equivalent to saying that, at any floor $n < 81$, any singly indexed puzzle Q can be reached by exactly $(81 - n)!$ different paths from the top (all of which start necessarily from the complete grid defined as the first index of Q). These paths are the $(81 - n)!$ different ways of deleting one by one its missing $81-n$ clues from its solution grid.

Notice that this would not be true for non-indexed puzzles that have multiple solutions. This is where the first index is useful.

Let N be the number of complete grids (N is known to be close to 6.67×10^{21}, but this is pointless here). At each floor n, there are $N \times 81! / n!$ doubly indexed puzzles and $N \times 81! / (81-n)! / n!$ singly indexed puzzles. For each n, there is therefore a uniform probability $P(n) = 1/N \times 1/81! \times (81-n)! \times n!$ that a singly indexed puzzle Q at floor n is reached by a random (uniform) search starting from one of the complete grids. What is important here is the ratio: $P(n+1) / P(n) = (n + 1) / (81 - n)$, giving the relative probability of being reached by the generation process, for two singly indexed puzzles with respectively n+1 and n clues.

The above formula is valid globally if we start from all the complete grids, as above, but it is also valid for all the single solution puzzles if we start from a single complete grid (just forget N in the proof above). (Notice however that it is not valid if we start from a subgrid instead of a complete grid.)

Now, call B the set of (non indexed) minimal puzzles. On B, all the puzzles are minimal. Any puzzle strictly above B has redundant clues and a single solution. Notice that, for all the puzzles on B and above B, singly indexed and non-indexed puzzles are in one-to-one correspondence. Therefore, the relative probability of two minimal puzzles is given by the above formula.

On the set B of minimal puzzles, there is thus a probability Pr naturally induced by the different Pn's and it is the probability that a minimal puzzle Q is output by our controlled-bias generator. It depends only on the number of clues and it is defined by $Pr(Q) = P(n)$ if Q has n clues.

The most important here is that, by construction of Pr on B (a construction which models the workings of the virtual controlled bias generator), the fundamental relation: $Pr(n+1)/Pr(n) = (n+1)/(81-n)$ holds for any two minimal puzzles, with respectively n+1 and n clues.

For n < 41, this relation means that a minimal puzzle with n clues is more likely to be reached from the top than a minimal puzzle with n+1 clues. More precisely, we have: $Pr(40) = Pr(41)$, $Pr(39) = 42/40 \times Pr(40)$, $Pr(38) = 43/39 \times Pr(39)$. Repeated application of the formula gives $Pr(24) = 61.11 \times Pr(30)$: *a puzzle with 24 clues has about 61 times more chances of being output by the controlled-bias generator than a puzzle with 30 clues.* This is indeed a very strong bias.

A non-biased generator would give the same probability to all the minimal puzzles. The above analysis shows that *the controlled bias generator:*
- is unbiased when restricted (by filtering its output) to n-clue puzzles, for any fixed n,
- is strongly biased towards puzzles with fewer clues,

- this bias is well known and given by Pr(n+1) / Pr(n) = (n + 1) / (81 – n),
- the puzzles produced are uncorrelated, provided that the complete grids are chosen in an uncorrelated way.

As we know precisely the bias with respect to uniformity, we can correct it easily by applying correction factors cf(n) to the probabilities on B. Only the relative values of the cf(n) is important: they satisfy cf(n+1) / cf(n) = (81-n)/(n+1). Mathematically, after normalisation, cf is just the relative density of the uniform distribution on B with respect to the probability distribution Pr.

This analysis also shows that a classical top-down generator is still more strongly biased towards puzzles with fewer clues because, instead of discarding the current path when it meets a multi-solution puzzle, it backtracks to the previous floor and tries again to go deeper.

6.2.3. Computing unbiased means and standard deviations using a controlled-bias generator

In practice, how can one compute unbiased statistics of minimal puzzles based on a (large) sample produced by a controlled-bias generator? Consider any random variable X defined (at least) on the set of minimal puzzles. Define: on(n) = the number of n-clue puzzles in the sample, E(X, n) = the mean value of X for n-clue puzzles in the sample and $\sigma(X, n)$ = the standard deviation of X for n-clue puzzles in the sample.

The mean and standard-deviation of X on a sample are classically computed as:
$$\text{mean}(X) = \sum_n [E(X, n) \times on(n)] / \sum_n on(n)$$
$$\sigma(X) = \sqrt{\{\sum_n [\sigma(X, n)^2 \times on(n)] / \sum_n [on(n)]\}}.$$

The unbiased mean and standard deviation of X must then be estimated as (this is merely the mean and standard deviation for a weighted average):
unbiased-mean(X) = $\sum_n [E(X, n) \times on(n) \times cf(n)] / \sum_n [on(n) \times cf(n)]$;
unbiased-$\sigma(X) = \sqrt{\{\sum_n [\sigma(X, n)^2 \times on(n) \times cf(n)] / \sum_n [on(n) \times cf(n)]\}}$.

These formulæ show that the cf(n) sequence needs be defined only modulo a multiplicative factor. It is convenient to choose cf(26) = 1. This gives the following sequence of correction factors (in the range n = 19-31, which includes all the puzzles of all the samples we have obtained with all the random generators considered here):
[0.00134 0.00415 0.0120 0.0329 0.0843 0.204 0.464 1 2.037 3.929 7.180 12.445 20.474]

It may be shocking to consider that 30-clue puzzles in a sample must be given a weight 61 times greater than 24-clue puzzles, but it is a fact. As a result of this strong bias of the controlled-bias generator (strong but known and much smaller than the other generators), unbiased statistics for the mean number of clues of minimal puzzles (and any variable correlated with this number) must rely on extremely large samples with sufficiently many 29-clue and 30-clue puzzles.

6.3. The real distribution of clues and the number of minimal puzzles

The above formulæ show that the number-of-clue distribution of the controlled-bias generator is the key for computing unbiased statistics.

6.3.1. The number-of-clue distribution as a function of the generator

Generator → sample size → #clues↓	bottom-up 1,000,000 % (sample)	top-down 1,000,000 % (sample)	ctr-bias 5,926,343 % (sample)	real % (estimated)
20	0.028	0.0044	0.0	0.0
21	0.856	0.24	0.0030	0.000034
22	8.24	3.45	0.11	0.0034
23	27.67	17.25	1.87	0.149
24	36.38	34.23	11.85	2.28
25	20.59	29.78	30.59	13.42
26	5.45	12.21	33.82	31.94
27	0.72	2.53	17.01	32.74
28	0.054	0.27	4.17	15.48
29	0.0024	0.017	0.52	3.56
30	0	0.001	0.035	0.41
31	0	0	0.0012	0.022
mean	**23.87**	**24.38**	**25.667**	**26.577**
std-dev	1.08	1.12	1.116	1.116

Table 6.1: The experimental number-of-clue distribution (%) for the bottom-up, top-down and controlled-bias generators and the estimated real distribution.

After applying the above formulæ to estimate the real number-of-clue distribution, Table 6.1 shows that the bias with respect to the number of clues is very strong in all the generators we have considered; moreover, *controlled-bias, top-down and bottom-up are increasingly biased towards puzzles with fewer clues.* Graphically, the estimated number-of-clue distribution is very close to Gaussian.

Table 6.1 partially explains Tables 6.3 and 6.4 in section 6.4. More precisely, it explains why there can be a noticeable W rating bias in the samples produced by the bottom-up and top-down generators, in spite of the weak correlation coefficient between the number of clues and the W rating of a puzzle: the bias with respect to the number of clues is very strong in these generators.

6.3.2. Collateral result: the number of minimal puzzles

The number of minimal Sudoku puzzles has been a longstanding open question. We can now provide precise estimates for the distribution of the mean number of n-clue minimal puzzles per complete grid (mean and standard deviation in the second and third columns of Table 6.2).

number of clues	number of n-clue minimal puzzles per complete grid: mean	number of n-clue minimal puzzles per complete grid: relative error (~ 1 std dev)	mean number of tries
20	6.152×10^6	70.7%	7.6306×10^{11}
21	1.4654×10^9	7.81%	9.3056×10^9
22	1.6208×10^{12}	1.23%	2.2946×10^8
23	6.8827×10^{12}	0.30%	1.3861×10^7
24	1.0637×10^{14}	0.12%	2.1675×10^6
25	6.2495×10^{14}	0.074%	8.4111×10^5
26	1.4855×10^{15}	0.071%	7.6216×10^5
27	1.5228×10^{15}	0.10%	1.5145×10^6
28	7.2063×10^{14}	0.20%	6.1721×10^6
29	1.6751×10^{14}	0.56%	4.8527×10^7
30	1.9277×10^{13}	2.2%	7.3090×10^8
31	1.1240×10^{12}	11.6%	2.0623×10^{10}
32	4.7465×10^{10}	70.7%	7.6306×10^{11}
Total	$\mathbf{4.6655 \times 10^{15}}$	**0.065%**	

Table 6.2: Mean number of n-clue minimal puzzles per complete grid. Last column: inverse of the proportion of n-clue minimal puzzles among n-clue sub-grids

Another number of interest (e.g. for the first naïve algorithm given in section 6.1) is the mean number of tries one must do to find an n-clue minimal puzzle by randomly deleting 81-n clues from a complete grid. It is the inverse of the proportion of n-clue minimal puzzles among n-clue sub-grids, given by the last column in Table 6.2.

One can also get:

– after multiplying the total mean by the number of complete grids (known to be 6,670,903,752,021,072,936,960 [Felgenhauer et al. 2005]), *the total number of minimal Sudoku puzzles: 3.1055×10³⁷, with 0.065% relative error;*

– after multiplying the total mean by the number of non isomorphic complete grids (known to be 5,472,730,538 [Russell et al. 2006]), *the total number of non isomorphic minimal Sudoku puzzles: 2.5477×10²⁵, also with 0.065% relative error.*

6.4. The W-rating distribution as a function of the generator

We can now apply the bias correction formulæ of section 6.2.3 to estimate the W rating distribution. Table 6.3 shows that the mean W rating of the minimal puzzles in a sample depends noticeably on the type of generator used to produce them and that all the generators give rise to mean complexity below the real values.

Generator	bottom-up	top-down	ctr-bias	real
sample size	10,000	50,000	5,926,343	
W rating : mean	**1.80**	**1.94**	**2.22**	**2.45**
W rating : std dev	1.24	1.29	1.35	1.39
max W found in sample	11	13	16	

Table 6.3: The W-rating means and standard deviations for bottom-up, top-down and controlled-bias generators, compared with the estimated real values.

The mean W rating gives only a very pale idea of what really happens, because the first two levels, W_0 and W_1, concentrate a large part of the distribution, for any of the generators. With the full distributions, Table 6.4 provides more detail about the bias in the W rating for the three kinds of generators (with the same sample sizes as in Table 6.3). All these distributions have the same two modes as the real distribution, at levels W_0 and W_3. But, when one moves from bottom-up to top-down to controlled-bias to real, the mass of the distribution moves progressively to the right. This displacement towards higher complexity occurs mainly at the first W levels, after which it is only slight, but still visible.

More detailed analyses (available on our website), in particular with skewness and kurtosis, seem to show that there is a (non absolute) barrier of complexity, such that, when we consider n-clue puzzles and when the number n of clues increases:
- the n-clue mean W rating increases;
- the proportion of puzzles with W rating away from the n-clue mean increases;
but:
- the proportion of puzzles with W rating far below the n-clue mean increases;

- the proportion of puzzles with W rating far above the n-clue mean decreases.

Graphically, the W rating distribution of n-clue puzzles looks like a wave. When n increases, the wave moves to the right, with a longer tail on its left and a steeper front on its right. The same remarks apply if the W rating is replaced by the SER.

Generator → W-rating ↓	bottom-up % (sample)	top-down % (sample)	ctr-bias % (sample)	real % (estimated)
0 (first mode →)	46.27	41.76	35.08	29.17
1	13.32	12.06	9.82	8.44
2	12.36	13.84	13.05	12.61
3 (second mode →)	15.17	16.86	20.03	22.26
4	10.18	12.29	17.37	21.39
5	1.98	2.42	3.56	4.67
6	0.49	0.55	0.79	1.07
7	0.19	0.15	0.21	0.29
8	0.020	0.047	0.055	0.072
9	0.010	0.013	0.015	0.020
10	0*	$3.8\ 10^{-3}$	$4.4\ 10^{-3}$	$5.5\ 10^{-3}$
11	0.01*	$1.5\ 10^{-3}$	$1.2\ 10^{-3}$	$1.5\ 10^{-3}$
12-16	0*	$1.1\ 10^{-3}$	$4.3\ 10^{-4}$	$5.4\ 10^{-4}$

*Table 6.4: The W-rating distribution (in %) for bottom-up, top-down and controlled-bias generators, compared with the estimated real distribution. A * sign on a result means that the number of puzzles justifying it is too small to allow a precise value.*

6.5. Stability of the classification results

6.5.1. Insensivity of the controlled-bias generator wrt the source of complete grids

There remains a final question: do the above results depend on the source of complete grids? Until now, we have done as if this was not a problem. Nevertheless, producing the unbiased and uncorrelated collections of complete grids, necessary in the first step of all the puzzle generators, is all but obvious. It is known that there are 6.67×10^{21} complete grids; it is therefore impossible to have a generator scan them all. Up to isomorphisms, there are "only" 5.47×10^9 complete grids, but this remains a very large number and storing them in uncompressed format would require about half a terabyte.

In 2009, Glenn Fowler provided both a collection of all the (equivalence classes of) complete grids in a compressed format (only 6 gigabytes) and a real time decompressor. All the results reported above for the controlled bias generator were obtained with this *a priori* unbiased source of complete grids. (Notice that, due to the normalisation and compression of grids, it is unbiased only when one does full scans of its grids, whence the queer sizes of some of our samples of controlled-bias minimal puzzles).

Before this, all the generators we tried had a first phase consisting of creating a complete grid and this is where some type of bias could slip in at this level. Nevertheless, we tested several sources of complete grids based on very different generation principles and the classification results remained very stable.

This insensitivity of the controlled-bias generator to the source of complete grids can be understood intuitively: it deletes in the mean two thirds of the initial grid data and any structure that might be present in the complete grids and cause a bias is washed away by the deletion phase.

6.4.2. Insensivity of the classification results wrt the generators implementation

As can be seen from additional results on our website, we have tested several independent implementations of the bottom-up and top-down generators, using in particular various pseudo-random number generators for the selection of clue deletions (or additions in the bottom-up case); they all lead to the same conclusions.

6.6 The W rating is a good approximation of the B rating

The above statistical results are unchanged when the W rating is replaced by the B rating. Indeed, in 10,000 puzzles tested, only 20 have different W and B ratings. Moreover, *in spite of non-confluence of the whip resolution theories, the maximum length of whips in a single resolution path using only loopless whips and obtained by the "simplest first" strategy (defined in section 5.5.2 for the B rating) is a good approximation of both the W and B ratings*.

7. g-labels, g-candidates, g-whips and g-braids

After introducing the purely structural notion of a "grouped-label" or "g-label", we give a new description of whips of length one. Having g-labels (or, equivalently, whips of length one) is an intrinsic property of a CSP that has deep consequences for its resolution theories. When a CSP has g-labels, one can define two new families of resolution rules: g-whips and g-braids, extending the resolution power obtained by whips and braids.

7.1. g-labels, g-links, g-candidates and whips[1]

7.1.1. g-labels and g-links

7.1.1.1. General definition of a grouped label (g-label) in a CSP

Definition: a *g-label* is a pair <V, g>, where V is a CSP variable and g is a set of labels for V, such that:

– the cardinality of g is greater than one (but g is not the full set of labels for V);

– there is at least one label l such that l is not a label for V and l is linked (possibly by different constraints) to all the labels in g;

– for any label l that is not a label for V, l is linked (possibly by different constraints) to all the labels in g as soon as there are two of them to which it is linked (possibly by different constraints);

– g is maximal with these properties.

Miscellaneous remarks:

– when CSP variable V is clear, we often speak of g-label g, but one must be careful with this abuse of language; (see the Sudoku discussion in section 7.1.1.3);

– with this definition, a label is not a g-label (due to the first condition);

– as a result of this first condition, there are CSPs with no g-label;

– the maximality condition has the effect of minimising the number of g-labels one must consider when looking for chain patterns built on them; accepting non maximal g-labels would increase the computational complexity of the corresponding resolution rules without providing any more generality (as can easily be checked from the definitions of g-whips and g-braids below);

– one can introduce a new, auxiliary sort: g-Label, with a constant symbol for every g-label and with variable symbols g, g', g_1, g_2, …;

– in the LatinSquare CSP, there are no g-labels;

– in Sudoku, all the elements of a g-label g are linked to l by the same constraint;

– in the n-Queens CSP, there are g-labels but their different elements are always linked to l by two or three different constraints; see section 7.8.1.

7.1.1.2. g-links

Definitions: we say that a g-label <V, g> and a label l are g-linked if l is not a label for V and l is linked to all the elements of g; and we define an auxiliary predicate *g-linked* with signature (g-Label, Label) by:
g-linked(<V, g>, l) ≡ $\forall v$ ¬label(l, V, v) ∧ $\forall l' \in g$ linked(l', l); we say that a g-label <V, g> and a label l are *compatible* if they are not g-linked.

Definition: we say that a g-label <V, g> is compatible with a g-label <V', g'> if g contains some label l compatible with <V', g'> . Notice that this is a symmetric relation, in spite of the non symmetric definition (most of the time, we shall use this relation in its apparently non-symmetric form); it is equivalent to: there are some l ∈ g and some l' ∈ g' such that l and l' are not linked.

Definition: we say that a label l [respectively a g-label <V, g>] is compatible with a set S of labels and g-labels if l [resp. <V, g>] is compatible with each element of S.

7.1.1.3. Grouped labels (g-labels) in Sudoku

As an example, let us analyse the situation in Sudoku. Informally, a g-label could be defined as the set of labels for a given Number "in" the intersection of a row and a block or "in" the intersection of a column and a block (these are the only possibilities). These intersections are known respectively as row-segments and column-segments (sometimes also as mini-rows and mini-columns).

Then, g-label $(n°, r°, c_{ijk})$ would be the mediator of a symmetric conjugacy relationship between the set of labels $(n°, r°, c°_1)$ such that rc-cell $(r°, c°_1)$ is in row $r°$ but not in block $b°$ and the set of labels for <variable, value> pairs <$b°n°$, $s°_2$> such that rc-cell [$b°$, $s°_2$] is in block $b°$ but not in row $r°$. Similarly, if $(r_{ijk}, c°) =$ [$b°$, s_{pqr}], then g-label $(n°, r_{ijk}, c°)$ would be the mediator of a conjugacy between the set of labels $(n°, r°_1, c°)$ such that rc-cell $(r°_1, c°)$ is in column $c°$ but not in block $b°$ and the set of labels for <variable, value> pairs <$b°n°$, $s°_2$> such that rc-cell [$b°$, $s°_2$] is in block $b°$ but not in column $c°$.

"Conjugacy", in the above sentences, must be understood in the following sense. When two sets of labels are conjugated via a g-label as above, a proof that all the candidates from one set are impossible leads in an obvious way to a proof that all the candidates from the other set are also impossible. Thus, when one knows that, in row $r°$ [resp. in column $c°$], number $n°$ can only be in block $b°$, one can delete $n°$ from all the rc-cells in block $b°$ and not in row $r°$ [resp. and not in column $c°$]. Conversely, when one knows that, in block $b°$, number $n°$ can only be in row $r°$ [resp. in column $c°$], one can delete $n°$ from all the rc-cells in row $r°$ [resp. in column $c°$] and not in block $b°$. These rules are among the most basic ones in Sudoku; they are usually named row-block and column-block interactions (or "locked candidates"). In Sudoku, g-labels correspond to what is also sometimes called "hinges": they are hinges for the conjugacy. As shown in *HLS* (see also the end of section 7.1.2), these ***basic interactions are equivalent to whip[1]***.

Nevertheless, this kind of symmetric conjugacy between two CSP variables is specific to Sudoku. We have chosen to define the notion of a g-label in a much more general way, involving only one CSP variable, so that it can be applied when it is not the intersection of two CSP variables and there is no associated symmetric conjugacy relationship. In particular, g-labels in the n-Queens CSP (section 7.8.1) will not be defined by two CSP variables.

According to our formal definition, Sudoku has the following 972 "g-labels":

– for each Row $r°$, for each Number $n°$, three g-labels for CSP variable $Xr°n°$:
$<Xr°n°, r°n°c123>$, $<Xr°n°, r°n°c456>$ and $<Xr°n°, r°n°c789>$, where:
$r°n°c123$ is the set of three labels $\{(n°, r°, c1), (n°, r°, c2), (n°, r°, c3)\}$;
$r°n°c456$ is the set of three labels $\{(n°, r°, c4), (n°, r°, c5), (n°, r°, c6)\}$;
$r°n°c789$ is the set of three labels $\{(n°, r°, c7), (n°, r°, c8), (n°, r°, c9)\}$;

– for each Column $c°$, for each Number $n°$, three g-labels for CSP variable $Xc°n°$: $<Xc°n°, c°n°r123>$, $<Xc°n°, c°n°r456>$ and $<Xc°n°, c°n°r789>$, where:
$c°n°r123$ is the set of three labels $\{(n°, c°, r1), (n°, c°, r2), (n°, c°, r3)\}$;
$c°n°r456$ is the set of three labels $\{(n°, c°, r4), (n°, c°, r5), (n°, c°, r6)\}$;
$c°n°r789$ is the set of three labels $\{(n°, c°, r7), (n°, c°, r8), (n°, c°, r9)\}$;

– for each Block $b°$, for each Number $n°$, three g-labels for CSP variable $Xb°n°$:
$<Xb°n°, b°n°s123>$, $<Xb°n°, c°n°s456>$ and $<Xb°n°, c°n°s789>$, where:
$b°n°s123$ is the set of three labels $\{(n°, b°, s1), (n°, c°, s2), (n°, c°, s3)\}$;
$b°n°s456$ is the set of three labels $\{(n°, b°, s4), (n°, c°, s5), (n°, c°, s6)\}$;
$b°n°s789$ is the set of three labels $\{(n°, b°, s7), (n°, c°, s8), (n°, c°, s9)\}$;

– for each Block $b°$, for each Number $n°$, three g-labels for CSP variable $Xb°n°$:
$<Xb°n°, b°n°s147>$, $<Xb°n°, c°n°s258>$ and $<Xb°n°, c°n°s369>$, where:
$b°n°s147$ is the set of three labels $\{(n°, b°, s1), (n°, c°, s4), (n°, c°, s7)\}$;
$b°n°s258$ is the set of three labels $\{(n°, b°, s2), (n°, c°, s5), (n°, c°, s8)\}$;
$b°n°s369$ is the set of three labels $\{(n°, b°, s3), (n°, c°, s6), (n°, c°, s9)\}$.

The two groups of g-labels for the Xbn CSP variables may seem redundant with respect to the first two groups: their sets of label triplets are the same as the sets of label triplets related to rows and columns. But they are not considered as g-labels for the same CSP variables.

In Sudoku, this difference has always been in implicit existence with the classical distinction between the rules of interaction from blocks to rows (or columns) and rules of interaction from rows (or columns) to blocks; these rules had even been given different names: pointing and claiming (that are now falling into oblivion).

Contrary to what we did for labels (considering them as equivalence classes of pre-labels), *we do not consider two g-labels as being essentially the same if they have the same sets of labels but different underlying CSP variables*. The reason for this will be clear after the SudoQueens example in section 7.8.3.

7.1.2. g-candidates and their correspondence with whips of length one

Definitions: we say that a g-label <V, g> for a CSP variable V is a *g-candidate* for V in a resolution state RS if there are at least two different labels l_1 and l_2 in g such that l_1 and l_2 are present as candidates in RS, i.e. RS |= candidate(l_1) and RS |= candidate(l_2). Thus, in the same spirit as in the definition of a g-label, we consider that an ordinary candidate is not a g-candidate. The above defined notion of "g-linked" can be extended straightforwardly from g-labels to g-candidates, by considering the complete g-labels underlying the g-candidates. As for "compatibility" between a candidate l and a g-candidate g, there is the condition that g must contain at least two candidates compatible with l.

g-labels act like the logical "or" of several candidates (but not any combination of any candidates, only structurally fixed combinations for the same CSP variable, predefined by the set of g-labels): once one knows that the true value of V is one of those in the g-candidate, it is not necessary to know precisely which of them is true; one can always conclude that any candidate g-linked to the g-label can be eliminated.

It can also be noticed that g-labels could be used to define two kinds of extended elementary resolution rules (which could be called g-resolution rules, as they deal with g-labels, g-links, g-values and g-candidates in addition to labels, links, values and candidates): gS would assert a g-value predicate for a g-label <V, g> and gECP would eliminate any candidate g-linked to a g-label <V, g>.

But the following remark will lead us farther and will require no extension of the notion of a resolution theory. If <V, g> is a g-label for V, Z is a candidate g-linked to it and l = <V, x> is any candidate in g, then V{x .} is a whip[1] with target Z. Conversely, for any whip[1]: V{x .} with target Z, there must be at least another value x' for V such that <V, x'> is in g, is still a candidate and is linked to Z (otherwise, the whip would degenerate into a Single, a possibility we have excluded from the definition of a whip); if one defines g as the set of labels for V that are linked to Z, then $<V, g>$ is a g-label for variable V and Z is g-linked to it.

7.2. g-bivalue chains, g-whips and g-braids

We now introduce extensions of bivalue chains, whips and braids by allowing the right-linking (but not the left-linking) objects to be either candidates or g-candidates.

Definition: *a g-regular sequence of length n associated with a sequence (V1, ... Vn) of CSP variables* is a sequence of length 2n [or 2n-1] $(L_1, R_1, L_2, R_2, ... L_n, [R_n])$, such that:

– for $1{\leq}k{\leq}$ n, L_k is a candidate,

– for $1{\leq}k{\leq}$ n [or $1{\leq}k{<}n$], R_k is a candidate or a g-candidate,

– for each k, L_k has a representative $<V_k, l_k>$ with V_k and R_k is a candidate or a g-candidate $<V_k, r_k>$ for V_k; this "strong continuity" or "strong g-continuity" (depending on what R_k is) from L_k to R_k implies "continuity" or "g-continuity" (i.e. link or g-link) from L_k to R_k.
The L_k are called the *left-linking candidates* of the sequence and the R_k the *right-linking candidates or g-candidates*.

Definition: A *g-regular chain* is a g-regular sequence that satisfies all the additional R_{k-1} to L_k g-continuity conditions: L_k is linked or g-linked to R_{k-1} for all k.

7.2.1. Definition of g-bivalue chains

Definition: in any CSP, given a candidate Z (which will be a target), a *g-bivalue-chain of length n* $(n \geq 1)$ is a *g-regular chain* $(L_1, R_1, L_2, R_2, ... L_n, R_n)$ associated with a sequence $(V_1, ... V_n)$ of CSP variables, such that:

– Z is neither equal to any candidate in $\{L_1, R_1, L_2, R_2, ... L_n, R_n\}$, nor a member of any g-candidate in this set, for any $1{\leq}k{<}n$;

– Z is linked to L_1;

– R_1 is the only candidate or g-candidate for V_1 compatible with Z;

– for any $1 < k \leq n$, R_k is the only candidate or g-candidate for V_k compatible with R_{k-1};

– Z is not a label for V_m;

– Z is linked or g-linked to R_m.

Notice that these conditions imply that Z cannot be a label for any of the CSP variables V_k.

Theorem 7.1 (g-bivalue-chain rule for a general CSP): in any resolution state of any CSP, if Z is a target of a g-bivalue-chain, then it can be eliminated (formally, this rule concludes ¬candidate(Z)).

Proof: the proof is short and obvious but it will be the basis for the proof of all our forthcoming chain, whip and braid rules including g-labels.

If Z was True, then L_1 and all the other candidates for V_1 linked to Z would be eliminated by ECP; therefore R_1 would have to be or to contain the true value of V_1; but then L_2 and all the candidates for V_2 linked or g-linked to R_1 would be eliminated by ECP or W_1 and R_2 would have to be or to contain the true value of V_2....; finally R_n would have to be or to contain the true value of V_n; which would contradict the hypothesis that Z was True. Therefore Z can only be False. qed.

Notation: a g-bivalue-chain of length n, together with a potential target elimination, is written symbolically as:
g-bivalue-chain[n]: $\{L_1\ R_1\} - \{L_2\ R_2\} - ...... - \{L_n\ R_n\} \Rightarrow \neg candidate(Z)$,
where the curly brackets recall that the two candidates or g-candidates inside have representatives with the same CSP variable.

Re-writing the candidates or g-candidates as <variable, value> or <variable, g-value> pairs and "factoring" the CSP variables out of the pairs, a bivalue chain will also be written symbolically in either of the more explicit forms:
g-bivalue-chain[n]: $V_1\{l_1\ r_1\} - V_2\{l_2\ r_2\} - ...... - V_n\{l_n\ r_n\} \Rightarrow \neg candidate(Z)$, or:
g-bivalue-chain[n]: $V_1\{l_1\ r_1\} - V_2\{l_2\ r_2\} - ...... - V_n\{l_n\ r_n\} \Rightarrow V_Z \neq v_Z$.

In spite of the apparently non reversible definition, one has:

Theorem 7.2: a g-bivalue-chain is reversible.

Proof: the main point of the proof is the construction of the reversed chain (L'_1, R'_1, L'_2, R'_2, ... L'_n, R'_n). It is based on the reversed sequence of CSP variables and defined as follows (for a similar theorem, see section 9.2.2):

– $L'_k = R_{n-k+1}$ if R_{n-k+1} is a candidate; L'_k = any element in R_{n-k+1} if R_{n-k+1} is a g-candidate; thus, L'_k is always a candidate;

– $R'_k = L_{n-k+1}$ plus all the candidates for V_{n-k+1} that are linked to R_{n-k}; thus, R'_k can be a candidate or a g-candidate.

7.2.2. Definition of g-whips

Definition: given a candidate Z (which will be the target), a *g-whip* of length n ($n \geq 1$) built on Z is a g-regular sequence (L_1, R_1, L_2, R_2, ... L_n) [notice that there is no R_n] associated with a sequence (V_1, ... V_n) of CSP variables, such that:

– Z is neither equal to any candidate in $\{L_1, R_1, L_2, R_2, ... L_n\}$ nor a member of any g-candidate in this set;

– L_1 is linked to Z;

– for each $1 < k \leq n$, L_k is linked or g-linked to R_{k-1}; this is what we call g-continuity from R_{k-1} to L_k;

– for any $1 \leq k < n$, R_k is the only candidate or g-candidate for V_k compatible with Z and with all the previous right-linking candidates and g-candidates (i.e. with Z and with all the R_i, $1 \leq i < k$);

– Z is not a label for V_n;

– V_n has no candidate compatible with Z and with all the previous right-linking candidates and g-candidates (but V_n has more than one candidate).

Notice that left-linking candidates are labels, as in the case of whips; they are not g-labels. Accepting g-labels instead of labels would lead to no added generality but it would entail unnecessary complications. This is the main reason for our restrictive definition of a g-label (i.e. a label is not a g-label).

Theorem 7.3 (g-whip rule for a general CSP): in any resolution state of any CSP, if Z is a target of a g-whip, then it can be eliminated (formally, this rule concludes ¬candidate(Z)).

Proof: the proof is a simple adaptation of that for g-bivalue-chains, adding the elimination of all the z-candidates by ECP and, at each step, the elimination of all the next t-candidates by ECP or W_1. The end is slightly different: the last condition on the g-whip entails that, if the target Z was True, there would be no possible value for the last variable V_n (because it is not a CSP-Variable for Z).

7.2.3. Definition of g-braids

Definition: given a candidate Z (which will be the target), a *g-braid* of length n ($n \geq 1$) built on Z is a g-regular sequence (L_1, R_1, L_2, R_2, ... L_n) [notice that there is no R_n] associated with a sequence (V_1, ... V_n) of CSP variables, such that:

– Z is neither equal to any candidate in {L_1, R_1, L_2, R_2, ... L_n} nor a member of any g-candidate in this set;

– L_1 is linked to Z;

– for any $1 < k \leq n$, L_k is either linked to a previous right-linking candidate or to the target or g-linked to a previous right-linking g-candidate; this is the only (but major) structural difference with g-whips (for which the only linking possibility is R_{k-1}); the "g-continuity" condition of g-whips is not satisfied by g-braids;

– for any $1 \leq k < n$, R_k is the only candidate or g-candidate for V_k compatible with Z and with all the previous right-linking candidates and g-candidates (i.e. with Z and with all the R_i, $1 \leq i < k$);

– Z is not a label for V_n;

– V_n has no candidate compatible with Z and with all the previous right-linking candidates and g-candidates (but V_n has more than one candidate).

As in g-whips, left-linking candidates are labels, not g-labels. Here also, accepting g-labels instead of labels would lead to no added generality but it would entail unnecessary complications.

Theorem 7.4 (g-braids rule for a general CSP): in any resolution state of any CSP, if Z is a target of a g-braid, then it can be eliminated (formally, this rule concludes ¬candidate(Z)).

Proof: obvious (almost the same as in the g-whips case).

Definition: in any of the above defined g-bivalue chains, g-whips or g-braids, a candidate other than L_k for a CSP variable V_k is called a t- [respectively a z-] candidate if it is incompatible with a previous right-linking candidate or g-candidate [resp. with the target]. Notice that a candidate can be z- and t- at the same time and that the t- and z- candidates are not considered as being part of the pattern. Notice also that a right-linking g-candidate can contain z- and/or t-candidates, as long as it has more than one non-z and non-t candidate (otherwise, the only compatible candidate is considered as a mere right-linking candidate).

7.2.4. Properties of g-whips and g-braids

g-whips and g-braids have properties very similar to those of whips and braids, namely: linearity, g-continuity (for g-whips), non anticipativeness, left-composability,... In the next sections, we shall see that g-braids also have the two strongest properties of braids: confluence and relationship with gT&E, i.e. $T\&E(W_1)$.

7.3. g-whip and g-braid resolution theories; the gW and gB ratings

One can now define two new families of resolution theories and two new ratings, in a way that strictly parallels what was done for whips and braids in chapter 5. As was the case for the W and B ratings, the gW and gB ratings of an instance will be measures of the hardest step in its simplest resolution path with g-whips or g-braids; they will not take into account the whole path.

7.3.1. g-whip resolution theories in a general CSP; the gW rating

Recall that BRT(CSP) is the Basic Resolution Theory of the CSP defined in section 4.3.

Definition: for any $n \geq 0$, let gW_n be the following resolution theory:

– $gW_0 = BRT(CSP) = W_0 = B_0$,

– $gW_1 = gW_0 \cup$ {rules for g-whips of length 1} $= W_1$,

– $gW_2 = gW_1 \cup$ {rules for g-whips of length 2},

– ....

– $gW_n = gW_{n-1} \cup$ {rules for g-whips of length n},

– $gW_\infty = \cup_{n \geq 0} gW_n$.

Definition : the **gW-rating** of an instance P of the CSP, noted $gW(P)$, is the smallest $n \leq \infty$ such that P can be solved within gW_n. An instance P has gW rating n if it can be solved using only g-whips of length no more than n but it cannot be solved using only g-whips of length strictly smaller than n. By convention, $gW(P) = \infty$ means that P cannot be solved by g-whips.

The gW rating has some good properties one can expect of a rating:

– it is defined in a purely logical way, independent of any implementation; the gW rating of an instance is an intrinsic property of this instance;

– in the Sudoku case, it is invariant under symmetry and supersymmetry; similar symmetry properties will be true for any CSP, if it has symmetries of any kind and they are properly formalised.

7.3.2. g-braid resolution theories in a general CSP; the gB rating

Definition: for any $n \geq 0$, let gB_n be the following resolution theory:

– $gB_0 = BRT(CSP) = gW_0 = W_0 = B_0$,

– $gB_1 = gB_0 \cup$ {rules for g-braids of length 1} $= gW_1 = W_1 = B_1$,

– $gB_2 = gB_1 \cup$ {rules for g-braids of length 2},

– ....

– $gB_n = T_n(g\text{-braids}) : gB_{n-1} \cup$ {rules for g-braids of length n},

– $gB_\infty = \cup_{n \geq 0} gB_n$.

Definition : the **gB-rating** of an instance P of the CSP, noted $gB(P)$, is the smallest $n \leq \infty$ such that P can be solved within gB_n. An instance P has gB rating n if it can be solved using only g-braids of length no more than n but it cannot be solved using only g-braids of length strictly smaller than n. By convention, $gB(P) = \infty$ means that P cannot be solved by g-braids.

The gB rating has all the good properties one can expect of a rating:

– it is defined in a purely logical way, independent of any implementation; the gB rating of an instance is an intrinsic property of this instance;

– as will be shown in the second next section, it is based on an increasing sequence of theories (gB$_n$) with the confluence property; this insures *a priori* better computational properties ; in particular, one can define a "simplest first" resolution strategy able to provide the gB rating after following a single resolution path;

– in the Sudoku case, it is invariant under symmetry and supersymmetry ; similar properties will be true for any CSP with symmetries properly formalised.

7.4. Comparison of the ratings based on whips, braids, g-whips and g-braids

The first natural question is: how do these two new ratings differ from the W and B ratings associated with ordinary whips and braids? For any CSP, any instance P and any $1 \leq n \leq \infty$, it is obvious that $gB_n(P) \leq \{gW_n(P), B_n(P)\} \leq W_n(P)$, but the relationship between $gW_n(P)$ and $B_n(P)$ is not obvious at all.

Statistically, in Sudoku, there is surprisingly little difference between the four ratings for instances with finite W ratings. Based on 21,371 puzzles generated by the controlled bias generator, only 49 cases with $gW(P) < W(P)$ were found. This is a proportion of 0.23%. In most of these cases, the difference was 1. In 3 cases, the difference was 2. In 1 case, the difference was 5 (see section 7.7.1).

In what follows, as there can be no confusion, we use the same symbol to name a resolution theory T and the set of instances of the CSP solvable in it, i.e. we use T to mean {P / P solvable in T}.

7.4.1. In any CSP, $W_2 = B_2 \subseteq gW_2 = gB_2$

Theorem 5.5 has shown that $W_2 = B_2$. The proof below will show that $gW_2 = gB_2$. As a result, one has $W_2 = B_2 \subseteq gW_2 = gB_2$.

This is the most one can hope in general: the inclusion $B_2 \subset gW_2$ is strict in Sudoku ($gW_2 \not\subset B_2$), as shown by the counter-example to equality in section 7.7.2. The example in section 7.7.3 will even show that $gW_2 \not\subset B_\infty$.

Theorem 7.5: In any CSP, any elimination done by a g-braid of length 2 can be done by a g-whip of same or shorter length; as a result, $gB_2 = gW_2$.

Proof: Let $B = V_1\{l_1\ r_1\} - V_2\{l_2\ .\} \Rightarrow V_z \neq v_z$ be a g-braid[2] with target $Z = <V_z, r_z>$ in some resolution state RS. If variable V_2 has a candidate $<V_2, v'>$ (it may be $<V_2, l_2>$) such that $<V_2, v'>$ is linked or g-linked to $<V_1, r_1>$, then $V_1\{l_1\ r_1\}$ $- V_2\{v'\ .\} \Rightarrow V_z \neq v_z$ is a g-whip[2] with target Z. Otherwise, $<V_2, l_2>$ is linked to $<V_z, v_z>$ and $V_2\{l_2\ .\} \Rightarrow V_z \neq v_z$ is a shorter g-whip[1] with target Z.

7.4.2. In any CSP, $gW_3 = gB_3$

Theorem 7.6: In any CSP, any elimination done by a g-braid of length 3 can be done by a whip or a g-whip of same or shorter length; as a result, $gB_3 = gW_3$.

Proof: The proof is a little harder than that of $gB_2 = gW_2$. It involves three kinds of changes: 1) re-ordering the various cells; exchanging the roles of left-linking, right-linking and t- objects; exchanging candidates with g-candidates. It is a very good exercise on the manipulation of these notions.

Let $B = V_1\{l_1 \ r_1\} - V_2\{l_2 \ r_2\} - V_3\{l_3 \ .\} \Rightarrow V_z \neq v_z$ be a g-braid[3] in some resolution state RS. We can always suppose that is has been pruned of its useless branches, i.e. of any part $V_k\{l_k \ r_k\}$ such that no candidate for any posterior CSP variable is linked or g-linked to r_k. This entails in particular that CSP variable V_3 has a candidate linked or g-linked to $<V_2, r_2>$; by modifying B if necessary, we can always suppose it is $<V_3, l_3>$. Then all the other candidates for V_3 are linked or g-linked to (at least) one of $<V_z, v_z>$, $<V_1, r_1>$ or $<V_2, r_2>$. We now consider two subcases.

1) If CSP variable V_2 has at least one candidate linked or g-linked to $<V_1, r_1>$, we can always suppose it is $<V_2, l_2>$ (otherwise, we modify the l_2 of the original g-braid). Then B is a g-whip[3] built on target $<V_z, v_z>$.

2) Otherwise, V_2 has only candidates linked or g-linked to $<V_z, v_z>$; then "$V_2\{l_2 \ r_2\} - V_3\{l_3$ " is a possible beginning for a g-whip built on target $<V_z, v_z>$. Moreover, in this case, V_3 must have at least one candidate $<V_3, t_3>$ linked or g-linked to $<V_1, r_1>$ (otherwise $V_1\{l_1 \ r_1\}$ would be a useless branch of B and it would have been pruned). If V_3 has only one such t_3, let gt_3 be t_3; if V_3 has several such t_3, they can only belong to a same g-label, say gt_3, for V_3. Let r'_1 be r_1 if r_1 is a candidate and any candidate in r_1 if r_1 is a g-candidate. Then the following is a g-whip[3] built on target $<V_z, v_z>$: $V_2\{l_2 \ r_2\} - V_3\{l_3 \ gt_3\} - V_1\{r'_1 \ .\}$. qed.

Case 2 is where a tentative proof of $B_3 \subseteq W_3$ along similar lines would fail: in some subcases, we need a right-linking g-candidate gt3, even if B had only right-linking candidates. (Of course, this does not prove that $B_3 \not\subset W_3$.)

7.4.3. General comparisons

Getting eventually a lower rating is not the only advantage of having g-whips.

We already mentioned that the inclusion $B_2 \subset gW_2$ is strict in Sudoku (i.e. $gW_2 \not\subset B_2$). But we also have the much stronger (a priori very unexpected) result that gW_2 cannot be reduced in general to whips or braids of any length, i.e. $gW_2 \not\subset B_\infty$ (which obviously implies that $gW_\infty \not\subset B_\infty$). This will be shown by the example of section 7.7.3. Notice however that such instances will be very exceptional, at least for the Sudoku CSP, as "almost all" the randomly generated puzzles can be solved with whips (chapter 6).

The simple counter-example in section 7.7.3 is related to the presence in the puzzle of a Sudoku specific pattern, a Swordfish (see chapter 8). The example in section 7.7.4 is much more complex but it shows that even when the Sudoku specific Subset patterns are not involved, one can prove that $gW_\infty \not\subset B_\infty$ (indeed $gW_{18} \not\subset B_\infty$).

What about the converse? Is $B_\infty \subseteq gW_\infty$? With a puzzle that can be solved by braids (of maximal length 6) but not by g-whips, section 7.7.5 gives a negative answer: $B_\infty \not\subset gW_\infty$. What the smallest n such that $B_n \not\subset gW_n$ is remains an open question; we only know that $n \leq 6$.

Finally, none of B_∞ and gW_∞ is included in the other.

Now, as a g-whip is a particular case of a g-braid, one has $gW_n \subseteq gB_n$ for all n. But the converse is not true in general, except for n = 0, 1, 2 or 3. g-braids are a true generalisation of g-whips. Even in the Sudoku case (for which whips solve almost any puzzle), there seems to be (rare) examples of puzzles that can be solved with g-braids but (probably) not with g-whips: one will appear in section 7.7.6.

7.5. The confluence property of the gB_n resolution theories

7.5.1. The confluence property of g-braid resolution theories

Theorem 7.7: each of the gB_n resolution theories, $0 \leq n \leq \infty$, is stable for confluence; therefore it has the confluence property.

Let n be fixed. We must show that, if an elimination of a candidate Z could have been done in a resolution state RS_1 by a g-braid B of length $m \leq n$ and with target Z, it will always still be possible, starting from any further state RS_2 obtained from RS_1 by consistency preserving assertions and eliminations, if we use a sequence of rules from gB_n. Let B be: $\{L_1\ R_1\} - \{L_2\ R_2\} - \ldots - \{L_p\ R_p\} - \{L_{p+1}\ R_{p+1}\} - \ldots - \{L_m\ .\}$, with target Z, where the R_k's are candidates or g-candidates modulo Z and the previous R_i's.

The proof follows that for braids in section 5.5, with a few additional subtleties. Consider first the state RS_3 obtained from RS_2 by applying repeatedly the rules in BRT until quiescence. As BRT has the confluence property, this state is uniquely defined. (Notice that we could legitimately apply rules from W_1 instead of only BRT, but we would have no guarantee that they do all the eliminations needed in the following proof, because we do not know yet that W_1 has the confluence property).

If, in RS_3, target Z has been eliminated, there remains nothing to prove. If target Z has been asserted, then the instance of the CSP is contradictory; if not yet detected in RS_3, this contradiction can be detected by CD in a state posterior to RS_3, reached by a series of applications of rules from W_1, following the g-braid structure of B.

Otherwise, we must consider all the elementary events related to B that can have happened between RS_1 and RS_3 as well as those we must provoke in posterior resolution states RS. For this, we start from B' = what remains of B in RS_3 and we let RS = RS_3. At this point, B' may not be a g-braid in RS. We progressively update RS and B' by repeating the following procedure, for p = 1 to p = m, until it produces a new (possibly shorter) g-braid B' with target Z in RS – a situation that is bound to happen. (This is a difference with the braids case: we have to consider a state RS posterior to RS_3). We return from this procedure as soon as B' is a g-braid in RS. All the references below are to the current RS and B'.

a) If, in RS, the left-linking or any t- or z- candidate of CSP variable V_p has been asserted (this can only be between RS_1 and RS_3), then all the candidates linked to it have been eliminated by ECP in RS_3, in particular: Z and/or the candidate(s) R_k (k<p) to which it is linked and/or all the elements of the g-candidate(s) R_k (k<p) to which it is g-linked; if Z is among them, there remains nothing to prove; otherwise, the procedure has already been successfully terminated by case f1 or f2α of the first such k.

b) If, in RS, left-linking candidate L_p has been eliminated (but not asserted) (it can therefore no longer be used as a left-linking candidate in a g-braid) and if CSP variable V_p still has a z- or a t- candidate C_p (i.e. a candidate C_p linked or g-linked to Z or to some previous R_i), then replace L_p by C_p. Now, up to C_p, B' is a partial g-braid in RS with target Z. Notice that, even if L_p was linked or g-linked to R_{p-1} (e.g. if B was a g-whip) this may not be the case for C_p; therefore trying to prove a similar theorem for g-whips would fail here.

c) If, in RS, any t- or z- candidate of V_p has been eliminated (but not asserted), this does not change the basic structure of B (at stage p). Continue with the same B'.

d) If, in RS, right-linking candidate R_p or a candidate R_p' in right-linking g-candidate R_p has been asserted (p can therefore not be the last index of B'), R_p can no longer be used as an element of a g-braid, because it is no longer a candidate or a g-candidate. Contrary to the proof for braids, and only because of this d case, we cannot be sure that this assertion occurred in RS_3. We must palliate this. First eliminate by ECP or W_1 any left-linking or t- candidate for any CSP variable of B' after p that is incompatible with R_p, i.e. linked or g-linked to it, if it is still present in RS. Now, considering the g-braid structure of B upwards from p, more eliminations and assertions can been done by rules from W_1. (Notice that we are not trying to do more eliminations or assertions than needed to get a g-braid in RS; in particular, we continue to consider R_p, not R_p'; in any case, it will be excised from B'; but, most of all, we do not have to find the shortest possible g-braid!)

Let q be the smallest number strictly greater than p such that, in RS, CSP variable V_q still has a (left-linking, t- or z-) candidate C_q that is not linked or g-linked to any of the R_i for $p \leq i < q$ (by definition of a g-braid, C_q is therefore linked or g-linked to Z or to some R_i with $i < p$). Apply the following rules from W_1 (if they have not yet been applied between RS_2 and RS) for each of the CSP variables V_u of B with index u increasing from p+1 to q-1 included:
- eliminate its left-linking candidate L_u by ECP or W_1;
- at this stage, CSP variable V_u had no left-linking, t- or z- candidate;
- if R_u is a candidate, assert it by S and eliminate by ECP all the left-linking and t-candidates for CSP variables after u that are incompatible with R_u in the current RS;
- if R_u is a g-candidate, it cannot be asserted by S; eliminate by W_1 all the left-linking and t- candidates for CSP variables after u that are incompatible with R_u in the current RS.

In the new RS thus obtained, excise from B' the part related to CSP variables p to q-1 (included) and, if L_q has been eliminated in the passage from RS_2 to RS, replace it by C_q; for each integer $s \geq p$, decrease by q-p the index of CSP variable V_s and of its candidates and g-candidates in the new B'. In RS, B' is now, up to p (the ex q), a partial g-braid in gB_n with target Z.

e) If, in RS, left-linking candidate L_p has been eliminated (but not asserted), and if CSP variable V_p has no t- or z- candidate in RS (complementary to case b), then there are now two cases (V_p must have at least one candidate).

e1) If R_p is a candidate, then V_p has only one possible value, namely R_p. If R_p has not yet been asserted by S somewhere between RS_2 and RS, do it now; this case is now reducible to case d (because the assertion of R_p also entails the elimination of L_p); go back to case d for the same value of p (this does not introduce an infinite loop!). Otherwise, go to the next p.

e2) If R_p is a g-candidate, then R_p cannot be asserted by S; use it, for any CSP variable after p, to eliminate by W_1 any of its t-candidates that is g-linked to R_p. Let q be the smallest number strictly greater than p such that, in RS, CSP variable V_q still has a (left-linking, t- or z-) candidate C_q that is not linked or g-linked to any of the R_i for $p \leq i < q$. Replace RS by the state obtained after all the assertions and eliminations similar to those in case d above have been done. Then, in RS, excise the part of B' related to CSP variables p to q-1 (included), replace L_q by C_q (if L_q has been eliminated in the passage from RS_2 to RS_3) and re-number the posterior elements of B', as in case d. In RS, B' is now, up to p (the ex q), a partial g-braid in gB_n with target Z. If p is its last index, it is a g-braid; return it and stop.

f) Finally, consider eliminations occurring in a right-linking candidate or g-candidate R_p. This implies that p cannot be the last index of B'. There are two cases.

f1) If, in RS, right-linking candidate R_p of B has been eliminated (but not asserted) or marked (by f2γ) in a previous step (i.e. it has become a t-candidate), then replace B' by its initial part: $\{L_1 R_1\} - \{L_2 R_2\} - \ldots - \{L_p .\}$. At this stage, B' is in RS a (possibly shorter) g-braid with target Z. Return B' and stop.

f2) If, in RS, a candidate in right-linking g-candidate R_p has been eliminated (but not asserted) or marked in a previous step, then:

f2α) either there remains no unmarked candidate of R_p in RS; then replace B' by its initial part: $\{L_1 R_1\} - \{L_2 R_2\} - \ldots - \{L_p .\}$; at this stage, B' is in RS a (possibly shorter) g-braid with target Z; return B' and stop;

f2β) or the remaining unmarked candidates of R_p in RS still make a g-candidate and B' does not have to be changed;

f2γ) or there remains only one unmarked candidate R_p' of R_p; replace R_p by R_p' in B'. We must also prepare the next steps by putting marks. Any t-candidate of B that was g-linked to R_p, if it is still present in RS, can still be considered as a t-candidate in B', where it is now linked to R_p' instead of being g-linked to R_p; this does not raise any problem. However, this substitution may entail that candidates that were not t-candidates in B become t-candidates in B'; if they are left-linking candidates of B', this is not a problem either; but if any of them is a right-linking candidate or an element of a right-linking g-candidate for B', then mark it so that the same procedure (i.e. f1 or f2) can be applied to it in a later step.

Notice that, as was the case for braids, this proof works only because the notions of being linked and g-linked do not depend on the resolution state.

7.5.2. g-braid resolution strategies consistent with the gB rating

As explained in section 4.5.3 and in exactly the same way as in the braids case, we can take advantage of the confluence property of g-braid resolution theories to define a "simplest firth" strategy that will always find the simplest (in terms of the length of the g-braids it will use) solution after following a single resolution path. As a result, it will also compute the gB rating of an instance. The following order satisfies this requirement:

ECP < S <
bivalue-chain[1] < whip[1] < g-whip[1] < braid[1] < g-braid[1] <
... < ...
bivalue-chain[k] < whip[k] < g-whip[k] < braid[k] < g-braid[k] <
bivalue-chain[k+1] < whip[k+1] < g-whip[k+1] < braid[k+1] < g-braid[k+1] < ...

Notice that bivalue-chains, whips, g-whips and braids being special cases of g-braids of same length, their explicit presence in the set of rules does not change the final result (z-chains and t-whips could also be added in the landscape). We put them here because when we look at a resolution path, it may be nicer to see simple patterns appear instead of more complex ones (g-braids). Also, it allows to see (in the Sudoku case) that, in practice, g-braids that are neither g-whips nor braids do not appear very often in the resolution paths.

Here, we have put g-whips before braids of same length, because they are structurally simpler and experiments confirm this complexity hierarchy (in terms of computation times and memory requirements). This choice has no impact on the gB rating.

As in the case of ordinary braids, the above ordering does not completely define a deterministic procedure: it does not set any precedence between different chains of same type and length. This could be done by using an ordering of the candidates instantiating them, based e.g. on their lexicographic order. But one can also decide that, for all practical purposes, which of these equally prioritised rule instantiations should be "fired" first should be chosen randomly.

7.6. The "gT&E vs g-braids" theorem

In section 5.6.1, we defined the procedure T&E(T, Z, RS) for any candidate Z, any resolution state RS and any resolution theory T with the confluence property. In this section, we consider $T = W_1 = B_1$ and we set gT&E = T&E(W_1). It is obvious that any elimination that can be done by a g-braid B can be done by gT&E, using a sequence of rules from $B_1 = W_1$, following the structure of B. The converse is more interesting:

Theorem 7.8: for any instance of any CSP, any elimination that can be done by gT&E can be done by a g-braid. Any instance of a CSP that can be solved by gT&E can be solved by g-braids.

Proof: Let RS be a resolution state and let Z be a candidate eliminated by gT&E(Z, RS) using some auxiliary resolution state RS'. Following the steps of resolution theory B_1 in RS', we progressively build a g-braid in RS with target Z. But we must do this in a way a little smarter than in our proof for mere braids. First, remember that B_1 contains only four types of rules: ECP (which eliminates candidates), S (which asserts a value for a CSP variable), W_1 (whips of length 1, which eliminates candidates) and CD (which detects a contradiction on a CSP variable).

Consider the sequence $(P_1, P_2, ..., P_k, ...P_n)$ of rule applications in RS' based on rules from W_1 different from ECP and suppose that P_n is the first occurrence of CD

(there must be at least one occurrence of CD if Z is eliminated by gT&E). We first define the R_k and V_k sequences, for $k < n$:
- if P_k is of type S, then it asserts a value R_k for some CSP variable V_k; add R_k and V_k at the end of the appropriate sequences;
- if P_k is of type whip[1]: $\{M_k .\} \Rightarrow \neg$candidate($C_k$) for some CSP variable V_k, then define R_k as the g-candidate for V_k that contains M_k and is g-linked to C_k; (notice that C_k will not necessarily be L_{k+1}); add R_k and V_k to the sequences.

We shall build a g-braid[n] in RS with target Z, with the R_k's as its sequence of right-linking candidates or g-candidates and with the V_k's as its sequence of CSP variables. We only have to define properly the L_k's. We do this successively for $k = 1, ..., k = n$. As the proofs for $k = 1$ and for the passage from k to k+1 are almost identical, we skip the case $k = 1$. Suppose we have done it until k and consider CSP variable V_{k+1}.

Whatever rule P_{k+1} is (S or whip[1]), the fact that it can be applied means that, apart from R_{k+1} (if it is a candidate) or the labels contained in R_{k+1} (if it is a g-candidate), all the other labels for CSP variable V_{k+1} that were still candidates for V_{k+1} in RS (and there must be at least one, say L_{k+1}) have been eliminated in RS' by the assertion of Z and the previous rule applications. But these previous eliminations can only result from being linked or g-linked to Z or to some R_i, $i \le k$. $\{L_{k+1} \ R_{k+1}\}$ is therefore a legitimate extension for our partial g-braid.

End of the procedure: at step n, a contradiction is obtained by CD for some variable V_n. It means that all the candidates for V_n that were still candidates for V_n in RS (and there must be at least one, say L_n) have been eliminated in RS' by the assertion of Z and the previous rule applications. But these previous eliminations can only result from being linked or g-linked to Z or to some R_i, $i < n$. L_n is thus the last left-linking candidate of the g-braid we were looking for in RS.

Here again (as in the proof of confluence), this proof works only because the existence of a link or a g-link between two candidates does not depend on the resolution state. Again, it is very unlikely that the gT&E procedure followed by the construction in this proof would produce the shortest available g-braid in RS.

7.7. Exceptional examples

This section provides the proofs by examples announced in section 7.4.

7.7.1. A puzzle with W=B=7 and gW=2

In section 7.4, we mentioned the rare case of a puzzle P with finite W rating but with very different W and gW ratings: $gW(P) = W(P) - 5$. One might think that this can happen only for hard puzzles, but the example in Figure 7.1 shows that it can also happen with relatively simple ones: here, we have $gW(P) = 2$ and $W(P) = 7$.

1				6				9
	8		1			5	6	
	3		6			8		1
								7
				8	2	4	6	
	4		9				2	
5	6		8	4	1			
		7			3			

1	2	3	4	5	6	7	8	9
4	5	6	7	8	9	1	3	2
7	8	9	1	3	2	5	6	4
2	3	4	6	9	7	8	5	1
6	1	8	5	2	4	3	9	7
9	7	5	3	1	8	2	4	6
3	4	1	9	7	5	6	2	8
5	6	2	8	4	1	9	7	3
8	9	7	2	6	3	4	1	5

Figure 7.1. *Puzzle P (cb# xxxx) with W(P)=B(P)=7 and gW(P)=2*

   1) If we accept g-whips, there is a very short resolution path:

***** SudoRules version 15b.1.8-gW *****
26 givens and 206 candidates and 1339 nrc-links
singles ==> r8c9 = 3, r8c3 = 2
whip[1]: c9n5{r9 .} ==> r9c8 ≠ 5
whip[1]: r8n9{c7 .} ==> r9c8 ≠ 9, r9c7 ≠ 9,
whip[1]: r8n7{c8 .} ==> r7c7 ≠ 7
whip[1]: r6n3{c5 .} ==> r5c4 ≠ 3, r5c5 ≠ 3
;;; Resolution state RS$_1$
whip[2]: b4n6{r5c1 r5c3} - b4n8{r5c3 .} ==> r5c1 ≠ 4,r5c1 ≠ 2
whip[2]: b4n8{r5c1 r5c3} - b4n6{r5c3 .} ==> r5c1 ≠ 9
whip[2]: b4n6{r5c3 r5c1} - b4n8{r5c1 .} ==> r5c3 ≠ 5, r5c3 ≠ 4
whip[1] : r5n4{c6 .} ==> r4c6 ≠ 4
whip[2]: b4n6{r5c3 r5c1} - b4n8{r5c1 .} ==> r5c3 ≠ 1
whip[2]: b4n8{r5c3 r5c1} - b4n6{r5c1 .} ==> r5c3 ≠ 9
;;; Resolution state RS$_2$
g-whip[2]: r3n7{c1 c456} - c4n7{r2 .} ==> r6c1 ≠ 7
singles to the end

   2) If we accept only whips, the resolution path is much longer:

***** SudoRules version 15b.1.8-W *****
;;; same path up to RS$_2$
whip[3]: c4n7{r1 r6} - c2n7{r6 r2} - r3n7{c1 .} ==> r1c5 ≠ 7
whip[3]: c2n7{r1 r6} - c4n7{r6 r1} - r3n7{c6 .} ==> r2c1 ≠ 7
whip[3]: c4n7{r2 r6} - c2n7{r6 r1} - r3n7{c1 .} ==> r2c5 ≠ 7, r2c6 ≠ 7
whip[3]: r5n2{c6 c2} - r1n2{c2 c4} - b8n2{r9c4 .} ==> r4c5 ≠ 2
whip[3]: b4n7{r6c2 r4c1} - r3n7{c1 c6} - b8n7{r7c6 .} ==> r6c5 ≠ 7
whip[3]: r9c2{n1 n9} - r9c1{n9 n8} - r9c8{n8 .} ==> r9c7 ≠ 1

whip[3]: r9c8{n8 n1} - r9c2{n1 n9} - r9c1{n9 .} ==> r9c9 ≠ 8
whip[5]: r4n2{c1 c6} - r4n7{c6 c5} - b8n7{r7c5 r7c6} - r3n7{c6 c1} - r6c1{n7 .} ==> r4c1 ≠ 9
whip[5]: r4c8{n9 n5} - r4c5{n5 n7} - b8n7{r7c5 r7c6} - r3n7{c6 c1} - r6c1{n7 .} ==> r4c3 ≠ 9
whip[3]: r9c2{n9 n1} - c3n1{r7 r6} - c3n9{r6 .} ==> r2c2 ≠ 9
whip[6]: b3n1{r2c7 r2c8} - r9c8{n1 n8} - r9c1{n8 n9} - r6c1{n9 n7} - b1n7{r3c1 r1c2} - c4n7{r1 .}
==> r2c7 ≠ 7
**whip[7]: c3n6{r2 r5} - c1n6{r5 r2} - c1n4{r2 r4} - c1n2{r4 r3} - b3n2{r3c9 r2c9} -
c9n8{r2 r7} - c3n8{r7 .} ==> r2c3 ≠ 4**
whip[7]: r3n3{c1 c5} - c3n3{r3 r7} - b7n1{r7c3 r9c2} - b7n9{r9c2 r9c1} - r6c1{n9 n7} - r3n7{c1 c6}
- c4n7{r1 .} ==> r2c1 ≠ 3
whip[7]: r1n2{c5 c2} - c1n2{r2 r4} - c6n2{r4 r5} - r2n2{c6 c9} - r3c9{n2 n4} - c6n4{r3 r2} -
c1n4{r2 .} ==> r3c5 ≠ 2
whip[7]: r4n2{c1 c6} - r4n7{c6 c5} - b8n7{r7c5 r7c6} - r3n7{c6 c1} - c1n2{r3 r2} - r1c2{n2 n5} -
r2c2{n5 .} ==> r4c1 ≠ 4
hidden-single-in-a-block ==> r4c3 = 4
whip[4]: b5n4{r5c6 r5c4} - r1n4{c4 c7} - c7n7{r1 r8} - c7n9{r8 .} ==> r5c6 ≠ 9
whip[4]: r3c3{n9 n3} - r3c5{n3 n7} - c4n7{r1 r6} - r6c1{n7 .} ==> r3c1 ≠ 9
whip[5]: b3n7{r1c7 r2c8} - c8n1{r2 r9} - r7c7{n1 n6} - r9c7{n6 n4} - r1n4{c7 .} ==> r1c4 ≠ 7
;;; only now do we get the crucial elimination with a whip[2]:
whip[2]: c4n7{r6 r2} - r3n7{c6 .} ==> r6c1 ≠ 7
singles to the end

3) Interestingly (anticipating on chapter 8), this puzzle can also be solved with Subset rules, but it gets a higher rating than with g-whips; i.e. g-whips are better than Subset in this case.

***** SudoRules version 15b.1.8-WS *****
;;; same path up to RS$_1$
hidden-pairs-in-a-row: r5{n6 n8}{c1 c3} ==> r5c3 ≠ 9, r5c3 ≠ 5, r5c3 ≠ 4, r5c3 ≠ 1, r5c1 ≠ 9, r5c1 ≠4
whip[1] : r5n4{c6 .} ==> r4c6 ≠ 4
hidden-pairs-in-a-row: r5{n6 n8}{c1 c3} ==> r5c1 ≠ 2
;;; same situation as RS$_2$ (all the whips[2] in the W or gW resolution paths are hidden pairs)
naked-triplets-in-a-row: r9{c1 c2 c8}{n8 n9 n1} ==> r9c9 ≠ 8, r9c7 ≠ 1
swordfish-in-rows: n7{r3 r4 r7}{c6 c1 c5} ==> r6c5 ≠ 7
;;; The crucial elimination is now obtained with a swordfish:
swordfish-in-rows: n7{r3 r4 r7}{c6 c1 c5} ==> r6c1 ≠ 7
singles to the end

7.7.2. *gW$_2$ ⊄ B$_2$: a puzzle with W=3, B=3, gW=2, gB=2*

Our second example (puzzle cb#1249 in Figure 7.2) proves that the obvious inclusion B$_2$ ⊂ gW$_2$ is not an equality in general ("obvious" because B$_2$ = W$_2$).

1) The resolution path with g-whips gives gW(P) = 2:

***** SudoRules version 15b.1.12-gW *****
27 givens, 200 candidates and 1254 nrc-links

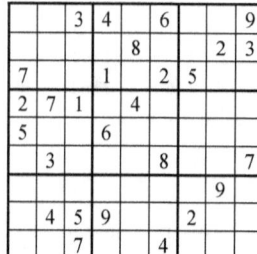

singles ==> r7c3 = 2, r1c2 = 2, r2c2 = 5, r2c4 = 7, r1c5 = 5, r2c6 = 9, r3c5 = 3, r4c7 = 9, r5c9 = 2
whip[1]: c9n1{r9 .} ==> r7c7 ≠ 1, r8c8 ≠ 1, r9c7 ≠ 1, r9c8 ≠ 1
whip[1]: c2n1{r9 .} ==> r7c1 ≠ 1, r8c1 ≠ 1, r9c1 ≠ 1
whip[1]: r4n6{c8 .} ==> r6c8 ≠ 6, r6c7 ≠ 6
whip[1]: r4n8{c8 .} ==> r5c8 ≠ 8, r5c7 ≠ 8
whip[2]: r4c6{n3 n5} - r4c4{n5 .} ==> r4c8 ≠ 3
whip[1]: r4n3{c4 .} ==> r5c6 ≠ 3
whip[2]: r4c6{n5 n3} - r4c4{n3 .} ==> r4c8 ≠ 5, r4c9 ≠ 5
singles ==> r6c8 = 5, r6c4 = 2, r9c5 = 2
;;; Resolution state RS$_1$
g-whip[2]: c7n8{r1 r789} - r8n8{c9 .} ==> r1c1 ≠ 8
singles to the end

		3	4		6			9
				8			2	3
7			1		2	5		
2	7	1		4				
5			6					
	3				8			7
						9		
	4	5	9			2		
		7			4			

1	2	3	4	5	6	7	8	9
4	5	6	7	8	9	1	2	3
7	8	9	1	3	2	5	4	6
2	7	1	3	4	5	9	6	8
5	8	4	6	9	7	3	1	2
6	3	9	2	1	8	4	5	7
3	6	2	5	7	1	8	9	4
8	4	5	9	6	3	2	7	1
9	1	7	8	2	4	6	3	5

Figure 7.2. *A puzzle P (cb #1249) with gW(P)=2 and W(P)=B(P)=3*

2) The resolution path with whips gives W(P) = 3; the resolution path with braids is exactly the same, i.e. no non-whip braid appears in it, and B(P) = 3:

***** SudoRules version 15b.1.12-W *****
;;; same path up to RS$_1$
whip[3]: b1n1{r1c1 r2c1} - c1n4{r2 r6} - r6c7{n4 .} ==> r1c7 ≠ 1
whip[3]: c7n6{r7 r2} - b3n1{r2c7 r1c8} - c8n7{r1 .} ==> r8c8 ≠ 6
whip[3]: c1n4{r2 r6} - r6c7{n4 n1} - r2n1{c7 .} ==> r2c1 ≠ 6
whip[3]: r2c3{n4 n6} - b4n6{r6c3 r6c1} - c1n4{r6 .} ==> r2c7 ≠ 4
whip[1]: r2n4{c1 .} ==> r3c3 ≠ 4
whip[3]: r2c7{n6 n1} - r6c7{n1 n4} - c8n4{r5 .} ==> r3c8 ≠ 6
whip[3]: b6n8{r4c9 r4c8} - r3c8{n8 n4} - c9n4{r3 .} ==> r7c9 ≠ 8
whip[3]: r2c3{n4 n6} - r2c7{n6 n1} - r6c7{n1 .} ==> r6c3 ≠ 4
whip[3]: b1n9{r3c2 r3c3} - b1n6{r3c3 r2c3} - r6c3{n6 .} ==> r3c2 ≠ 8
whip[3]: b1n8{r1c1 r3c3} - b1n9{r3c3 r3c2} - b7n9{r9c2 .} ==> r9c1 ≠ 8
whip[3]: c1n8{r8 r1} - c7n8{r1 r7} - r8n8{c9 .} ==> r9c2 ≠ 8
whip[3]: b7n8{r7c1 r7c2} - c7n8{r7 r9} - r8n8{c8 .} ==> r1c1 ≠ 8
singles to the end

7.7.3. $gW_2 \not\subset B_\infty$: a puzzle not solvable by braids of any length but solvable in gW_2

The example in Figure 7.3 (a puzzle from Mauricio's swordfish collection) goes much further: it proves that $gW_2 \not\subset B_\infty$ and therefore $gW_\infty \not\subset B_\infty$.

		1			2			3
				1			4	
2			4			5		
		6			7			8
	5						2	
9			3			4		
		8			1			5
	9			6				
1			9			7		

6	4	1	5	9	2	8	7	3
8	7	5	6	1	3	2	4	9
2	3	9	4	7	8	5	6	1
3	1	6	2	4	7	9	5	8
7	5	4	1	8	9	3	2	6
9	8	2	3	5	6	4	1	7
4	2	8	7	3	1	6	9	5
5	9	7	8	6	4	1	3	2
1	6	3	9	2	5	7	8	4

Figure 7.3. A puzzle P with B(P)=∞ but gW(P)=2

Using the T&E procedure and the "T&E vs braids" theorem, it is easy to check that this puzzle is not solvable by braids, let alone by whips. But it is in gT&E and it can therefore be solved by g-braids. Let us try to do better and solve it by g-whips.

```
***** SudoRules version 15b.1.12-gW *****
24 givens, 214 candidates and 1289 nrc-links
g-whip[2]: c3n4{r5 r789} - r7n4{c2 .} ==> r5c5 ≠ 4
g-whip[2]: r1n9{c5 c789} - c9n9{r3 .} ==> r5c5 ≠ 9
singles to the end
```

Anticipating on chapter 8, this puzzle can also be solved by Subsets of size 3, more precisely by Swordfish; actually, we find two Swordfish (for two different numbers) in the same three columns, a very exceptional situation. This puzzle will also count as a very rare example of a Swordfish not completely subsumed by whips.

```
***** SudoRules version 15b.1.12-B-S *****
24 givens, 214 candidates and 1289 nrc-links
swordfish-in-columns n4{c3 c6 c9}{r9 r5 r8} ==> r9c5 ≠ 4, r9c2 ≠ 4, r8c1 ≠ 4, r5c5 ≠ 4, r5c1 ≠ 4
swordfish-in-columns n9{c3 c6 c9}{r3 r2 r5} ==> r5c7 ≠ 9, r5c5 ≠ 9
singles to the end
```

7.7.4. $gW_\infty \not\subset B_\infty$: a puzzle not solvable by braids of any length but solvable in gW_{18}

Even without invoking puzzles, as in section 7.7.3, involving the rare case of a Subset pattern that is not subsumed by whips or braids, there are examples that can be solved by g-whips but not by braids. Consider the puzzle shown in Figure 7.4 (known as "AI Broken Brick").

Using the T&E procedure and the "T&E vs braids" theorem, it is easy to check that this puzzle is not solvable by T&E and it has therefore no chance of being solvable by braids, let alone by whips. But it is solvable by gT&E and it can therefore be solved by g-braids. Let us try to do better and solve it by g-whips.

4				6			7	
						6		
	3				2			1
7					8	5		
	1		4					
	2		9	5				
						7		5
		9	1				3	
		3		4			8	

4	5	1	8	6	3	9	7	2
9	8	2	7	1	4	6	5	3
6	3	7	5	9	2	8	4	1
7	9	6	3	2	8	5	1	4
3	1	5	4	7	6	2	9	8
8	2	4	9	5	1	3	6	7
1	4	8	6	3	9	7	2	5
2	7	9	1	8	5	4	3	6
5	6	3	2	4	7	1	8	9

Figure 7.4. *Puzzle "AI Broken Brick" with B(P)=∞ and gW(P)=18*

***** SudoRules version 15b.1.8-gW *****
23 givens, 219 candidates and 1366 nrc-links
hidden-single-in-a-column ==> r2c6 = 4
whip[1] : c3n7{r3 .} ==> r2c2 ≠ 7
whip[4]: b4n5{r5c3 r5c1} - b4n9{r5c1 r4c2} - r2c2{n9 n8} - r1c2{n8 .} ==> r3c3 ≠ 5, r2c3 ≠ 5, r1c3 ≠ 5
hidden-single-in-a-column ==> r5c3 = 5
whip[9]: r6n7{c9 c6} - b5n1{r6c6 r4c5} - r4n3{c5 c4} - b5n6{r4c4 r5c6} - r8c6{n6 n5} - r9c6{n5 n9} - r7n9{c6 c8} - r5c8{n9 n2} - r4n2{c8 .} ==> r6c9 ≠ 3
whip[10]: b1n6{r3c1 r3c3} - r4c3{n6 n4} - r6c3{n4 n8} - r6c1{n8 n3} - b4n6{r6c1 r4c2} - c4n6{r4 r7} - b9n6{r7c8 r8c9} - c9n4{r8 r6} - r6c7{n4 n1} - r9n1{c7 .} ==> r9c1 ≠ 6
whip[13]: b1n6{r3c1 r3c3} - r4c3{n6 n4} - r6c3{n4 n8} - r6c1{n8 n3} - b4n6{r6c1 r4c2} - c4n6{r4 r9} - r8n6{c6 c9} - c9n4{r8 r6} - r6c7{n4 n1} - c6n1{r6 r1} - r1c3{n1 n2} - r7c3{n2 n1} - r9n1{c1 .} ==> r7c1 ≠ 6
whip[13]: c6n5{r8 r1} - r1n1{c6 c3} - r2n1{c1 c5} - b5n1{r4c5 r6c6} - c7n1{r6 r9} - r9c1{n1 n2} - b1n2{r2c1 r2c3} - b1n7{r2c3 r3c3} - b1n6{r3c3 r3c1} - r3n5{c1 c8} - r2c8{n5 n9} - c1n9{r2 r5} - c7n9{r5 .} ==> r9c4 ≠ 5
whip[1]: c4n5{r1 .} ==> r1c6 ≠ 5
whip[15]: r4c3{n6 n4} - r6c3{n4 n8} - r3c3{n8 n7} - b1n6{r3c3 r3c1} - b4n6{r5c1 r4c2} - c4n6{r4 r9} - r8n6{c6 c9} - c9n4{r8 r6} - r6n7{c9 c6} - r8c6{n7 n5} - r9c6{n5 n9} - r7n9{c6 c8} - b9n1{r7c8 r9c7} - r6n1{c7 c8} - r6n6{c8 .} ==> r7c3 ≠ 6
g-whip[15]: c7n1{r9 r6} - c6n1{r6 r1} - c5n1{r2 r4} - c8n1{r4 r7} - b9n9{r7c8 r9c9} - c6n9{r9 r7} - c6n3{r7 r456} - r4n3{c4 c9} - b3n3{r2c9 r1c7} - r1n9{c7 c2} - r4n9{c2 c8} - r2n9{c8 c5} - b2n3{r2c5 r2c4} - r2n7{c4 c3} - c3n1{r2 .} ==> r9c7 ≠ 2

g-whip[18]: r4n1{c8 c5} - b2n1{r2c5 r1c6} - c6n9{r1 r789} - r7n9{c5 c6} - c6n3{r7 r456} - r4n3{c4 c9} - r4n2{c9 c4} - r5c5{n2 n7} - b6n7{r5c9 r6c9} - c9n4{r6 r8} - r8c7{n4 n2} - r8c5{n2 n8} - r3c5{n8 n9} - r2c5{n9 n3} - r1n3{c6 c7} - c7n9{r1 r9} - r9n1{c7 c1} - r9n2{c1 .} **==> r4c8 ≠ 9**

whip[5]: r4n9{c9 c2} - r1n9{c2 c6} - c6n1{r1 r6} - c7n1{r6 r9} - r9n9{c7 .} ==> r2c9 ≠ 9

g-whip[10]: c5n1{r2 r4} - c6n1{r6 r1} - b2n3{r1c6 r123c4} - r4n3{c4 c9} - r4n9{c9 c2} - r2c2{n9 n5} - r1c2{n5 n8} - r1n9{c2 c789} - r2c8{n9 n2} - r2c9{n2 .} ==> r2c5 ≠ 8

g-whip[14]: b9n1{r9c7 r7c8} - r4n1{c8 c5} - b2n1{r2c5 r1c6} - c6n9{r1 r7} - c6n3{r7 r456} - r4n3{c4 c9} - r4n9{c9 c2} - c2n4{r4 r789} - r7n4{c3 c2} - r7n6{c2 c4} - r4c4{n6 n2} - r9c4{n2 n7} - r9c6{n7 n5} - r8c6{n5 .} ==> r9c7 ≠ 9

naked-single ==> r9c7 = 1

whip[5]: c6n5{r8 r9} - r9n9{c6 c9} - r4n9{c9 c2} - r1c2{n9 n8} - r2c2{n8 .} ==> r8c2 ≠ 5

whip[7]: r4n9{c2 c9} - b9n9{r9c9 r7c8} - r2n9{c8 c5} - c5n1{r2 r4} - r4n3{c5 c4} - b2n3{r1c4 r1c6} - b2n1{r1c6 .} ==> r1c2 ≠ 9

whip[5]: r1c2{n8 n5} - r2c2{n5 n9} - b4n9{r4c2 r5c1} - b4n3{r5c1 r6c1} - b4n8{r6c1 .} ==> r1c3 ≠ 8, r2c3 ≠ 8, r3c3 ≠ 8

whip[8]: r1c2{n5 n8} - r2c2{n8 n9} - b4n9{r4c2 r5c1} - b4n3{r5c1 r6c1} - b4n8{r6c1 r6c3} - r6c7{n8 n4} - b3n4{r3c7 r3c8} - b3n5{r3c8 .} ==> r2c1 ≠ 5

whip[9]: r1c2{n8 n5} - r2c2{n5 n9} - b4n9{r4c2 r5c1} - b4n3{r5c1 r6c1} - b4n8{r6c1 r6c3} - r6c7{n8 n4} - r3c7{n4 n9} - c8n9{r3 r7} - c5n9{r7 .} ==> r3c1 ≠ 8

whip[10]: b7n5{r9c1 r9c2} - r1c2{n5 n8} - r2c2{n8 n9} - r4n9{c2 c9} - b9n9{r9c9 r7c8} - c5n9{r7 r3} - c6n9{r1 r9} - r9n7{c6 c4} - r3n7{c4 c3} - b1n6{r3c3 .} ==> r3c1 ≠ 5

whip[1]: c1n5{r9 .} ==> r9c2 ≠ 5

whip[3]: c2n6{r7 r4} - c2n9{r4 r2} - r3c1{n9 .} ==> r8c1 ≠ 6

whip[1]: b7n6{r9c2 .} ==> r4c2 ≠ 6

whip[5]: b1n8{r2c1 r1c2} - b3n8{r1c9 r3c7} - b3n4{r3c7 r3c8} - b3n5{r3c8 r2c8} - c2n5{r2 .} ==> r2c4 ≠ 8

whip[8]: r3c1{n9 n6} - r3c3{n6 n7} - r3c5{n7 n8} - b8n8{r8c5 r7c4} - c3n8{r7 r6} - r6c1{n8 n3} - r6c7{n3 n4} - b3n4{r3c7 .} ==> r3c8 ≠ 9

whip[10]: r1c2{n8 n5} - r2c2{n5 n9} - b4n9{r4c2 r5c1} - b4n3{r5c1 r6c1} - b4n8{r6c1 r6c3} - r6c7{n8 n4} - r8n4{c7 c9} - r8n6{c9 c6} - c4n6{r9 r4} - b4n6{r4c3 .} ==> r8c2 ≠ 8

whip[9]: r3c1{n9 n6} - r3c3{n6 n7} - r3c5{n7 n8} - r8n8{c5 c1} - r6c1{n8 n3} - r5c1{n3 n9} - b1n9{r2c1 r2c2} - c8n9{r2 r7} - c5n9{r7 .} ==> r3c7 ≠ 9

whip[8]: r8c7{n2 n4} - r3c7{n4 n8} - c5n8{r3 r7} - c3n8{r7 r6} - b6n8{r6c9 r5c9} - b6n7{r5c9 r6c9} - c9n4{r6 r4} - r6n4{c8 .} ==> r8c5 ≠ 2

whip[9]: r1c2{n8 n5} - r1c4{n5 n3} - r2n3{c5 c9} - r4n3{c9 c5} - c5n1{r4 r2} - r1c6{n1 n9} - r1c7{n9 n2} - r8c7{n2 n4} - r3c7{n4 .} ==> r1c9 ≠ 8

whip[9]: b3n3{r1c7 r2c9} - r4n3{c9 c5} - r4n1{c5 c8} - r6n1{c8 c6} - r1c6{n1 n9} - r1c9{n9 n2} - r4n2{c9 c4} - b8n2{r9c4 r7c5} - b8n9{r7c5 .} ==> r1c4 ≠ 3

whip[2]: r1c2{n8 n5} - r1c4{n5 .} ==> r1c7 ≠ 8

whip[4]: r1c4{n5 n8} - r3n8{c5 c7} - b3n4{r3c7 r3c8} - b3n5{r3c8 .} ==> r2c4 ≠ 5

whip[8]: c7n9{r1 r5} - c1n9{r5 r3} - b1n6{r3c1 r3c3} - r4c3{n6 n4} - r6c3{n4 n8} - c7n8{r6 r3} - b3n4{r3c7 r3c8} - b3n5{r3c8 .} ==> r2c8 ≠ 9

whip[1]: b3n9{r1c7 .} ==> r1c6 ≠ 9

whip[1]: c6n9{r9 .} ==> r7c5 ≠ 9

whip[4]: r1c3{n2 n1} - r2c3{n1 n7} - r2c4{n7 n3} - r1c6{n3 .} ==> r2c1 ≠ 2

whip[1]: c1n2{r9 .} ==> r7c3 ≠ 2

whip[5]: r6n1{c8 c6} - r1c6{n1 n3} - r2n3{c5 c9} - b3n8{r2c9 r3c7} - b3n4{r3c7 .} ==> r6c8 ≠ 4
whip[7]: r6c8{n6 n1} - c6n1{r6 r1} - c5n1{r2 r4} - r4n3{c5 c4} - c6n3{r6 r7} - r7n9{c6 c8} -
b9n6{r7c8 .} ==> r4c9 ≠ 6
whip[8]: r8c7{n2 n4} - r8c9{n4 n6} - r8c2{n6 n7} - r9c2{n7 n6} - r9c4{n6 n7} - r2c4{n7 n3} -
r2c9{n3 n8} - r3c7{n8 .} ==> r9c9 ≠ 2
whip[8]: c2n5{r2 r1} - b1n8{r1c2 r2c1} - c1n9{r2 r5} - c8n9{r5 r7} - r9n9{c9 c6} - r9n5{c6 c1} -
r8c1{n5 n2} - b9n2{r8c7 .} ==> r2c2 ≠ 9
hidden-single-in-a-column ==> r4c2 = 9
whip[1]: c2n4{r8 .} ==> r7c3 ≠ 4
whip[2]: r1c2{n8 n5} - r2c2{n5 .} ==> r7c2 ≠ 8
whip[1]: c2n8{r1 .} ==> r2c1 ≠ 8
whip[5]: r8c5{n8 n7} - r3c5{n7 n9} - c1n9{r3 r2} - c1n1{r2 r7} - r7c3{n1 .} ==> r7c5 ≠ 8
whip[4]: b8n9{r7c6 r9c6} - r9n5{c6 c1} - r9n2{c1 c4} - r7c5{n2 .} ==> r7c6 ≠ 3
g-whip[2]: r4n3{c9 c456} - c6n3{r5 .} ==> r1c9 ≠ 3
whip[3]: b4n3{r5c1 r6c1} - c7n3{r6 r1} - c6n3{r1 .} ==> r5c5 ≠ 3
whip[3]: b4n3{r5c1 r6c1} - c7n3{r6 r1} - c6n3{r1 .} ==> r5c9 ≠ 3
whip[3]: c9n8{r6 r2} - b3n3{r2c9 r1c7} - c7n9{r1 .} ==> r5c7 ≠ 8
whip[3]: c4n8{r3 r7} - c3n8{r7 r6} - c7n8{r6 .} ==> r3c5 ≠ 8
hidden-single-in-a-column ==> r8c5 = 8
whip[2]: b2n5{r3c4 r1c4} - c4n8{r1 .} ==> r3c4 ≠ 7
whip[2]: r7n1{c1 c3} - r7n8{c3 .} ==> r7c1 ≠ 2
whip[3]: r9c2{n6 n7} - b8n7{r9c6 r8c6} - b8n5{r8c6 .} ==> r9c6 ≠ 6
whip[4]: r9c9{n9 n6} - r9c2{n6 n7} - r9c4{n7 n2} - r7n2{c5 .} ==> r7c8 ≠ 9
singles to the end

7.7.5. $B_\infty \not\subset gW_\infty$: a puzzle P with $gW(P) = \infty$ but $B(P) = 6$

With Figure 7.5, we now have a puzzle P (of moderate difficulty) not solvable
by g-whips but solvable by braids: $B(P) = gB(P) = 6$ but $W(P) = gW(P) = \infty$.

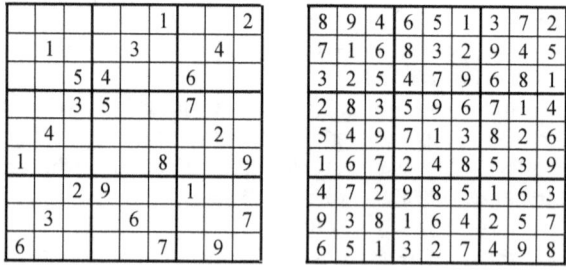

Figure 7.5. *A puzzle P with $B(P) = 6$ but $gW(P) = W(P) = \infty$*

Not only is this puzzle not solvable by whips or g-whips, it allows no
elimination at all by whips or g-whips. Let us try with braids :

***** SudoRules version 15b.1.12-B *****

25 givens, 204 candidates and 1214 nrc-links

braid[5]: b9n6{r7c9 r7c8} - r7n3{c8 c6} - r2c9{n8 n5} - c6n5{r2 r8} - r8c8{n8 .} ==> r7c9 ≠ 8

braid[5]: b9n2{r9c7 r8c7} - c7n4{r8 r6} - r8c8{n8 n5} - r6n5{c7 c2} - r9c2{n8 .} ==> r9c7 ≠ 8

braid[6]: r9c2{n5 n8} - r8c8{n5 n8} - r2c9{n5 n8} - c7n8{r1 r5} - c3n8{r2 r1} - c4n8{r9 .} ==>
r9c9 ≠ 5

whip[6]: b4n5{r5c1 r6c2} - r9n5{c2 c5} - r1n5{c5 c8} - r2c9{n5 n8} - b6n8{r5c9 r4c8} - r8c8{n8 .}
==> r5c7 ≠ 5

braid[5]: r5c7{n8 n3} - r8c8{n8 n5} - c6n3{r5 r7} - r7c8{n8 n6} - r6c8{n6 .} ==> r4c8 ≠ 8

whip[3]: r5n5{c1 c9} - r2c9{n5 n8} - b6n8{r5c9 .} ==> r5c1 ≠ 8

braid[5]: r5c7{n8 n3} - r8c8{n8 n5} - c6n3{r5 r7} - r7c8{n8 n6} - r6c8{n6 .} ==> r8c7 ≠ 8

whip[6]: b2n5{r1c5 r2c6} - c9n5{r2 r5} - c1n5{r5 r8} - r9c2{n5 n8} - b8n8{r9c5 r8c4} - r8c8{n8 .}
==> r7c5 ≠ 5

braid[5]: r7c5{n8 n4} - r8c8{n8 n5} - r6n4{c5 c7} - r8c7{n5 n2} - r8c6{n5 .} ==> r7c8 ≠ 8

whip[3]: c5n5{r1 r9} - r9c2{n5 n8} - r7n8{c1 .} ==> r1c5 ≠ 8

whip[4]: r7c5{n8 n4} - r6n4{c5 c7} - b9n4{r9c7 r9c9} - b9n8{r9c9 .} ==> r8c4 ≠ 8

braid[6]: b8n5{r8c6 r9c5} - r2c9{n5 n8} - r9c2{n5 n8} - r4n8{c9 c1} - b8n8{r9c5 r7c5} -
r3n8{c9 .} ==> r2c6 ≠ 5

hidden-single-in-a-block ==> r1c5 = 5

whip[2]: r9n5{c2 c7} - c8n5{r7 .} ==> r6c2 ≠ 5

hidden-single-in-a-block ==> r5c1 = 5

whip[6]: b8n3{r7c6 r9c4} - r6n3{c4 c7} - b6n4{r6c7 r4c9} - r9c9{n4 n8} - r8c8{n8 n5} -
b6n5{r6c8 .} ==> r7c8 ≠ 3

whip[4]: r4c8{n1 n6} - r7c8{n6 n5} - b6n5{r6c8 r6c7} - b6n4{r6c7 .} ==> r4c9 ≠ 1

whip[5]: c8n3{r1 r6} - b6n5{r6c8 r6c7} - c7n3{r6 r9} - c7n4{r9 r8} - b9n2{r8c7 .} ==> r3c9 ≠ 3

whip[6]: r6n4{c7 c5} - c6n4{r4 r7} - b8n3{r7c6 r9c4} - r9c9{n3 n8} - r8c8{n8 n5} - b8n5{r8c6 .}
==> r8c7 ≠ 4

whip[4]: r8c4{n1 n2} - r8c7{n2 n5} - b8n5{r8c6 r7c6} - b8n3{r7c6 .} ==> r9c4 ≠ 1

whip[4]: b9n4{r9c9 r7c9} - r7n3{c9 c6} - b8n5{r7c6 r8c6} - b8n4{r8c6 .} ==> r9c3 ≠ 4

whip[6]: b8n5{r8c6 r7c6} - b8n3{r7c6 r9c4} - r9n2{c4 c7} - c7n4{r9 r6} - b6n5{r6c7 r6c8} -
r6n3{c8 .} ==> r8c6 ≠ 2

whip[6]: c7n4{r9 r6} - b6n5{r6c7 r6c8} - r6n3{c8 c4} - b8n3{r9c4 r7c6} - b8n5{r7c6 r8c6} -
r8c7{n5 .} ==> r9c7 ≠ 2

singles ==> r8c7 = 2, r8c4 = 1, r9c3 = 1

whip[5]: r2n7{c1 c4} - r5n7{c4 c5} - r3n7{c5 c8} - c8n1{r3 r4} - c5n1{r4 .} ==> r1c3 ≠ 7

whip[5]: r3c6{n2 n9} - r2c6{n9 n6} - r5c6{n6 n3} - b8n3{r7c6 r9c4} - b8n2{r9c4 .} ==> r3c5 ≠ 2

whip[5]: c2n9{r1 r4} - c5n9{r4 r5} - r5n1{c5 c9} - b3n1{r3c9 r3c8} - r3n3{c8 .} ==> r3c1 ≠ 9

whip[6]: b6n5{r6c8 r6c7} - r6n4{c7 c5} - r7c5{n4 n8} - r9c5{n8 n2} - r9c4{n2 n3} - r6n3{c4 .} ==>
r6c8 ≠ 6

whip[6]: r3c9{n8 n1} - r5n1{c9 c5} - c5n9{r5 r4} - r5n9{c6 c3} - r5n7{c3 c4} - c5n7{r6 .} ==>
r3c5 ≠ 8

whip[1]: c5n8{r9 .} ==> r9c4 ≠ 8

whip[4]: c4n8{r2 r1} - c7n8{r1 r5} - c3n8{r5 r8} - b9n8{r8c8 .} ==> r2c9 ≠ 8

naked-single ==> r2c9 = 5

whip[2]: c7n4{r6 r9} - c7n5{r9 .} ==> r6c7 ≠ 3

whip[2]: c7n4{r9 r6} - c7n5{r6 .} ==> r9c7 ≠ 3

whip[1]: b9n3{r9c9 .} ==> r5c9 ≠ 3

whip[4]: c7n4{r9 r6} - b6n5{r6c7 r6c8} - r6n3{c8 c4} - r9n3{c4 .} ==> r9c9 ≠ 4
whip[2]: r6n4{c5 c7} - c9n4{r4 .} ==> r7c5 ≠ 4
naked-single ==> r7c5 = 8
whip[2]: b6n4{r4c9 r6c7} - r9n4{c7 .} ==> r4c5 ≠ 4
whip[3]: r8c8{n5 n8} - r9n8{c9 c2} - b7n5{r9c2 .} ==> r7c8 ≠ 5
singles ==> r7c8 = 6, r4c8 = 1, r3c9 = 1, r5c5 = 1
whip[3]: b5n4{r4c6 r6c5} - r9c5{n4 n2} - r4c5{n2 .} ==> r4c6 ≠ 9
whip[3]: b9n4{r7c9 r9c7} - r9n5{c7 c2} - r7n5{c2 .} ==> r7c6 ≠ 4
whip[3]: r5c7{n3 n8} - c9n8{r5 r9} - r9n3{c9 .} ==> r5c4 ≠ 3
whip[3]: b4n7{r6c3 r5c3} - r5n9{c3 c6} - b5n3{r5c6 .} ==> r6c4 ≠ 7
whip[3]: r6n6{c2 c4} - b5n3{r6c4 r5c6} - r5n9{c6 .} ==> r5c3 ≠ 6
whip[4]: c3n4{r1 r8} - r8c6{n4 n5} - r8c8{n5 n8} - r3n8{c8 .} ==> r1c3 ≠ 8
whip[3]: b9n8{r9c9 r8c8} - c3n8{r8 r2} - r3n8{c2 .} ==> r5c9 ≠ 8
singles to the end

7.7.6. $gB_\infty \neq gW_\infty$: a puzzle solvable by g-braids but probably not by g-whips

Finding a Sudoku puzzle solvable by g-braids but not by g-whips is very hard: one can rely neither on random generators (all the puzzles we produced with them – about ten millions – were solvable by whips) nor on Subset rules that would not be subsumed by g-whips but would be by g-braids (see chapter 8 for comments on this). The following (Figure 7.6) gives the only such puzzle (#77) in the Magictour-top1465 collection. Using the "gT&E vs g-braids" theorem, it is easy to show that it can be solved by g-braids. But the following resolution path with g-whips shows that these are not enough to make substantial advances in the solution.

7					4			
	2			7			8	
		3			8			9
			5			3		
	6			2			9	
		1			7			6
			3			9		
	3			4			6	
		9			1			5

7	9	8	6	3	5	4	2	1
1	2	6	9	7	4	5	8	3
4	5	3	2	1	8	6	7	9
9	7	2	5	8	6	3	1	4
5	6	4	1	2	3	8	9	7
3	8	1	4	9	7	2	5	6
6	1	7	3	5	2	9	4	8
8	3	5	7	4	9	1	6	2
2	4	9	8	6	1	7	3	5

Figure 7.6. A puzzle (Magictour-top1465#77) solvable by g-braids but not by g-whips

***** SudoRules version 15.1.8-gW *****
24 givens and 219 candidates
hidden-single-in-a-row ==> r9c8 = 3
whip[1]: r9n4{c2 .} ==> r7c1 <> 4, r7c2 <> 4, r7c3 <> 4
g-whip[8]: c4n6{r2 r9} - r1n6{c4 c3} - b1n8{r1c3 r1c2} - b1n9{r1c2 r2c1} - b1n1{r2c1 r3c123} - r3c5{n1 n5} - r7c5{n5 n8} - r9c5{n8 .} ==> r2c6 <> 6

whip[11]: c3n4{r4 r2} - c6n4{r2 r4} - b5n6{r4c6 r4c5} - r9c5{n6 n8} - r7c5{n8 n5} - r3c5{n5 n1} - c4n1{r3 r5} - b5n8{r5c4 r6c4} - b5n9{r6c4 r6c5} - r6c2{n9 n5} - r3c2{n5 .} ==> r5c1 <> 4

whip[12]: r9c5{n8 n6} - r7c5{n6 n5} - r3c5{n5 n1} - r4c5{n1 n9} - r6c4{n9 n4} - r5c6{n4 n3} - b4n3{r5c1 r6c1} - c1n9{r6 r2} - r2c4{n9 n6} - r1n6{c6 c3} - b1n8{r1c3 r1c2} - b1n1{r1c2 .} ==> r6c5 <> 8

g-whip[14]: r3n4{c4 c123} - c3n4{r2 r4} - c6n4{r4 r2} - r5c6{n4 n3} - b4n3{r5c1 r6c1} - r6n4{c1 c8} - r6n2{c8 c7} - b4n2{r6c1 r4c1} - r9n2{c1 c4} - c6n2{r8 r1} - b2n3{r1c6 r1c5} - r1c9{n3 n1} - r5n1{c9 c7} - b6n5{r5c7 .} ==> r5c4 <> 4

whip[15]: b3n6{r2c7 r3c7} - b3n7{r3c7 r3c8} - r3n2{c8 c4} - c4n4{r3 r6} - r5c6{n4 n3} - r6c5{n3 n9} - r4c6{n9 n6} - r1n6{c6 c3} - b1n8{r1c3 r1c2} - r6c2{n8 n5} - c8n5{r6 r1} - r1c6{n5 n9} - r1c4{n9 n1} - r5c4{n1 n8} - r5c1{n8 .} ==> r2c4 <> 6

After this point, there is no whip or g-whip of length less than 18. While trying g-whips[18], the number of partial g-whips to be analysed suddenly gets so large that SudoRules encounters memory overflow problems. Given the poor partial results above (only 8 eliminations after the HS(row)), it is unlikely that a g-whip solution can be obtained. Exercise for the reader: write a better implementation of g-whips (less greedy for memory) and prove that there is indeed no g-whip solution.

7.8. g-labels and g-whips in n-Queens and in SudoQueens

7.8.1. g-labels in n-Queens

We have seen in section 5.11 that, in the n-Queens CSP, one can identify a label with a cell in the grid. From the various examples of whip[1] we have already seen there, we can understand that the g-labels of n-Queens are:

— for variable $Xr°$:

- all the symmetric sets of horizontal triplets of cells in row $r°$ that are separated by k other cells, $0 \leq k \leq IP((n-3)/2)$, provided that: 1) either the second diagonal passing though the leftmost cell, the first diagonal passing through the rightmost cell and the column passing through the inner cell meet in a cell above $r°$ and inside the grid; 2) or the first diagonal passing though the leftmost cell, the second diagonal passing through the rightmost cell and the column passing through the inner cell meet in a cell under $r°$ and inside the grid. The labels l g-linked to such a g-label correspond to the meeting points; (there are at most 2 such labels, symmetric with respect to $r°$);

- all the sets of horizontal pairs of cells in row $r°$ that are separated by k other cells ($0 \leq k \leq n-2$), provided that the column passing through one cell and one of the two diagonals passing through the other cell meet in a cell inside the grid, and provided that they are not part of some of the previous g-labels (maximality condition). The labels l g-linked to such a g-label correspond to the meeting points; (depending on $r°$, k and n, there are at most 2 or 4 such labels, symmetric with respect to $r°$ and the column containing l_2);

– for variable $X_{c°}$: similar g-labels obtained by 90° rotation.

Notice that, contrary to the Sudoku case, *any* label l g-linked to a g-label $<V, g>$ for a CSP variable V must use at least two different types of constraints (row, column, first diagonal or second diagonal) for its links with the various elements of g and at least one of these constraints is not defined by a CSP variable.

A simple case of a g-whip[3] can already be seen in the example of Figure 5.8, section 5.11.4. The first whip[4] elimination there can be replaced by a g-whip[3]:

g-whip[3]: r8{c5 c1} – r2{c1 c58} – r4{c8 .} $\Rightarrow$ ¬r2c9 (G eliminated)

7.8.2. A g-whip[3] example in 9-Queens

Accepting the same solution grid as that in Figure 5.8, the puzzle in Figure 7.7 is based on the same first two givens, but a different third one (r3c3, r6c2 and r8c5).

	c1	c2	c3	c4	c5	c6	c7	c8	c9
r1	o	o	o	+	o		o	B	A
r2	+	o	o	o	o	o			
r3	o	o	*	o	o	o	o	o	o
r4	o	o	o	o	o	+			o
r5	o	o	o		o		E	o	+
r6	o	*	o	o	o	o	o	o	o
r7	o	o	o	o	o	o	o	+	D
r8	o	o	o	o	*	o	o	o	o
r9	F	o	o	o	o	o	+	C	o

Figure 7.7. g-whips in a 9-Queens instance

***** Manual solution *****
whip[2]: c6{r1 r5} – c4{r5 .} $\Rightarrow$ ¬r1c9 (A eliminated)
g-whip[3]: r4{c8 c67} – r5{c6 c9} – r7{c9 .} $\Rightarrow$ ¬r1c8 (B eliminated)
g-whip[3]: r4{c8 c67} – r5{c6 c9} – r7{c9 .} $\Rightarrow$ ¬r9c8 (C eliminated)
whip[3]: r9{c7 c1} – r2{c1 c7}- r4{c7 .} $\Rightarrow$ ¬r7c9 (D eliminated)
single in r7: r7c8

whip[1]: r4{c7 .} ⇒ ¬r5c9 (E eliminated)
whip[2]: r4{c6 c7} – r2{c7 .} ⇒ ¬r9c1 (F eliminated)
single in r9: r9c7; single in c1: r2c1; single in r4: r4c6; single in r1: r1c4; single in r5: r5c9
Solution found in gW3.

7.8.3. g-labels in n-SudoQueens

n-SudoQueens was introduced in section 5.11.8. The g-labels of n-SudoQueens are both those of n-Sudoku (without their "n" coordinate) and those of n-Queens. As a result, the set of labels of a g-label can be included in the set of labels of another g-label (for a different CSP variable). For instance, consider 9-SudoQueens and the following two g-labels:
- <Xb1, g1> associated with CSP variable Xb1: <Xb1, r3c123>,
- <Xr3, g2> associated with CSP variable Xr3: <Xr3, r3c12>.

Let l be a label with respective representatives (r, c) and [b, s] in the two coordinate systems. Then:
- l is g-linked to <Xb1, g1> if and only if b = b1;
- l is g-linked to <Xr3, g2> if and only if (r = r2 or r = r4) and (c = 1 or c = 2).

This example shows that, although the set of labels in g2 is included in the set of labels in g1, none of the sets of labels linked to them is included in the other. This justifies that we keep the CSP variable as an explicit component of a g-label.

7.8.4. A g-whip[4] example in 9-SudoQueens

The example in Figure 7.8 shows an example of a g-whip[4] in 9-SudoQueens.

We shall also use it to illustrate how one can find instances of a CSP. When we introduced n-SudoQueens in section 5.11.8, we did not know for sure whether this CSP was not too constrained to have instances, at least for small values of n. So we tried to find instances for increasing values of n. As mentioned in that section, there are no instances for n = 2 or n = 4. But we found the instance in Figure 5.12 for n = 9, by a heuristic technique of adding queens progressively in the cell that is linked to the fewest other cells, so that we destroy fewer possibilities for the next ones. We started by cells in the two main diagonals, as close as possible to a corner (lesser destruction). When we reached the situation in Figure 5.13 (three queens given, in cells r1c1, r2c8 and r3c5), block b5 had only two possibilities left; r5c4 is linked to 11 available cells and r5c6 to 12; so we chose to put a queen in r5c4; but we were unable to find a solution. We then tried to prove that r5c4 was impossible; this is how we found the following g-whip[4] and a first resolution path showing that there is a unique solution. We keep this g-whip here for illustrative purposes:

g-whip[4]: c6{r7 r89} – b9{r9c7 r9c9} – b6{r4c9 r4c7} – r4{c2 .} ⇒ ¬r5c4 (A eliminated)

In this g-whip: r5c6 and r7c6 are z-candidates for Xr6 (1st cell); r8c7 is both a z- and a t-candidate for Xb9 (2nd cell); r5c6 is both a z- and a t- candidate for Xb6; r6c7, r4c9 and r6c9 and t-candidates for Xb6 (3rd cell); r4c3 and r5c2 are t-candidates for Xr4 (last cell).

	c1	c2	c3	c4	c5	c6	c7	c8	c9
r1	*	o	o	o	o	o	o	o	o
r2	o	o	o	o	o	o	o	*	o
r3	o	o	o	o	*	o	o	o	o
r4	o	-3	-0	o	o	o	+3	o	-2
r5	o	-0	o	A	o	-0	o	o	-0
r6	o	o	-0	o	o	o	-2	o	+2
r7	o	-0	o	-0	o	-0	o	o	o
r8	o	o		-0	o	+1	-0	o	+2
r9	o		B	-0	o	+1	-1	o	o

Figure 7.8. A partial grid for 9-SudoQueens

In Figure 5.13, in addition to our previous conventions, the characters in bold in a cell mean the following:
"-0" : the z-candidates of this g-whip;
"+n" the right-linking candidate or g-candidate for the n-th CSP variable;
"-n" the candidates linked to the n-th previous right-linking pattern in the n-th cell; they can be left-linking or t-candidates for the (n+1)-th CSP variables.

[Later, we found a simpler pattern, a whip[3], for the same elimination:

***** Manual solution *****
whip[3]: b4{r5c2 r4c2} - b9{r9c7 r8c9} - b6{r6c9 .} $\Rightarrow$ ¬r5c4 (A eliminated)
;;; the sequel has nothing noticeable:
single in block b5: r5c6
whip[1]: c4{r9 .} $\Rightarrow$ ¬r9c3 (B eliminated)
single in block b7: r7c2; single in column c3: r4c3; single in column c4: r8c4; single in row r6: r6c9; single in row r9: r9c7
Solution found in gW$_4$ (The solution is given in Figure 5.12.)]

Part Three

BEYOND G-WHIPS AND G-BRAIDS

8. Subset rules in a general CSP

This chapter has the two complementary goals of defining elementary Subset rules in any CSP and of showing that whips, g-whips, braids and g-braids subsume "almost all" the instances of these rules. This is not to mean that such elementary Subset rules (that are globally much weaker than whips) should not be preferred to chain rules when they can be applied; on the contrary, they may provide a shorter or a better understandable solution. But, when merely added to them, they do not bring much more resolution power; things are different when they are combined, as they will be in chapter 9, with the general "zt-ing" technique of whips and braids. Preparing the introduction of such combinations is the third goal of this chapter.

For the Subsets of size greater than two, we pay particular attention to the definitions: we want them to be comprehensive enough to get the broadest coverage but restrictive enough to exclude degenerated cases: for us, two Singles do not make a Pair, a Pair and a Single do not make a Triplet, a Triplet and a Single do not make a Quad, two Pairs do not make a Quad, … This modelling choice is consistent with what has already been done in the Sudoku case in *HLS1*, but it is now also closely related to how these patterns can be assigned a well defined "size" and ranked with respect to the W_n, B_n, gW_n and gB_n hierarchies; this will be essential in chapter 9 when we take them as building blocks of "S_p-whips" and "S_p-braids".

In sections 8.2 to 8.4, we define an S_p-subset rule in the general CSP framework (for $p = 2$, $p = 3$ and $p = 4$ – corresponding respectively to Pairs, Triplets and Quads) and we illustrate it by the classical form it takes in Sudoku, depending on which families of CSP variables one considers. For Sudoku, we write the Subsets in rows and leave it to the reader to write the corresponding Subsets in columns and in blocks (e.g. using meta-theorems 4.1 and 4.3 on symmetry and analogy). We give both the English and the formal logic statements and we insist once more on the symmetry and super-symmetry relationships between Naked, Hidden and Super-Hidden Subsets of same size (see Figure 8.1). Subsets are the simplest example of how the general CSP framework unifies, in a still stronger way than the mere symmetry relationships already present in *HLS1*, patterns that would otherwise be considered as different: ***in the CSP framework, Naked, Hidden and Super-Hidden Subset rules are not only related by symmetry relationships (for Subsets of given size), they are the very same rule.*** (Symmetry, super-symmetry and analogy of rules have already been illustrated in this book by whips and braids, but in a different, more powerful, way: they use only basic predicates having these properties.)

Though they were not formulated in CSP terms, all the classical Subset rules of sections 8.2 to 8.4 (except the Special Quads) were present in *HLS1*, in their Sudoku specific form. But our perspective here is different: we are less concerned with these patterns for themselves than with their relationship with whips and braids – whence the general subsumption theorems of section 8.6 and the choice of examples in section 8.7, mainly centred on showing rare cases not covered by subsumption.

8.1. Transversality, S_p-labels and S_p-links

In the same way as, in chapter 7, we had to introduce a distinction between g-labels (defined as maximal sets of labels) and g-candidates (that did not have to be maximal), we must now introduce a distinction between:

– S_p-labels, that can only refer to CSP variables and transversal sets of labels (which can be considered as a kind of maximality condition on S_p-labels),

– and S_p-subsets, in which considerations about mandatory and optional candidates will appear.

8.1.1. Set of labels transversal to a set of CSP variables

Definition: for p>1, given a set of p different CSP variables $\{V_1, V_2, ..., V_p\}$, we say that a non-empty set S of at most p different labels is *transversal* with respect to $\{V_1, V_2, ..., V_p\}$ for constraint c if:

– none of these labels has a representative with two of these CSP variables;

– all these labels are pairwise linked by c;

– S is maximal, in the sense that no label pertaining to one of these CSP variables could be added to it without contradicting the first two conditions.

Remarks:

– the first condition will always be true for pairwise strongly disjoint CSP variables, i.e. CSP variables such that no two of them share a label; but we do not adopt this stronger condition on CSP variables; adopting it would not change the general theory (for Subsets in the present chapter and for Reversible S_p-chains, S_p-whips and S_p-braids in chapter 9) and it would not restrict the applications to Sudoku; but it may restrict the applications to others CSPs; moreover, the corresponding definition for g-Subsets in chapter 10 would restrict the applications, even for Sudoku (see the example in section 10.3).

– the second condition could be generalised by allowing labels in the transversal set to be pairwise linked by different constraints. In LatinSquare or Sudoku, due to the theorems proven in chapter 11 of *HLS1*, such pairwise constraints can always be replaced by a global constraint as in the present definition; this is also obviously true in n-Queens. In case a CSP had a transversal set that could not be defined via a unique constraint, we think modelling choices should be investigated. Anyway, the apparently more general condition would not change the theory developed in this chapter and in chapter 9 (it is nowhere used in the proofs) – although it may have a noticeable negative impact on the complexity of any possible implementation.

Typical examples of transversal sets of labels occur when the CSP can be represented on a k dimensional grid and two candidates differing by only one coordinate are contradictory, as can be illustrated by the Sudoku or LatinSquare examples: given CSP variables Xrc1 and Xrc2, $\{<Xrc1, n°>, <Xrc2, n°>\}$ is a transversal set of labels, for any fixed Number n°; given CSP variables Xrn1 and Xrn2, $\{<Xrn1, c°>, <Xrn2, c°>\}$ is also a transversal set of labels for any fixed Column c°... But there is no reason to restrict the above definition to such "geometrical" cases. In particular, a transversal set of labels does not have to be associated with a "transversal" CSP variable (in the sense that, e.g. in Sudoku, variable Xc°n° could be called transversal to variable Xr°n°): in n-Queens, given two CSP variables Xr_1 and Xr_2 corresponding to different rows, the set of intersections of any diagonal (which is not associated with any CSP variable) with these rows defines a transversal set of labels (see section 8.8.1 for an example).

8.1.2. S_p-labels and S_p-links

Definitions: for any integer p>1, an S_p-label is a couple of data: {CSPVars, TransvSets}, where CSPVars is a set of p different CSP variables and TransvSets is a set of p different transversal sets of labels for these variables (each one for a well defined constraint). An S-label is an S_p-label for some p >1.

Definition: a label l is S_p-linked or simply S-linked to an S_p-label S = {CSPVars, TransvSets} if there is some k, 1≤k≤p, such that l is linked by the constraint c_k of $TransvSets_k$ to all the labels of $TransvSets_k$ (where $TransvSets_k$ is the k-th element of TransvSets). In these conditions, l is also called *a potential target of the S_p-label*.

Miscellaneous remarks:

– with this definition, a label and a g-label are not S_p-labels (due to the condition p>1); for labels, this is a mere matter of convention, but this choice is more convenient for the sequel;

– as a result of this condition, there may be CSPs with no S_p-labels;

– different transversal sets in the S_p-label are not required to be disjoint;

– in a sense, an S_p-label corresponds to the maximal extent of a possible S_p-subset (as defined below).

Notation: in the forthcoming definition of Subsets, we shall need a means of specifying that, in some transversal sets, some labels must exist while others may exist or not. We shall write this as e.g. $\{<V_1, v_1>, <V_2, v_2>, ..., (<V_k, v_k>), ...\}$.

This should be understood as follows: a label not surrounded with parentheses must exist; a pseudo-label surrounded with parentheses may exist or not; if it exists, then it is named $<V_k, v_k>$.

8.2. Pairs

8.2.1. Pairs in a general CSP

Definition: in any resolution state RS of any CSP, a *Pair* (or S_2-*subset*) is an S_2-label {CSPVars, TransvSets}, where:

– CSPVars = $\{V_1, V_2\}$,

– TransvSets is composed of the following transversal sets of labels:
  $\{<V_1, v_{11}>, <V_2, v_{21}>\}$ for constraint c_1,
  $\{<V_1, v_{12}>, <V_2, v_{22}>\}$ for constraint c_2,
such that:

– in RS, V_1 and V_2 are disjoint, i.e. they share no candidate;

– $<V_1, v_{11}> \neq <V_1, v_{12}>$ and $<V_2, v_{22}> \neq <V_2, v_{21}>$;

– in RS, V_1 has the two mandatory candidates $<V_1, v_{11}>$ and $<V_1, v_{12}>$ and no other candidate;

– in RS, V_2 has the two mandatory candidates $<V_2, v_{22}>$ and $<V_2, v_{21}>$ and no other candidate.

A *target of a Pair* is defined as a candidate S_2-linked to the underlying S_2-label.

Theorem 8.1 (S_2 rule): in any CSP, a target of a Pair can be eliminated.

Proof: as the two transversal sets play similar roles, we can suppose that Z is linked to both $<V_1, v_{11}>$ and $<V_2, v_{21}>$. If Z was True, these candidates would be eliminated by ECP. As V_1 and V_2 have only two candidates each, their other candidate ($<V_1, v_{12}>$, respectively $<V_2, v_{22}>$) would be asserted by S, which is impossible, as they are linked. Notice that the proof works only because V_1 and V_2 share no candidate in RS (and therefore in no posterior resolution state).

The rest of this section shows how, choosing pairs of variables in different sub-families of CSP variables, the familiar Naked Pairs, Hidden Pairs and Super-Hidden Pairs (X-Wing) of Sudoku (or LatinSquare) appear as mere Pairs in the above defined sense.

8.2.2. Naked Pairs in Sudoku

For the definition of Naked Pairs, there can be no ambiguity and we adopt the standard formulation. Naked Pairs in a row, or NT(row), is the following rule:

if there is a row r and there are two different columns c_1 and c_2 and two different numbers n_1 and n_2, such that:
- the candidates for cell (r, c_1) are exactly the two numbers n_1 and n_2,
- the candidates for cell (r, c_2) are exactly the two numbers n_1 and n_2,
then eliminate the two numbers n_1 and n_2 from the candidates for any other rc-cell in row r in rc-space.

Validity is very easy to prove directly from this (almost) standard formulation of the problem: in row r, each of the two cells defined by columns c_1 and c_2 must get a value and only two values (n_1 and n_2) are available for them, which entails that, whatever distribution is made between them of these two values, none of these two values remains available for the other cells in the same row.

The logical formulation strictly parallels the English one (except that, as is often the case, something which is formulated in natural language as "if there exists a row ...", which should apparently translate into an existential quantifier, must be written with a universal quantifier):

$\forall r \forall \neq (c_1, c_2) \forall \neq (n_1, n_2)$
    { candidate(n_1, r, c_1) $\wedge$ candidate(n_2, r, c_1) $\wedge$
      candidate(n_2, r, c_2) $\wedge$ candidate(n_1, r, c_2) $\wedge$
      $\forall c \in \{c_1, c_2\} \forall n \neq n_1, n_2$ ¬candidate(n, r, c)
$\Rightarrow$
      $\forall c \neq c_1, c_2 \ \forall n \in \{n_1, n_2\}$ ¬candidate(n, r, c) }.

Exercise: show that this is exactly what Pairs of the general definition give when applied to CSP variables Xrc_1 and Xrc_2, with transversal sets defined by CSP variables (considered as constraints) Xrn_1 and Xrn_2.

8.2.3. Hidden Pairs in Sudoku

If we apply meta-theorem 4.2 to Naked Pairs in a row, permuting the words "number" and "column", we obtain the rule for Hidden Pairs in a row (once transposed into rn-space, a Hidden Pairs in a row looks graphically like a Naked Pairs in a row would in rc-space):
if there is a row r and there are two different numbers n_1 and n_2 and two different columns c_1 and c_2, such that:
- the candidates (columns) of rn-cell (r, n_1) (in rn-space) are exactly c_1 and c_2,
- the candidates (columns) of rn-cell (r, n_2) (in rn-space) are exactly c_1 and c_2,
then eliminate the two columns c_1 and c_2 from the candidates for any other rn-cell (r, n) in row r in rn-space.

$\forall r \forall \neq (n_1, n_2) \forall \neq (c_1, c_2)$
    { candidate(n_1, r, c_1) $\wedge$ candidate(n_1, r, c_2) $\wedge$
        candidate(n_2, r, c_2) $\wedge$ candidate(n_2, r, c_1) $\wedge$

$$\forall n \in \{n_1, n_2\} \forall c \neq c_1, c_2 \ \neg candidate(n, r, c)$$
$$\Rightarrow$$
$$\forall n \neq n_1, n_2 \forall c \in \{c_1, c_2\} \ \neg candidate(n, r, c) \}.$$

Exercise: show that this is exactly what Pairs of the general definition give when applied to CSP variables Xrn_1 and Xrn_2, with transversal sets defined by CSP variables (considered as constraints) Xrc_1 and Xrc_2.

8.2.4. Super Hidden Pairs in Sudoku (X-Wing)

This is not yet the full story: one can iterate the application of meta-theorem 4.2 and a rule SHP(row) can be obtained from rule HP(row) by permuting the words "row" and "number". Let us first do this permutation formally, i.e. by applying the S_m transform to HP(row) = S_{cn}(NP(row)). We get the logical formulation for Super Hidden Pairs in rows, or SHP(row):

$\forall n \forall \neq (r_1, r_2) \forall \neq (c_1, c_2)$
   { candidate$(n, r_1, c_1) \wedge$ candidate$(n, r_1, c_2) \wedge$
   candidate$(n, r_2, c_2) \wedge$ candidate$(n, r_2, c_1) \wedge$
   $\forall r \in \{r_1, r_2\} \forall c \neq c_1, c_2 \ \neg candidate(n, r, c)$
   $\Rightarrow$
      $\forall r \neq r_1, r_2 \forall c \in \{c_1, c_2\} \ \neg candidate(n, r, c) \}.$

Let us now try to understand the result, with a strict English transcription:
if there is a number n and there are two different rows r_1 and r_2 and two different columns c_1 and c_2 such that:
- the candidates (columns) of rn-cell (r_1, n) (in rn-space) are c_1 and c_2 and no other column,
- the candidates (columns) of rn-cell (r_2, n) (in rn-space) are c_1 and c_2 and no other column,
then eliminate the two columns c_1 and c_2 from the candidates (columns) for any other rn-cell (r, n) in column n in rn-space.

Exercise: show that this is exactly what Pairs of the general definition give when applied to CSP variables Xr_1n and Xr_2n, with transversal sets defined by CSP variables (considered as constraints) Xc_1n and Xc_2n.

As the meaning of this rule is not absolutely clear in rc-space, let us make it more explicit with a new equivalent formulation based on rc-space: if there is a number n and there are two different rows r_1 and r_2, such that, in these rows, n appears as a candidate only in columns c_1 and c_2, then, in any of these two columns, eliminate n from the candidates for any row other than r_1 and r_2. We find the usual formulation of X-Wing in rows. Finally, we have shown that the familiar X-Wing in rows is the super-hidden version of Naked Pairs in a row: SHP(row) $\equiv S_m$(HP(row)) $\equiv S_m(S_{cn}$(NP(row))) = X-Wing(row).

8.3. Triplets

8.3.1. Triplets in a general CSP

There may be several formulations of Triplets. Here, we adopt one that is neither too restrictive (the presence of some of the candidates potentially involved is not mandatory) nor too comprehensive (i.e., by making mandatory the presence of some of the candidates involved, it does not allow degenerated cases). The justification was done in *HLS1* for Sudoku, but it is valid for the general CSP.

Definition: in any resolution state RS of any CSP, a *Triplet* (or S_3-*subset*) is an S_3-label {CSPVars, TransvSets}, where:

– CSPVars = $\{V_1, V_2, V_3\}$,

– TransvSets is composed of the following transversal sets of labels:

   $\{<V_1, v_{11}>, (<V_2, v_{21}>), <V_3, v_{31}>\}$ for constraint c_1,
   $\{<V_1, v_{12}>, <V_2, v_{22}>, (<V_3, v_{32}>)\}$ for constraint c_2,
   $\{(<V_1, v_{13}>), <V_2, v_{23}>, <V_3, v_{33}>\}$ for constraint c_3,

such that:

– in RS, V_1, V_2 and V_3 are pairwise disjoint, i.e. no two of these variables share a candidate;

– $<V_1, v_{11}> \neq <V_1, v_{12}>$, $<V_2, v_{22}> \neq <V_2, v_{23}>$ and $<V_3, v_{33}> \neq <V_3, v_{31}>$;

– in RS, V_1 has the two mandatory candidates $<V_1, v_{11}>$ and $<V_1, v_{12}>$, one optional candidate $<V_1, v_{13}>$ (supposing this label exists) and no other candidate;

– in RS, V_2 has the two mandatory candidates $<V_2, v_{22}>$ and $<V_2, v_{23}>$, one optional candidate $<V_2, v_{21}>$ (supposing this label exists) and no other candidate;

– in RS, V_3 has the two mandatory candidates $<V_3, v_{33}>$ and $<V_3, v_{31}>$, one optional candidate $<V_3, v_{32}>$ (supposing this label exists) and no other candidate.

A *target of a Triplet* is defined as a candidate S_3-linked to the underlying S_3-label.

Theorem 8.2 (S_3 rule): in any CSP, a target of a Triplet can be eliminated.

Proof: as the three transversal sets play similar roles, we can suppose that Z is linked to the first, i.e. to $<V_1, v_{11}>$, $<V_2, v_{21}>$ (and $<V_3, v_{31}>$ if it exists). If Z was True, these candidates (if they are present) would be eliminated by ECP. Each of V_1, V_2 and V_3 would have at most two candidates left. Any choice for V_1 would reduce to at most one the number of possibilities for each of V_2 and V_3 (due to the pairwise contradictions between members of each transversal set). Finally, the unique choice for V_2, if any, would in turn reduce to zero the number of possibilities for V_3.

The rest of this section shows how, choosing sets of three variables in different sub-families of CSP variables, the familiar Naked Triplets, Hidden Triplets and

Super-Hidden Triplets (Swordfish) of Sudoku appear as mere Triplets of the general CSP.

8.3.2. Naked Triplets in Sudoku

There may be several definitions of Naked Triplets (see *HLS1* for a discussion). Here, we adopt the same as in *HLS1*, neither too restrictive nor too comprehensive (i.e. it does not allow degenerated cases). Naked Triplets in a row or NT(row):
if there is a row r and there are three different columns c_1, c_2 and c_3 and three different numbers n_1, n_2 and n_3, such that:
- cell (r, c_1) has n_1 and n_2 among its candidates,
- cell (r, c_2) has n_2 and n_3 among its candidates,
- cell (r, c_3) has n_3 and n_1 among its candidates,
- none of the cells (r, c_1), (r, c_2) and (r, c_3) has any candidate other than n_1, n_2 or n_3,
then eliminate the three numbers n_1, n_2 and n_3 from the candidates for any other cell in row r in rc-space.

$\forall r \forall \neq (c_1,c_2,c_3) \forall \neq (n_1,n_2,n_3)$
  { candidate$(n_1, r, c_1) \wedge$ candidate$(n_2, r, c_1) \wedge$
    candidate$(n_2, r, c_2) \wedge$ candidate$(n_3, r, c_2) \wedge$
    candidate$(n_3, r, c_3) \wedge$ candidate$(n_1, r, c_3) \wedge$
    $\forall c \in \{c_1, c_2, c_3\} \forall n \neq n_1,n_2,n_3 \neg$candidate$(n, r, c)$
  $\Rightarrow$
    $\forall c \neq c_1,c_2,c_3 \ \forall n \in \{n_1, n_2, n_3\} \neg$candidate$(n, r, c)$ }.

Exercise: show that this is exactly what Triplets of the general definition give when applied to CSP variables Xrc_1, Xrc_2 and Xrc_3, with transversal sets defined by CSP variables (considered as constraints) Xrn_1, Xrn_2 and Xrn_3.

8.3.3. Hidden Triplets in Sudoku

If we apply meta-theorem 4.2 to Naked Triplets in a row, permuting the words "number" and "column", we obtain the rule for Hidden Triplets in a row, or HT(row):
if there is a row r, and there are three different numbers n_1, n_2 and n_3 and three different columns c_1, c_2 and c_3, such that:
- rn-cell (r, n_1) (in in rn-space) has c_1 and c_2 among its candidates (columns),
- rn-cell (r, n_2) (in in rn-space) has c_2 and c_3 among its candidates (columns),
- rn-cell (r, n_3) (in in rn-space) has c_3 and c_1 among its candidates (columns),
- none of the rn-cells (r, n_1), (r, n_2) and (r, n_3) (in in rn-space) has any remaining candidate (column) other than c_1, c_2 and c_3,
then eliminate the three columns c_1, c_2 and c_3 from the candidates for any other rn-cell (r, n) in row r in rn-space.

$\forall r \forall \neq (n_1,n_2,n_3) \forall \neq (c_1,c_2,c_3)$
    { candidate(n_1, r, c_1) $\wedge$ candidate(n_1, r, c_2) $\wedge$
    candidate(n_2, r, c_2) $\wedge$ candidate(n_2, r, c_3) $\wedge$
    candidate(n_3, r, c_3) $\wedge$ candidate(n_3, r, c_1) $\wedge$
    $\forall n \in \{n_1, n_2, n_3\} \forall c \neq c_1,c_2,c_3$ ¬candidate(n, r, c)
$\Rightarrow$
    $\forall n \neq n_1,n_2,n_3 \forall c \in \{c_1, c_2, c_3\}$ ¬candidate(n, r, c) }.

Exercise: show that this is exactly what Triplets of the general definition give when applied to CSP variables Xrn_1, Xrn_2 and Xrn_3, with transversal sets defined by CSP variables (considered as constraints) Xrc_1, Xrc_2 and Xrc_3.

8.3.4. Super Hidden Triplets in Sudoku (Swordfish)

As in the case of Pairs, one can iterate the application of meta-theorem 4.2 and a rule SHT(row) rule can be obtained from rule HT(row) by permuting the words "row" and "number". If we apply the S_{rn} transform to HT(row) = S_{cn}(NT(row)), we get the logical formulation of Super Hidden Triplets in rows, or SHT(row):

$\forall n \forall \neq (r_1,r_2,r_3) \forall \neq (c_1,c_2,c_3)$
    { candidate(n, r_1, c_1) $\wedge$ candidate(n, r_1, c_2) $\wedge$
    candidate(n, r_2, c_2) $\wedge$ candidate(n, r_2, c_3) $\wedge$
    candidate(n, r_3, c_3) $\wedge$ candidate(n, r_3, c_1) $\wedge$
    $\forall r \in \{r_1, r_2, r_3\} \forall c \neq c_1,c_2,c_3$ ¬candidate(n, r, c)
$\Rightarrow$
    $\forall r \neq r_1,r_2,r_3 \forall c \in \{c_1, c_2, c_3\}$ ¬candidate(n, r, c) }.

Let us now try to understand the result, first with a direct English transliteration: if there is a number n, and there are three different rows r_1, r_2 and r_3 and three different columns c_1, c_2 and c_3, such that:
- rn-cell (r_1, n) (in rn-space) has c_1 and c_2 among its candidates (columns),
- rn-cell (r_2, n) (in rn-space) has c_2 and c_3 among its candidates (columns),
- rn-cell (r_3, n) (in rn-space) has c_3 and c_1 among its candidates (columns),
- none of the rn-cells (r_1, n), (r_2, n) and (r_3, n) (in rn-space) has any candidate (column) other than c_1, c_2 and c_3,
then eliminate the three columns c_1, c_2 and c_3 from the candidates (columns) for any other rn-cell (r, n) in column n in rn-space in rn-space .

Exercise: show that this is exactly what Triplets of the general definition give when applied to CSP variables Xr_1n, Xr_2n and Xr_3n, with transversal sets defined by CSP variables (considered as constraints) Xc_1n, Xc_2n and Xc_3n.

As this is not yet very explicit, let us try to clarify it by expressing it in rc-space and by temporarily forgetting part of the conditions: if there is a number n and there are three different rows r_1, r_2 and r_3 and three different columns c_1, c_2 and c_3, such that for each of the three rows the instance of number n that must be somewhere in each of these rows can actually only be in either of the three columns, then in any of the three columns eliminate n from the candidates for any row different from the given three.

What we find is the usual formulation of the rule for Swordfish in rows. There remains one point: the part of the conditions we have temporarily discarded. It is precisely what prevents Swordfish in rows from reducing to X-Wing in rows.

8.4. Quads

8.4.1. Quads in a general CSP

Finding the proper formulation for Quads, guaranteeing that it covers no degenerated case, is less obvious than for Triplets. Indeed, the simplest way is to introduce two types of Quads: Cyclic and Special. (In order to avoid technicalities, we shall show that there can only be these two types for the Sudoku CSP, but the analysis can be transposed to the general framework.) We choose to write the Special Quad in such a way that it is does not cover any case already covered by the Cyclic Quad. If we wanted to introduce larger Subsets, though one could always write a general formula expressing non-degeneracy (which would lead to computationally very inefficient implementations), it would get harder and harder to write an explicit (more efficient) list of non-degenerated subcases. As we shall see soon, in the 9×9 Sudoku case, this would be useless.

Definition: in any resolution state RS of any CSP, a *Cyclic Quad* (or *Cyclic S_4-subset*) is an S_4-label {CSPVars, TransvSets}, where:

– CSPVars = $\{V_1, V_2, V_3, V_4\}$,

– TransvSets is composed of the following transversal sets of labels:
$\{<V_1, v_{11}>, (<V_2, v_{21}>), (<V_3, v_{31}>), <V_4, v_{41}>\}$ for constraint c_1,
$\{<V_1, v_{12}>, <V_2, v_{22}>, (<V_3, v_{32}>), (<V_4, v_{42}>)\}$ for constraint c_2,
$\{(<V_1, v_{13}>), <V_2, v_{23}>, <V_3, v_{33}>, (<V_4, v_{43}>)\}$ for constraint c_3,
$\{(<V_1, v_{14}>), (<V_2, v_{24}>), <V_3, v_{34}>, <V_4, v_{44}>\}$ for constraint c_4,
such that:

– in RS, V_1, V_2, V_3 and V_4 are pairwise disjoint, i.e. no two of these variables share a candidate;

– $<V_1, v_{11}> \neq <V_1, v_{12}>$, $<V_2, v_{22}> \neq <V_2, v_{23}>$, $<V_3, v_{33}> \neq <V_3, v_{34}>$ and $<V_4, v_{44}> \neq <V_4, v_{41}>$;

– in RS, V_1 has the two mandatory candidates $<V_1, v_{11}>$ and $<V_1, v_{12}>$, two optional candidates $<V_1, v_{13}>$ and $<V_1, v_{14}>$ (supposing any of these labels exists) and no other candidate,

– in RS, V_2 has the two mandatory candidates $<V_2, v_{22}>$ and $<V_2, v_{23}>$, two optional candidates $<V_2, v_{24}>$ and $<V_2, v_{21}>$ (supposing any of these labels exists) and no other candidate,

– in RS, V_3 has the two mandatory candidates $<V_3, v_{33}>$ and $<V_3, v_{34}>$, two optional candidates $<V_3, v_{31}>$ and $<V_3, v_{32}>$ (supposing any of these labels exists) and no other candidate,

– in RS, V_4 has the two mandatory candidates $<V_4, v_{44}>$ and $<V_4, v_{41}>$, two optional candidates $<V_4, v_{42}>$ and $<V_4, v_{43}>$ (supposing any of these labels exists) and no other candidate.

Definition: in any resolution state RS of any CSP, a *Special Quad* (or *Special S_4-subset*) is an S_4-label {CSPVars, TransvSets}, where:

– CSPVars = $\{V_1, V_2, V_3, V_4\}$,

– TransvSets is composed of the following transversal sets of labels:

$\{<V_1, v_{11}>, <V_2, v_{21}>, <V_3, v_{31}>, (<V_4, v_{41}>\})$ for constraint c_1,
$\{<V_1, v_{12}>, (<V_2, v_{22}>), (<V_3, v_{32}>), <V_4, v_{42}>\}$ for constraint c_2,
$\{(<V_1, v_{13}>), <V_2, v_{23}>, (<V_3, v_{33}>), <V_4, v_{43}>\}$ for constraint c_3,
$\{(<V_1, v_{14}>), (<V_2, v_{24}>), (<V_3, v_{34}>), <V_4, v_{44}>\}$ for constraint c_4,

such that:

– in RS, V_1, V_2, V_3 and V_4 are pairwise disjoint, i.e. no two of these variables share a candidate;

– $<V_1, v_{11}> \neq <V_1, v_{12}>$, $<V_2, v_{21}> \neq <V_2, v_{23}>$ and $<V_3, v_{31}> \neq <V_3, v_{34}>$; moreover $<V_4, v_{42}>$, $<V_4, v_{43}>$ and $<V_4, v_{44}>$ are pairwise different;

– in RS, V_1 has the two mandatory candidates $<V_1, v_{11}>$ and $<V_1, v_{12}>$ and no other candidate;

– in RS, V_2 has the two mandatory candidates $<V_2, v_{21}>$ and $<V_2, v_{23}>$ and no other candidate;

– in RS, V_3 has the two mandatory candidates $<V_3, v_{31}>$ and $<V_3, v_{34}>$ and no other candidate;

– in RS, V_4 has the three mandatory candidates $<V_4, v_{42}>$, $<V_4, v_{43}>$ and $<V_4, v_{44}>$ and no other candidate.

In both cases, a *target of a Quad* is defined as a candidate S_4-linked to the underlying S_4-label.

Theorem 8.3 (S_4 rule): in any CSP, a target of a Quad can be eliminated.

Proof for the cyclic case: as the four transversal sets play similar roles, we can suppose that Z is linked to all of $<V_1, v_{11}>$, $<V_2, v_{21}>$, $(<V_3, v_{31}>)$ and $(<V_4, v_{41}>)$. If Z was True, these candidates (if they are present) would be eliminated by ECP. Each of V_1, V_2, V_3 and V_4 would have at most three candidates left. Any choice for V_1 would reduce to at most two the number of possibilities for V_2, V_3 and V_4. Any further choice among the remaining candidates for V_2 would reduce to at most one the number of possibilities for V_3 and V_4. Finally the unique choice left for V_3, if any, would reduce to zero the number of possibilities for V_4.

Proof for the special case: there are four subcases (the last two of which are similar to the second):
- suppose Z is linked to all of $<V_1, v_{11}>$, $<V_2, v_{21}>$, $<V_3, v_{31}>$ (and $<V_4, v_{41}>$). If Z was True, these candidates (if they are present) would be eliminated by ECP. Each of V_1, V_2, V_3, would have only one candidate left; choosing these as values would reduce to zero the number of possibilities for V_4.
- suppose Z is linked to all of $<V_1, v_{12}>$, $(<V_2, v_{22}>)$, $(<V_3, v_{32}>)$ and $<V_4, v_{42}>$. If Z was True, $<V_1, v_{12}>$ and $<V_4, v_{42}>$ would be eliminated by ECP; $<V_1, v_{11}>$ would then be asserted by S, which would eliminate $<V_2, v_{21}>$ and $<V_3, v_{31}>$. Then $<V_2, v_{23}>$ and $<V_3, v_{34}>$ would be asserted. This would leave no possibility for V_4.

The rest of this section shows how, choosing sets of four variables in different sub-families of CSP variables, the familiar Naked Quads, Hidden Quads and Super-Hidden Quads (Jellyfish) of Sudoku appear as mere Quads of the general CSP.

8.4.2. Naked Quads in Sudoku

The good formulation for Naked Quads is a little harder to find than for Triplets.

Naked Quads in a row (first tentative formulation, sometimes called Strict Naked Quads or Complete Naked Quads): if there is a row and there are four numbers and four cells in this row whose remaining candidates are exactly these four numbers, then remove these four numbers from the candidates for the other cells in this row. But there is a major problem: it is unnecessarily restrictive and situations where it can be applied are extremely rare (actually, in 10,000,000 randomly generated minimal puzzles, we have found no example that would use this form of Quads if simpler rules, i.e. Subsets and whips of size strictly less than four, are allowed).

Naked Quads in a row (second tentative formulation, sometimes called Comprehensive Naked Quads): if there is a row and there are four numbers and four cells in this row such that all their candidates are among these four numbers, then remove these four numbers from the candidates for all the other cells in this row. But, again, it has a major problem: it includes Naked Triplets in a row, Naked Pairs in a row and even Naked Single in a row as special cases.

So, neither of the usual two formulations of the Naked Quads rule is correct according to our guiding principles. How then can one formulate it so that it is comprehensive but does not subsume any of the rules for Naked Subsets of smaller size? It is enough to make certain that the four cells have no candidate other than the four given numbers (say n_1, n_2, n_3 and n_4), that each of them has more than one candidate (it is not a Naked-Single), that no two of them have exactly the same two candidates (which would make a Naked Pairs in a row) and that no three of them form a Naked Triplets in a row. There are only two ways to satisfy these conditions.

The first, most general way is to impose candidates n_1 and n_2 for cell 1, candidates n_2 and n_3 for cell 2, candidates n_3 and n_4 for cell 3 and candidates n_4 and n_1 for cell 4. This is the "Cyclic Naked Quads". We get the final formulation of this first case, more complex than usual but with its full natural scope:
if there is a row r and there are four different columns c_1, c_2, c_3 and c_4, and four different numbers n_1, n_2, n_3 and n_4, such that:
- cell (r, c_1) has n_1 and n_2 among its candidates,
- cell (r, c_2) has n_2 and n_3 among its candidates,
- cell (r, c_3) has n_3 and n_4 among its candidates,
- cell (r, c_4) has n_4 and n_1 among its candidates,
- none of the cells (r, c_1), (r, c_2), (r, c_3) and (r, c_4) has any candidate other than n_1, n_2, n_3 or n_4,
then eliminate the four numbers n_1, n_2, n_3 and n_4 from the candidates for any other cell in row r in rc-space.

$\forall r \forall \neq (c_1,c_2,c_3,c_4) \forall \neq (n_1,n_2,n_3,n_4)$
    { candidate(n_1, r, c_1) $\wedge$ candidate(n_2, r, c_1) $\wedge$
    candidate(n_2, r, c_2) $\wedge$ candidate(n_3, r, c_2) $\wedge$
    candidate(n_3, r, c_3) $\wedge$ candidate(n_4, r, c_3) $\wedge$
    candidate(n_4, r, c_4) $\wedge$ candidate(n_1, r, c_4) $\wedge$
    $\forall c \in \{c_1, c_2, c_3, c_4\} \forall n \neq n_1,n_2,n_3,n_4$ ¬candidate(n, r, c)
$\Rightarrow$
    $\forall c \neq c_1,c_2,c_3,c_4 \forall n \in \{n_1, n_2, n_3, n_4\}$ ¬candidate(n, r, c) }.

Exercise: show that this is exactly what Cyclic Quads of the general definition give when applied to CSP variables Xrc_1, Xrc_2, Xrc_3 and Xrc_4, with transversal sets defined by CSP variables (considered as constraints) Xrn_1, Xrn_2, Xrn_3 and Xrn_4.

The second way will be called Special Naked Quads in a row, a very rare pattern, with the following contents for its cells: $\{n_1\ n_2\}\ \{n_1\ n_3\}\ \{n_1\ n_4\}\ \{n_2\ n_3\ n_4\}$:

$\forall r \forall \neq (c_1,c_2,c_3,c_4) \forall \neq (n_1,n_2,n_3,n_4)$
    { candidate(n_1, r, c_1) $\wedge$ candidate(n_2, r, c_1) $\wedge$ $\forall n \neq n_1,n_2$ ¬candidate(n, r, c_1) $\wedge$
    candidate(n_1, r, c_2) $\wedge$ candidate(n_3, r, c_2) $\wedge$ $\forall n \neq n_1,n_3$ ¬candidate(n, r, c_2) $\wedge$
    candidate(n_1, r, c_3) $\wedge$ candidate(n_4, r, c_3) $\wedge$ $\forall n \neq n_1,n_4$ ¬candidate(n, r, c_3) $\wedge$
    candidate(n_2, r, c_4) $\wedge$ candidate(n_3, r, c_4) $\wedge$ candidate(n_4, r, c_4)
        $\wedge$ $\forall n \neq n_2,n_3,n_4$ ¬candidate(n, r, c_4)
$\Rightarrow$
    $\forall c \neq c_1,c_2,c_3,c_4 \forall n \in \{n_1, n_2, n_3, n_4\}$ ¬candidate(n, r, c) }.

Exercise: show that this is exactly what Special Quads of the general definition give when applied to CSP variables Xrc_1, Xrc_2, Xrc_3 and Xrc_4, with transversal sets defined by CSP variables (considered as constraints) Xrn_1, Xrn_2, Xrn_3 and Xrn_4.

Exercise: Transpose the above justification for the two definitions of Quads in Sudoku to the general CSP framework. (Show that there are no other possibilities than the Cyclic and Special Quads.)

8.4.3. Hidden Quads in Sudoku

The proper formulation of rules for Hidden Quads would not be obvious if we could not rely on super-symmetries and meta-theorem 4.2. But, if we apply meta-theorem 4.2 to Cyclic Naked Quads in a row and to Special Naked Quads in a row, permuting the words "number" and "column", we immediately obtain two rules, corresponding to what is known as "Hidden Quads in a row" in the Sudoku world:

Cyclic Hidden Quads in a row, or Cyclic HQ(row):
if there is a row r, and there are four different numbers n_1, n_2, n_3 and n_4 and four different columns c_1, c_2, c_3 and c_4, such that:
- rn-cell (r, n_1) (in rn-space) has c_1 and c_2 among its candidates (columns),
- rn-cell (r, n_2) (in in rn-space) has c_2 and c_3 among its candidates (columns),
- rn-cell (r, n_3) (in in rn-space) has c_3 and c_4 among its candidates (columns),
- rn-cell (r, n_4) (in in rn-space) has c_4 and c_1 among its candidates (columns),
- none of the rn-cells (r, n_1), (r, n_2), (r, n_3) and (r, n_4) (in in rn-space) has any remaining candidate (column) other than c_1, c_2, c_3 and c_4,
then eliminate the four columns c_1, c_2, c_3 and c_4 from the candidates for any other rn-cell (r, n) in row r in rn-space.

$\forall r \forall \neq (n_1,n_2,n_3,n_4) \forall \neq (c_1,c_2,c_3,c_4)$
    { candidate(n_1, r, c_1) $\wedge$ candidate(n_1, r, c_2) $\wedge$
      candidate(n_2, r, c_2) $\wedge$ candidate(n_2, r, c_3) $\wedge$
      candidate(n_3, r, c_3) $\wedge$ candidate(n_3, r, c_4) $\wedge$
      candidate(n_4, r, c_4) $\wedge$ candidate(n_4, r, c_1) $\wedge$
      $\forall n \in \{n_1, n_2, n_3, n_4\} \forall c \neq c_1, c_2, c_3, c_4$ $\neg$candidate(n, r, c)
  $\Rightarrow$
      $\forall n \neq n_1, n_2, n_3, n_4 \forall c \in \{ c_1, c_2, c_3, c_4\}$ $\neg$candidate(n, r, c) }.

And Special Hidden Quads in a row, or Special HQ(row):

$\forall r \forall \neq (n_1,n_2,n_3,n_4) \forall \neq (c_1,c_2,c_3,c_4)$
    { candidate(n_1, r, c_1) $\wedge$ candidate(n_1, r, c_2) $\wedge$ $\forall c \neq c_1, c_2$ $\neg$candidate(n_1, r, c) $\wedge$
      candidate(n_2, r, c_1) $\wedge$ candidate(n_2, r, c_3) $\wedge$ $\forall c \neq c_1, c_3$ $\neg$candidate(n_2, r, c) $\wedge$
      candidate(n_3, r, c_1) $\wedge$ candidate(n_3, r, c_4) $\wedge$ $\forall n \neq c_1, c_4$ $\neg$candidate(n_3, r, c) $\wedge$
      candidate(n_4, r, c_2) $\wedge$ candidate(n_4, r, c_3) $\wedge$ candidate(n_4, r, c_4) $\wedge$
        $\wedge$ $\forall c \neq c_2, c_3, c_4$ $\neg$candidate(n_4, r, c)

$\Rightarrow$

$\forall n \neq n_1,n_2,n_3,n_4 \forall c \in \{ c_1, c_2, c_3, c_4 \} \neg candidate(n, r, c) \}.$

Exercise: show that this is exactly what Cyclic and Special Quads of the general definition give when applied to CSP variables Xrn_1, Xrn_2, Xrn_3 and Xrn_4, with transversal sets defined by CSP variables (considered as constraints) Xrc_1, Xrc_2, Xrc_3 and Xrc_4.

8.4.4. Super Hidden Quads in Sudoku (Jellyfish)

Finally, there remains to consider a rule that should be called Cyclic Super Hidden Quads in rows, or SHQ(row), obtained from Cyclic Hidden Quads in a row by permuting the words "row" and "number", according to meta-theorem 4.2. Let us first do this formally, i.e. by applying the S_{rn} transform to HQ(row) = S_{cn}(NQ(row)):

$\forall n \forall \neq (r_1,r_2,r_3,r_4) \forall \neq (c_1,c_2,c_3,c_4)$
    $\{$ candidate(n, r_1, c_1) $\land$ candidate(n, r_1, c_2) $\land$
    candidate(n, r_2, c_2) $\land$ candidate(n, r_2, c_3) $\land$
    candidate(n, r_3, c_3) $\land$ candidate(n, r_3, c_4) $\land$
    candidate(n, r_4, c_4) $\land$ candidate(n, r_4, c_1) $\land$
    $\forall r \in \{r_1, r_2, r_3, r_4\} \forall c \neq c_1,c_2,c_3,c_4 \neg candidate(n, r, c)$
$\Rightarrow$
    $\forall r \neq r_1,r_2,r_3,r_4 \forall c \in \{c_1, c_2, c_3, c_4\} \neg candidate(n, r, c) \}.$

Exercise: show that this is exactly what Cyclic Quads of the general definition give when applied to CSP variables Xr_1n, Xr_2n, Xr_3n and Xr_4n, with transversal sets defined by CSP variables (considered as constraints) Xc_1n, Xc_2n, Xc_3n and Xc_4n.

In the same way as in the Triplets case, we can clarify this rule by temporarily forgetting part of the conditions: if there is a number n and there are four different rows r_1, r_2, r_3 and r_4 and four different columns c_1, c_2, c_3 and c_4, such that for each of the four rows the instance of number n that must be somewhere in each of these rows can actually only be in either of the four columns, then in any of the four columns eliminate n from the candidates for any row different from the given four.

This is the usual formulation of the rule for Jellyfish in rows. The part we have temporarily discarded corresponds to the conditions we have added to Comprehensive Cyclic Naked Quads in a row; it is just what prevents Jellyfish in rows from reducing to X-Wing in rows or to Swordfish in rows. Finally, we have not only shown that the familiar Jellyfish in rows is the supersymmetric version of Cyclic Naked Quads in a row, but we have also found the proper way to write this rule according to our guiding principles, in as comprehensive a way as possible.

We leave it to the reader to write the rule for Special Super Hidden Quads or Special Jellyfish.

8.5. Relations between Naked, Hidden and Super Hidden Subsets in Sudoku

The so-called "fishy patterns" (X-Wing, Swordfish, Jellyfish, ...) are very popular in the Sudoku micro-world, even the non-existent ones (such as Squirmbag, a would be Super Hidden Quintuplets in our vocabulary) and there are many very specific extensions of these patterns (such as "finned fish", "sashimi fish", ... See also chapter 10 for another kind of extension).

As can be seen by looking at the logical formulæ in the previous sections, a graph similar to that in Figure 4.2 for Singles would not be enough to describe all the rules available for Subsets of size greater than one. Moreover, there is a major difference between Singles and larger Subsets: in the latter, there are different numbers of quantified variables of different sorts: Numbers, Rows and Columns. Building on these differences, the question now is, how far can one go in the iteration of theorem 4.2 and in the definition of Subset rules: Naked, Hidden, Super-Hidden, Super-Super-Hidden, ...?

As for the Naked and Hidden Subsets, a well-known (and obvious) property of Subsets shows that we have found all of them: for any subset S of Numbers of size p ($1 \leq p < 9$), there is a complementary subset S^c of size 9-p (with $1 \leq 9-p < 9$). And S forms a Naked Subset of size p on p cells in a row [respectively a column, a block], if and only if S^c forms a Hidden Subset of size 9-p on the remaining 9-p cells in this row [resp. this column, this block]. As a result, no Naked or Hidden Subset rule for subsets of size greater than four is needed. For instance, Naked Quintuplets in a row is just Hidden Quads in the same row and Hidden Quintuplets in a row is just Naked Quads in the same row.

What was less known before *HLS1*, because super-symmetries had not been explicited, the mythical Super Hidden Quintuplets in a row (alias Squirmbag) is just Hidden Quads in a column (as shown by Figure 8.1 and the remarks above). This is a very interesting example of a named thing that had no independent existence.

Indeed, after the previous sections, several natural questions may arise, such as:

– what if, instead of applying symmetry S_{cn} to NP(row), we apply symmetry S_m?

– what if we formulate a rule analogous to X-Wing in rows but in rn-space – i.e. a rule that should be called Hidden X-Wing in rows or HXW(row) or HSHP(row)?

Do we get new unknown rules? The answer is no; the previous set of rules is strongly closed under symmetry and supersymmetry. More specifically, the full story is to be found in Figure 8.1. The first practical consequence of this for the sequel is that it exempts us from looking for new types of Subset rules (but see chapter 10 for g-Subset rules). Checking the assertions of Figure 8.1 is an easy exercise about the S_{rc}, S_m and S_{cn} transforms (one must just be very careful with the indices). As a detailed proof is available in *HLS*, we do not reproduce it here.

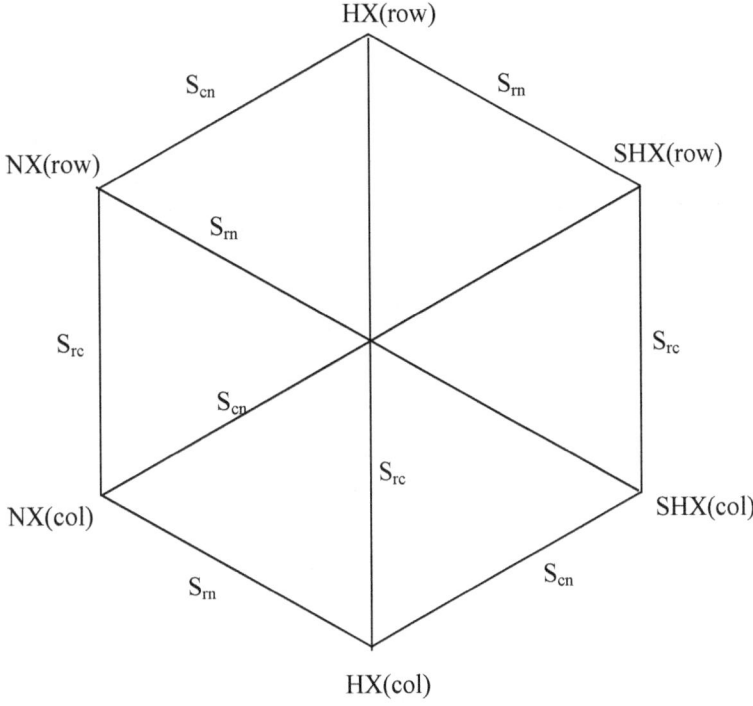

Figure 8.1. *Sudoku symmetries and supersymmetries (X = Pairs, Triplets or Quads – or Subsets of size ≤ IP(n/2) for Sudoku on n×n grids)*

[Historical note: after the first edition of *HLS*, we were informed that the idea of another view of Fish (i.e. of X-Wings, Swordfish and Jellyfish) had already been expressed by "Arcilla" on the late Sudoku Player's Forum, in the thread "a new (?) view of fish (naked or hidden)", November 3rd, 2006. The same thread also shows that similar ideas had been mentioned even before, still in informal ways or in programmers jargon (e.g. "the same program can be used to find Naked Subsets and Fish"). All this was very smart, though it missed the mathematical notions of symmetry and supersymmetry and the closely related idea (first presented in *HLS1*) of introducing the four 2D (rc, rn, cn and bn) spaces and cells as first class concepts, with their associated representations in an Extended Sudoku Board. As a result, it did not develop into a global framework and it led neither to the meta-theorems of chapter 4, nor to the systematic relationships displayed in Figure 8.1 (some of which are not obvious at all), nor to the idea of hidden chains introduced in *HLS1*.]

8.6. Subset resolution theories in a general CSP; confluence

8.6.1. Definition of the Subset resolution theories

The principle of the definitions for Pairs, Triplets and Quads can easily be extended to larger Subsets, although, as we mentioned above, the conditions for non-degeneracy may be tedious to write explicitly. Given a non-degenerated Subset pattern, we define its size to be the number of CSP variables (or transversal sets) in its definition: Pairs have size 2, Quads size 4, ... As should now be expected from our previous rules, we can define an increasing sequence of resolution theories.

Definition: In any CSP, the Subset resolution theories are defined as follows:
- $S_0 = BSRT$,
- $S_1 = W_1$,
- $S_2 = S_1 \cup \{\text{rules for non degenerated Pairs}\}$,
- $S_3 = S_2 \cup \{\text{rules for non degenerated Triplets}\}$,
- ....
- $S_{n+1} = S_n \cup \{\text{rules for non degenerated Subsets of size } n\}$,
- $S_\infty = \cup_{n \geq 0} S_n$.

Notice that, in this hierarchy of resolution theories, we put W_1 before Pairs; this is not only a matter of convention: as already noticed, whips of length 1 (when they exist) are the most basic pattern after Singles and it would not make much sense to define any resolution theory, apart from BSRT, without them.

In 9×9 Sudoku or Latin Squares, $S_\infty = S_4$. More generally, in n×n Sudoku or Latin Squares, $S_\infty = S_p$, with $p = IP(n/2)$ (IP is the integer part).

Theorem 8.4: in any CSP, each of the S_n resolution theories is stable for confluence; therefore, it has the confluence property.

Proof: let S be an S_p-subset ($p \leq n$), for CSP variables $\{V_1, ..., V_p\}$ and transversal sets (some of the labels below may be missing):
$\{<V_1, v_{11}>, <V_2, v_{21}>, ...<V_p, v_{p1}>\}$
....
$\{<V_1, v_{1p}>, <V_2, v_{2p}>, ...<V_p, v_{pp}>\}$

If Z is a target for S, it is linked to all the elements in some of these sets. There may happen two different events:
- if some optional candidate is eliminated from the transversal sets, what remains is still an S_p-subset and Z is still linked to it via the same transversal set;
- if a mandatory candidate is eliminated from a transversal set, either what remains is still an S_p-subset (due to the presence of the optional candidates) or what remains

can be split into two (or more) smaller Subsets or Singles and Z is still linked to one of them.

In any case, Z can still be eliminated by rules in S_n.

8.6.2. Complexity considerations (in Sudoku)

When we increase the size p of Subsets (p goes up from 2 to 4 as we pass from Pairs to Quads, via Triplets), the number of possible cases in each row (forgetting the Special Quads) increases from $(9 \times 8)^2 = 5184$ to $(9 \times 8 \times 7 \times 6)^2 = 9,144,576$ (4 different columns and 4 different numbers). Multiplying this by 9 rows and by 8 patterns (3 Naked, 3 Hidden and 2 Super-Hidden), i.e. by 72, gives an idea of the increase in complexity (from 373,248 to 658,409,472). These figures can be significantly improved by ordering the columns and/or numbers (and this is essential for an effective implementation), but the order of magnitude remains the same. Programming Triplets and Quads as rules in a knowledge based system is a very good exercise for AI students: they can see the importance of having a precise logical formulation before they start to code them in the specific formalism of their inference engine, they can be shown different techniques of rule optimisation and finally they can see at work Newell's famous distinction [Newell 1982] between the "knowledge level" (here a non-ambiguous English or MS-FOL formulation) and the "symbol level" (the rule in the syntax of the inference engine, where different logical conditions may have to be ordered, control facts may have to be added, different saliences, i.e. priorities of rules, may have to be introduced, ….).

8.6.3. Definition of the W_n+S_n, gW_n+S_n, B_n+S_n and gB_n+S_n theories and ratings

One can define the increasing sequence $(W_n+S_n, n \geq 0)$ of resolution theories:

- $W_0+S_0 = BSRT$,
- $W_1+S_1 = W_1$,
- …
- $W_{n+1}+S_{n+1} = W_n+S_n \cup W_{n+1} \cup S_{n+1}$,
- …
- $W_\infty+S_\infty = \cup_{n \geq 0} W_n+S_n$.

One can define in similar ways the increasing sequences $(gW_n+S_n, n \geq 0)$, $(B_n+S_n, n \geq 0)$ and $(gB_n+S_n, n \geq 0)$ of resolution theories. And with each of these sequences, one can associate a rating.

Theorem 8.5: in any CSP, each of the B_n+S_n and gB_n+S_n resolution theories is stable for confluence; therefore, it has the confluence property.

This is an immediate corollary to theorems 5.6 and 7.4 and lemma 4.1.

8.7. Whip subsumption results for Subset rules

After the previous definitions, this section describes the main relationships between Subsets and whips. In the Sudoku case, additional subsumption results can be found on our website for extended Subset patterns ("finned fish" and "sashimi fish"). In our opinion, this is the main section of this chapter; it establishes a strong link between the length of a whip or braid and the size of a Subset. For consistency reasons, patterns that can be seen either as whips [resp. g-whips, braids or g-braids] or as Subsets must be assigned the same W [resp. gW, B, gB] and S ratings. Moreover, the results proven here justify the *a priori* combinations (with the same n) of the S_n and W_n, gW_n, B_n or gB_n theories used in the above definitions.

8.7.1. Subsumption theorems in a general CSP

8.7.1.1. Pairs

Theorem 8.6: $S_2 \subseteq W_2$ (whips of length 2 subsume all the Pairs).

Proof: keeping the notations of theorem 8.1 and considering a target Z of the Pair that is linked to the first transversal set, the following whip eliminates Z:
whip[2]: $V_1\{v_{11} \ v_{12}\} - V_2\{v_{22} .\} \Rightarrow \neg candidate(Z)$.

The converse of the above theorem is false: $W_2 \not\subset S_2$. For a deep understanding of whips, this is as interesting as the theorem itself. The Sudoku example in section 8.8.1 has W(P) = 2 but S(P) = 3. It also has three very instructive examples of whip[2] that cannot be considered as Pairs. [For Sudoku experts: they cannot even be considered as generalised Pairs, e.g. as "finned X-Wings".] (See also section 10.1.6.1.)

8.7.1.2. Triplets

Theorem 8.7: W_3 subsumes "almost all" the Triplets.

Proof: keeping the notations of theorem 8.2 and considering a target Z of the Triplet that is linked to the first transversal set (the three of them play similar roles), the following whip eliminates Z in any CSP:
whip[3]: $V_1\{v_{11} \ v_{12}\} - V_2\{v_{22} \ v_{23}\} - V_3\{v_{33} .\} \Rightarrow \neg candidate(Z)$,
provided that $<V_1, v_{13}>$ is not a candidate for V_1.
The optional candidates of the Triplet appear in the whip as z or t candidates.

Considering that, in the above situation, the three CSP variables play symmetrical roles, there is only one case of a Triplet elimination that cannot be replaced by a whip[3] elimination. It occurs when the optional candidates for variables V_1, V_2 and V_3 in the transversal set to which the target is S_3-linked correspond to existing labels and are all effectively present in the resolution state.

This theorem is illustrated by the same Sudoku example as above (in section 8.8.1), whereas a Sudoku example of non-subsumption is given in section 8.8.2; it even shows that $S_3 \not\subset B_\infty$.

Replacing whips by braids would not change the above results.

8.7.1.3. Quads

Theorem 8.8: W_4 subsumes "almost all" the Cyclic Quads.

Keeping the notations of theorem 8.3, the following whip eliminates a target Z of the Cyclic Quad in any CSP:
whip[4]: $V_1\{v_{11}\ v_{12}\} - V_2\{v_{22}\ v_{23}\} - V_3\{v_{33}\ v_{34}\} - V_4\{v_{41}\ .\} \Rightarrow \neg\text{candidate}(Z)$,
provided that $<V_1, v_{13}>$ and $<V_1, v_{14}>$ (if they exist) are not candidates for V_1 and $<V_2, v_{23}>$ (if it exists) is not a candidate for V_2.
The optional candidates of the Quad appear in the whip as z or t candidates.

An exceptional example of non-subsumption for a Naked Quad elimination is given in section 8.8.3.

Theorem 8.9: B_4 subsumes all the Special Quads.

Keeping the notations of theorem 8.3, let Z be a target of the Special Quad:
- if Z is linked to the first transversal set, the following braid eliminates Z:
braid[4]: $V_1\{v_{11}\ v_{12}\} - V_2\{v_{21}\ v_{23}\} - V_3\{v_{31}\ v_{34}\} - V_4\{v_{44}\ .\} \Rightarrow \neg\text{candidate}(Z)$,
in which the first three left-linking candidates are linked to Z;
- if Z is linked to another transversal set, say the second, the following whip eliminates Z:
whip[4]: $V_1\{v_{12}\ v_{11}\} - V_2\{v_{21}\ v_{23}\} - V_4\{v_{43}\ v_{44}\} - V_3\{v_{34}\ .\} \Rightarrow \neg\text{candidate}(Z)$,
in which candidate $<V_4, v_{42}>$ appears as a z candidate for the third CSP variable.

8.7.2. Statistical almost subsumption results in Sudoku

The theorems in the previous subsection show that "almost all" of the eliminations done by Subsets can be done by whips. Can this "almost all", until now only specified by logical conditions, be given any numerical meaning? One has $W+S(P) \leq W(P)$ for any instance P and the question can be reformulated as: how frequently can the two ratings be different? Notice that this is not exactly an answer to our initial question, because equality of the ratings does not mean that the same eliminations were done; another resolution path may have been followed. Anyway, experiments with the first 10,000 random minimal puzzles in the Sudogen0 collection show that the W+S and the W ratings differ in only 8 cases: either non-subsumption cases are statistically very rare (as suggested by the above theorems) or they are well compensated by other eliminations.

8.7.3. Comparison of the resolution power of whips and Subsets of same length

Subsets are "almost" subsumed by whips of same length; but is there any reciprocal almost subsumption, so that both would have approximately the same resolution power? The answer is negative. The classification results in Table 8.1 show that, even with W_1 included in all the S_n theories, Subsets have a very weak resolution power compared to whips. The W line comes from the "ctr-bias" column of Table 6.4; the S line is based on a series of 275,867 puzzles from the controlled-bias generator. Only the part of the Table in bold is meaningful for this comparison.

rating →	0 (BRT)	1 (S_1=W_1)	2	3	4	4≤n<∞
S	35.08%	9.82%	**5.44%**	**0.36%**	**0.011%**	**0%**
W	35.08%	9.82%	**13.05%**	**20.03%**	**17.37%**	**qsp 100%**

Table 8.1: S and W distributions for the controlled-bias generator

One way of understanding these results is that the definition of Subsets is much more restrictive than the definition of whips. In Subsets, transversal sets are defined by a single constraint. In whips, the fact of being linked to the target or a given previous right-linking candidate plays a role very similar to each of these transversal sets. But being linked to a candidate is much less restrictive than being linked to it via a pre-assigned constraint. As shown by the almost subsumption results, the few Subset cases not covered by whips because of the restrictions on them related to sequentiality are too rarely met in practice to be able to compensate for this.

8.8. Subsumption and non-subsumption examples from Sudoku

This final section illustrates both subsumption and non-subsumption cases. It also shows concretely how Super Hidden Subsets can look like Naked ones in the appropriate 2D space.

8.8.1. $W_2 \not\subset S_2$; also an example of Swordfish subsumption by a whip[3]

Let us first prove that $W_2 \not\subset S_2$. For the puzzle P in Figure 8.2 (Royle17#18966), we shall show that $W(P) = 2$ and $S(P) = 3$.

After an initial sequence of 36 Hidden Singles, leading to the puzzle in the middle of Figure 8.2, we consider two resolution paths.

	5		4					
				3		8		
								1
3				8		7		
	6							5
			2					
			5		6		4	
1		8				3		

	5	1	4		8		3	6
6			1	3	5	8		4
4	8	3		6			5	1
3		5	6	8	4	7	1	
8	6		3		1	4	5	
	1	4	2	5		6	8	3
3			5		6		4	8
1		8				3	6	5
5		6	8		3			7

7	5	1	4	2	8	9	3	6
6	9	2	1	3	5	8	7	4
4	8	3	7	6	9	5	2	1
3	2	5	6	8	4	7	1	9
8	6	7	3	9	1	4	5	2
9	1	4	2	5	7	6	8	3
2	3	9	5	7	6	1	4	8
1	7	8	9	4	2	3	6	5
5	4	6	8	1	3	2	9	7

Figure 8.2. Puzzle Royle17#18966: 1) original, 2) after initial Singles, 3) solution

In the first path, using only the Subset theories, the simplest rule applicable is a Swordfish in columns (Figure 8.3); it allows four eliminations; after three have been done, Singles and ECP are enough to solve the puzzle, showing that S(P) = 3.

***** SudoRules version 15b.1.12-S *****
swordfish-in-columns: n7{c2 c4 c8}{r2 r3 r8} ==> r3c6 ≠ 7, r8c5 ≠ 7, r8c6 ≠ 7
singles to the end

In the second path, using only the whip theories, the simplest applicable rules are ***three very instructive cases of whip[2] that cannot be considered as Pairs [or even as g-Pairs*** (see 10.1.6.1)], leading to eliminations unrelated to the above Swordfish; these are enough to solve the puzzle with Singles and ECP, showing that W(P) = 2.

***** SudoRules version 15b.1.12-W *****
whip[2]: r1n7{c1 c5} - r5n7{c5 .} ==> r2c3 <> 7
whip[2]: r1n7{c5 c1} - r6n7{c1 .} ==> r3c6 <> 7
whip[2]: b4n7{r5c3 r6c1} - r1n7{c1 .} ==> r5c5 <> 7
singles to the end

Now, forgetting the simple whip[2] eliminations, we can also use this example to show how a Swordfish looks like in the proper 2D space. Spotting this Swordfish in the standard representation (upper part of Figure 8.3) may be difficult because it seems to be very degenerated (three of the nine rc-cells on which it lies are even decided). However, in the cn-representation (lower part of Figure 8.3), it looks like a very incomplete Naked-Triplets, but still a non-degenerated one. Indeed, it is a hidden xy-chain[3] (defined in *HLS1* as a kind of bivalue-chain[3], but in rn- instead of rc- space, and therefore a whip[3]).

Exercise: based on the proof of theorem 8.7, write the four whips[3] allowing the eliminations of the four Swordfish targets.

	c1	c2	c3	c4	c5	c6	c7	c8	c9	
r1	n2 / n7 n9	n5	n1	n4	n2 / n7 n9	n8	n2 / n9	n3	n6	r1
r2	n6	n2 / n7 n9	n2 / n̲7̲ n9	n1	n3	n5	n8	n2 / n7 n9	n4	r2
r3	n4	n8	n3	/ n7 n9	n6	n2 / n̲7̲ n9	n5	n2 / n7 n9	n1	r3
r4	n3	n2 / n9	n5	n6	n8	n4	n7	n1	n2 / n9	r4
r5	n8	n6	n2 / n7 n9	n3	/ n7 n9	n1	n4	n5	n2 / n9	r5
r6	/ n7 n9	n1	n4	n2	n5	/ n7 n9	n6	n8	n3	r6
r7	n2 / n7 n9	n3	n2 / n7 n9	n5	n1 n2 / n7 n9	n6	n1 n2 / n9	n4	n8	r7
r8	n1	n2 n4 / n7 n9	n8	/ n7 n9	n2 n4 / n̲7̲ n9	n2 / n̲7̲	n3	n6	n5	r8
r9	n5	n2 n4 / n9	n6	n8	n1 n2 n4 / n9	n3	n1 n2 / n9	n2 / n9	n7	r9
	c1	c2	c3	c4	c5	c6	c7	c8	c9	

	c1	c2	c3	c4	c5	c6	c7	c8	c9	
n1	r8	r6	r1	r2	/ r7 r9	r5	/ r7 r9	r4	r3	n1
n2	r1 / r7	r2 r4 / r8	r2 r5 / r7	r6	r1 / r7 r8 r9	r3 / r8	r1 / r7 r9	r2 r3 / r9	r4 r5	n2
n3	r4	r7	r3	r5	r2	r9	r8	r1	r6	n3
n4	r3	/ r8 r9	r6	r1	/ r8 r9	r4	r5	r7	r2	n4
n5	r9	r1	r4	r7	r6	r2	r3	r5	r8	n5
n6	r2	r5	r9	r4	r3	r7	r6	r8	r1	n6
n7	r1 r6 / r7	r2 / r8	r̲2̲ r5 / r7	r3 / r8	r1 r5 / r7 r̲8̲	r̲3̲ r6 / r̲8̲	r4	r2 r3	r9	n7
n8	r5	r3	r8	r9	r4	r1	r2	r6	r7	n8
n9	r1 r6 / r7	r2 r4 / r8 r9	r2 r5 / r7	r3 / r8	r1 r5 / r7 r8 r9	r3 r6 / r8	r1 / r7 r9	r2 r3 / r9	r4 r5	n9
	c1	c2	c3	c4	c5	c6	c7	c8	c9	

Figure 8.3. Puzzle Royle17#18966, seen in rc and cn spaces, after initial Singles have been applied. The four eliminations allowed by the Swordfish (in grey cells) are underlined.

8.8.2. $S_3 \not\subset B_\infty$: a Swordfish not subsumed by whips or braids

	c1	c2	c3	c4	c5	c6	c7	c8	c9	
r1	n4 n5 n6 n7 n8	n4 n6 n7 n8	n1	n5 n6 n7 n8	n5 n7 n8 n9	n2	n6 n8 n9	n6 n7 n8 n9	n3	r1
r2	n3 n5 n6 n7 n8	n3 n6 n7 n8	n3 n5 n7 n9	n5 n6 n7 n8	n1	n3 n5 n6 n8 n9	n2 n6 n8 n9	n4	n2 n6 n7 n9	r2
r3	n2	n3 n6 n7 n8	n3 n7 n9	n4	n3 n7 n8 n9	n3 n6 n8 n9	n5	n1 n6 n7 n8 n9	n1 n6 n7 n9	r3
r4	n3 n4	n1 n2 n3 n4	n6	n1 n2 n5	n2 n4 n5 n9	n7	n1 n9	n1 n3 n5 n9	n8	r4
r5	n3 n4 n7 n8	n5	n3 n4 n7	n6 n8	n1 n4 n8 n9	n4 n6 n8 n9	n1 n3 n6 n9	n2	n1 n6 n7 n9	r5
r6	n9	n1 n2 n7 n8	n2 n7	n3	n2 n5 n8	n5 n6 n8	n4	n1 n5 n6 n7	n1 n6 n7	r6
r7	n3 n4 n6 n7	n2 n3 n4 n6 n7	n8	n2 n7	n2 n3 n4 n7	n1	n2 n3 n6 n9	n3 n9	n5	r7
r8	n3 n4 n5 n7	n9	n2 n3 n4 n5 n7	n2 n5 n7 n8	n6	n3 n4 n5 n8	n1 n2 n3 n8	n1 n3 n8	n1 n2 n4	r8
r9	n1	n2 n3 n4 n6	n2 n3 n4 n5	n9	n2 n3 n4 n5 n8	n2 n3 n4 n5 n8	n7	n3 n6 n8	n2 n4 n6	r9
	c1	c2	c3	c4	c5	c6	c7	c8	c9	

	c1	c2	c3	c4	c5	c6	c7	c8	c9	
n1	r9	r4 r6	r1	r4 r5	r2	r7	r4 r5 r8	r4 r6 r8	r3 r5 r6 r8	n1
n2	r3	r4 r6 r7 r9	r6 r7 r9	r4 r7 r8	r4 r6 r7 r9	r1	r2 r7 r8	r5	r2 r8 r9	n2
n3	r2 r4 r5 r7 r8	r2 r3 r4 r5 r7 r9	r2 r3 r5 r8 r9	r6	r3 r7 r9	r2 r3 r8 r9	r4 r5 r7 r8	r4 r7 r8 r9	r1	n3
n4	r1 r4 r5 r7 r8	r1 r4 r7 r9	r5 r8 r9	r3	r4 r5 r7 r9	r5 r8 r9	r6	r2	r8 r9	n4
n5	r1 r2 r8	r5	r2 r8 r9	r1 r2 r4 r8	r1 r4 r7 r9	r2 r6 r8 r9	r3	r4 r6	r7	n5
n6	r1 r2 r7	r1 r2 r3 r7 r9	r4	r1 r2 r4	r8	r2 r3 r5 r6 r7	r1 r2 r5 r7	r1 r3 r6 r7	r2 r3 r5 r6 r9	n6
n7	r1 r2 r5 r7 r8	r1 r2 r3 r6 r7	r2 r3 r5 r6 r8	r1 r2 r7 r8 r7	r1 r3 r7	r4	r9	r1 r3 r6	r2 r3 r5 r6	n7
n8	r1 r2 r5	r1 r2 r3 r6	r7	r1 r2 r5 r8	r1 r3 r5 r6 r9	r2 r3 r5 r6 r8 r9	r1 r2 r8	r1 r3 r8 r9	r4	n8
n9	r6	r8	r2 r3	r9	r1 r3 r4 r5	r2 r3 r5	r1 r2 r4 r5 r7	r1 r3 r4 r7	r2 r3 r5	n9
	c1	c2	c3	c4	c5	c6	c7	c8	c9	

Figure 8.4. *Two Swordfish in columns at the same time, in rc and cn representations*

We have already met in section 7.7.3 (Figure 7.3, reproduced as Figure 8.5) the puzzle we shall now use to illustrate a case of non-subsumption of a Swordfish in columns by whips. We know from section 7.7.3 that this puzzle cannot be solved by braids of any length, let alone by whips. However, it has a resolution path using only Swordfish (besides rules in BSRT), which proves that at least one of the Swordfish eliminations cannot be replaced by a whip or a braid elimination.

	1			2				3
				1			4	
2			4			5		
		6			7			8
	5					2		
9			3			4		
		8			1			5
	9			6				
1			9			7		

6	4	1	5	9	2	8	7	3
8	7	5	6	1	3	2	4	9
2	3	9	4	7	8	5	6	1
3	1	6	2	4	7	9	5	8
7	5	4	1	8	9	3	2	6
9	8	2	3	5	6	4	1	7
4	2	8	7	3	1	6	9	5
5	9	7	8	6	4	1	3	2
1	6	3	9	2	5	7	8	4

Figure 8.5. *A puzzle P with W(P)=B(P)=∞ but S(P)=3*

***** SudoRules version 15b.1.12-S *****
24 givens, 214 candidates and 1289 nrc-links
swordfish-in-columns: n4{c3 c6 c9}{r5 r8 r9} ==> r9c5 ≠ 4, r9c2 ≠ 4, r8c1 ≠ 4, r5c5 ≠ 4, r5c1 ≠ 4
swordfish-in-columns: n9{c3 c6 c9}{r2 r3 r5} ==> r5c7 ≠ 9, r5c5 ≠ 9
;;; this swordfish allows three more eliminations, but they are interrupted by singles
singles to the end

As for the advantages of considering the four 2D spaces, notice that in the upper part of Figure 8.4 (rc-space at the start of resolution), it is difficult to distinguish the two Swordfish, because they are in the same columns and they have three rc-cells in common. In the lower part (cn-space), it is obvious: they lie in different rows.

Exercise: use theorem 8.7 and its proof to show exactly which eliminations done (or allowed) by the two Swordfish are subsumed by whips and which are not.

As previously shown in section 7.7.3, this puzzle can be solved by g-whip[2], but this is irrelevant to our present purposes, because these g-whips are unrelated to the two Swordfish.

8.8.3. A Jellyfish not subsumed by whips but solved by braids or g-whips

After theorem 8.8, whips subsume most cases of Cyclic Quads. But there are rare examples in which this is not the case, such as the puzzle in Figure 8.6

(#017#Mauricio-002#8#1). Not only is there a Quad elimination that cannot be done by whips or braids of length 4, but there is no whip of length < 18 that could do it. We shall also use this puzzle to illustrate the fact that allowing/disallowing one more resolution rule can occasionally have dramatic effects on the classification of a puzzle, although the statistical effects seem to be minor.

			1					
	1	2	3		4	5	6	
	2	3				7	8	
	4	7		6		1	2	
	3	1	8		7	6	4	
	5	8				2	3	

4	9	5	7	8	6	3	1	2
3	7	6	5	1	2	8	9	4
8	1	2	3	9	4	5	6	7
1	8	9	2	7	3	4	5	6
6	2	3	4	5	1	7	8	9
5	4	7	9	6	8	1	2	3
2	6	4	1	3	5	9	7	8
9	3	1	8	2	7	6	4	5
7	5	8	6	4	9	2	3	1

Figure 8.6. *Puzzle P with W+S(P)=4, B(P) = 10, W(P) >18 and gW(P) = 4*

8.8.3.1. Solution with whips and subsets, W+S(P)=4

Let us first find a solution combining whips and Subsets:

***** SudoRules version 13.7wter2-WS *****
nrc-chain[2]: c8n5{r4 r7} - r8n5{c9 c5} ==> r4c5 ≠ 5 (a special case of whip[2])
xyz-chain[3]: r6c4{n9 n5} - r5c5{n5 n4} - r9c5{n4 n9} ==> r4c5 ≠ 9 (a special case of whip[3])
naked-quads-in-a-block: b5{r5c4 r5c5 r5c6 r6c4}{ n1 n4 n5 n9} ==> r4c4 ≠ 4, r4c5 ≠ 4

;;; here , the application of Naked Quad is "interrupted" by the availability of a simpler rule :
whip[1]: b4n4{r5c4 .} ==> r5c9 ≠ 4
;;; now the Quad continues:
naked-quads-in-a-block: b5{r5c4 r5c5 r5c6 r6c4}{n1 n4 n5 n9} ==> r4c4 ≠ 1, r4c4 ≠ 5, r4c4 ≠ 9, r4c6 ≠ 5, r4c6 ≠ 9, r6c6 ≠ 5, r6c6 ≠ 9
;;; Resolution state RS$_1$

;;; the resolution state RS$_1$ reached at this point is displayed in Figure 8.7; here we have artificially isolated the last elimination allowed by this Quad, for later reference, because the same resolution state will be reached by another resolution path using only braids.
;;; let us now continue past resolution state RS$_1$:
naked-quads-in-a-block: b5{r5c4 r5c5 r5c6 r6c4}{n1 n4 n5 n9} ==> r4c6 ≠ 1
hidden-single-in-row r4 ==> r4c1 = 1

;;; we now reach a resolution state RS$_2$ (Figure 8.8) in which there is a Jellyfish; notice that this Jellyfish was already present in resolution state RS$_1$.

	c1	c2	c3	c4	c5	c6	c7	c8	c9	
r1	n3 n4 n5 n6 n7 n8 n9	n6 n7 n8 n9	n4 n5 n6 n9	n2 n5 n6 n7	n2 n5 n9	n2 n5 n6 n7 n8 n9	n4 n8 n9	n3 n1 n7	n1 n2 n3 n4 n7 n8 n9	r1
r2	n3 n4 n5 n6 n7 n8 n9	n6 n7 n8 n9	n4 n5 n6 n9	n2 n5 n6 n7	n1	n2 n5 n6 n8 n9	n3 n4 n8 n9	n7	n2 n3 n4 n7 n8 n9	r2
r3	n7 n8 n9	n1	n2	n3		n7 n8 n9	n4	n5	n6 n7 n8 n9	r3
r4	n1 n5 n6 n8 n9	n6 n8 n9	n5 n6 n9	~~n1~~ n2 ~~n4 n5~~ n7	n2 n3 ~~n4~~ ~~n9~~ n7 n8	**n1** n2 n3 ~~n5~~ n8 ~~n9~~	n4 n9	n5 n9	n3 n4 n5 n6 n9	r4
r5	n1 n5 n6 n9	n2	n3	n1 n4 n5 n9	n4 n5 n9	n1 n5 n9	n7	n8	(n4) n5 n6 n9	r5
r6	n5 n8 n9	n4	n7	n5 n9	n6	n3 ~~n5~~ n8 ~~n9~~	n1	n2	n3 n5 n9	r6
r7	n2 n4 n6 n7 n9	n6 n7 n9	n4 n6 n9	n1 n2 n4 n5 n6 n9	n2 n3 n4 n5 n9	n1 n2 n3 n5 n6 n9	n1 n8 n9	n1 n5 n7	n1 n5 n7 n8 n9	r7
r8	n2 n9	n3	n1	n8	n2 n5 n9	n7	n6	n4	n5 n9	r8
r9	n4 n6 n7 n9	n5	n8	n1 n4 n6 n9	n4 n9	n1 n6 n9	n2	n3	n1 n7 n9	r9
	c1	c2	c3	c4	c5	c6	c7	c8	c9	

Figure 8.7. *resolution state RS₁: a Naked Quad in block b5 (in grey cells); the nine candidates eliminated by the Quad just before resolution state RS₁ is reached are barred; the candidate (n4r5c9) eliminated by the whip[1] is between parentheses; the next candidate (n1r4c6) the Quad could eliminate is underlined; it is the target of no whip or braid.*

;;; let us now continue past RS₂:
jellyfish-in-columns: n9{c2 c3 c7 c8}{r1 r2 r4 r7} ==> r1c6 ≠ 9, r1c9 ≠ 9, r2c1 ≠ 9, r2c4 ≠ 9, r2c6 ≠ 9, r2c9 ≠ 9, r4c9 ≠ 9, r7c1 ≠ 9, r7c4 ≠ 9, r7c5 ≠ 9, r7c6 ≠ 9, r7c9 ≠ 9
nrc-chain[3]: c6n9{r5 r9} - r9c5{n9 n4} - b5n4{r5c5 r5c4} ==> r5c4 ≠ 9 (a special kind of whip[3])
jellyfish-in-columns: n9{c2 c3 c7 c8}{r1 r2 r4 r7} ==> r1c4 ≠ 9, r1c5 ≠ 9
singles to the end

8.8.3.2. Using only braids, B(P)=10

Suppose we now want a pure braids solution and we do not allow Subset rules. Then we get B(P) = 10.

***** SudoRules version 15b.1.8-B *****
whip[2]: c8n5{r4 r7} - r8n5{c9 .} ==> r4c5 ≠ 5
whip[3]: r6c4{n9 n5} - r5c5{n5 n4} - r9c5{n4 .} ==> r4c5 ≠ 9
;;; the following whips[4] replace all but one of the eliminations allowed by the Naked Quad in the previous resolution path:

	c1	c2	c3	c4	c5	c6	c7	c8	c9	
r1	n3 n4 n5 n6 n7 n8 n9	n6 n7 n8 n9	n4 n5 n6 n9	n2 n5 n6 n7	n2 n5 n7 n8 n9	n2 n5 n6 n8 n9	n3 n4 n8 n9	n1 n7	n1 n2 n3 n4 n7 n8 n9	r1
r2	n3 n4 n5 n6 n7 n8 n9	n6 n7 n8 n9	n4 n5 n6 n9	n2 n5 n6 n7 n9	n1	n2 n5 n6 n8 n9	n3 n4 n8 n9	n7	n2 n3 n4 n7 n8 n9	r2
r3	n7 n8 n9	n1	n2	n3	n7 n8 n9	n4	n5	n6	n7 n8 n9	r3
r4	n1	n6 n8 n9	n5 n6 n9	n2 n7	n2 n3 n7 n8	n2 n3 n8	n3 n4 n9	n5 n9	n3 n4 n5 n6 n9	r4
r5	n5 n6 n9	n2	n3	n1 n4 n5 n9	n4 n5 n9	n1 n5 n9	n7	n8	n5 n6 n9	r5
r6	n5 n8 n9	n4	n7	n5 n9	n6	n3 n8	n1	n2	n3 n5 n9	r6
r7	n2 n4 n6 n7 n9	n6 n7 n9	n4 n6	n1 n2 n4 n5 n6 n9	n4 n5 n9	n2 n3 n5 n6 n9	n1 n2 n3 n9	n1 n5 n8 n9	n1 n5 n7 n8 n9	r7
r8	n2 n9	n3	n1	n8	n2 n5 n9	n7	n6	n4	n5 n9	r8
r9	n4 n6 n7 n9	n5	n8	n1 n4 n6 n9	n4 n9	n1 n6 n9	n2	n3	n1 n7 n9	r9
	c1	c2	c3	c4	c5	c6	c7	c8	c9	

Figure 8.8. *Resolution state RS_2: a Jellyfish not subsumed by whips or g-braids*

whip[4]: b5n7{r4c4 r4c5} - b5n2{r4c5 r4c6} - b5n3{r4c6 r6c6} - b5n8{r6c6 .} ==> r4c4 ≠ 1, r4c4 ≠ 4, r4c4 ≠ 5, r4c4 ≠ 9
whip[4]: b5n7{r4c5 r4c4} - b5n2{r4c4 r4c6} - b5n3{r4c6 r6c6} - b5n8{r6c6 .} ==> r4c5 ≠ 4
whip[1]: b4n4{r5c4 .} ==> r5c9 ≠ 4
whip[4]: r6c4{n5 n9} - r5c6{n9 n1} - r5c4{n1 n4} - r5c5{n4 .} ==> r4c6 ≠ 5
whip[4]: r6c4{n9 n5} - r5c6{n5 n1} - r5c4{n1 n4} - r5c5{n4 .} ==> r4c6 ≠ 9
whip[4]: r6c4{n5 n9} - r5c6{n9 n1} - r5c4{n1 n4} - r5c5{n4 .} ==> r6c6 ≠ 5
whip[4]: r6c4{n9 n5} - r5c6{n5 n1} - r5c4{n1 n4} - r5c5{n4 .} ==> r6c6 ≠ 9

Here, we have reached the same resolution state as RS_1. But now, candidate n1r4c6 (underlined in Figure 8.7), which could be eliminated by the Naked Quad in the previous resolution path, is the target of no whip or braid; it is a rare case of a Quad elimination not subsumed by whips, braids, g-whips or g-braids.

As a consequence of this missing elimination, r4c1 = 1 cannot be asserted. Nevertheless, this does not prevent the Jellyfish from being present (it was already present in state RS_1). But, what is really exceptional here is that *none of the candidates that could be eliminated by the Jellyfish can be eliminated by a whip[4].*

The resolution path with braids continues, much harder than with Subsets:

whip[5]: b4n1{r4c1 r5c1} - r5n6{c1 c9} - b6n5{r5c9 r6c9} - r6c4{n5 n9} - r5n9{c4 .} ==> r4c1 ≠ 5

whip[5]: b4n1{r4c1 r5c1} - r5n6{c1 c9} - b6n9{r5c9 r6c9} - r6c4{n9 n5} - r5n5{c4 .} ==> r4c1 ≠ 9

whip[5]: r4n1{c1 c6} - b5n8{r4c6 r6c6} - b5n3{r6c6 r4c5} - b5n2{r4c5 r4c4} - b5n7{r4c4 .} ==> r4c1 ≠ 8

whip[6]: r8c9{n9 n5} - c8n5{r7 r4} - c8n9{r4 r7} - c7n9{r7 r4} - c3n9{r4 r1} - c2n9{r2 .} ==> r2c9 ≠ 9

whip[6]: r8c9{n9 n5} - c8n5{r7 r4} - c8n9{r4 r7} - c7n9{r7 r4} - c3n9{r4 r2} - c2n9{r1 .} ==> r1c9 ≠ 9

whip[6]: r9c5{n9 n4} - r5c5{n4 n5} - r8n5{c5 c9} - b9n9{r8c9 r9c9} - b7n9{r9c1 r8c1} - r3n9{c1 .} ==> r7c5 ≠ 9

braid[6]: b5n5{r5c4 r6c4} - r8c9{n5 n9} - r6n9{c4 c1} - r3n9{c1 c5} - r5c5{n5 n4} - r9c5{n9 .} ==> r5c9 ≠ 5

whip[7]: r9c5{n4 n9} - r5c5{n9 n5} - r8n5{c5 c9} - r8n9{c9 c1} - r3n9{c1 c9} - r6n9{c9 c4} - r5n9{c4 .} ==> r7c5 ≠ 4

whip[7]: r9c5{n9 n4} - r5c5{n4 n5} - r8n5{c5 c9} - r8n9{c9 c1} - r3n9{c1 c9} - r6n9{c9 c4} - r5n9{c4 .} ==> r1c5 ≠ 9

braid[7]: r2c8{n7 n9} - r3c9{n9 n8} - c7n8{r2 r7} - c7n9{r7 r4} - r9n7{c9 c1} - r3c1{n7 n9} - b4n9{r6c1 .} ==> r1c9 ≠ 7

braid[7]: r2c8{n7 n9} - r3c9{n9 n8} - c7n8{r2 r7} - c7n9{r7 r4} - r9n7{c9 c1} - r3c1{n7 n9} - b4n9{r6c1 .} ==> r2c9 ≠ 7

braid[7]: r2c8{n7 n9} - r9n7{c1 c9} - r3c9{n7 n8} - c7n8{r1 r7} - c7n9{r1 r4} - r3c1{n7 n9} - b4n9{r6c1 .} ==> r2c1 ≠ 7

braid[7]: r6c4{n5 n9} - r8c9{n5 n9} - r5n9{c4 c1} - r3n9{c1 c5} - r8n5{c9 c5} - r5c5{n5 n4} - r9c5{n9 .} ==> r6c9 ≠ 5

whip[1]: b6n5{r4c8 .} ==> r4c3 ≠ 5

whip[1]: c3n5{r1 .} ==> r1c1 ≠ 5, r2c1 ≠ 5

whip[4]: c1n5{r5 r6} - r6c4{n5 n9} - r5n9{c4 c9} - r5n6{c9 .} ==> r5c1 ≠ 1

hidden-single-in-a-block ==> r4c1 = 1

braid[10]: b4n8{r4c2 r6c1} - r6n5{c1 c4} - c5n3{r4 r7} - r6n9{c4 c9} - r8c9{n9 n5} - c5n5{r8 r1} - c5n2{r1 r8} - c5n7{r1 r3} - r3c1{n7 n9} - r8n9{c9 .} ==> r4c5 ≠ 8

whip[1]: c5n8{r1 .}2 ==> r1c6 ≠ 8, r2c6 ≠ 8

whip[7]: b2n8{r1c5 r3c5} - c1n8{r3 r6} - r6n5{c1 c4} - r6n9{c4 c9} - r3n9{c9 c1} - r8n9{c1 c5} - r9n9{c6 .} ==> r1c2 ≠ 8

whip[8]: b2n8{r1c5 r3c5} - c5n7{r3 r4} - c5n3{r4 r7} - c5n2{r7 r8} - r8c1{n2 n9} - r3n9{c1 c9} - r6n9{c9 c4} - r5n9{c4 .} ==> r1c5 ≠ 5

braid[5]: r6n3{c9 c6} - c5n3{r4 r7} - r6c4{n9 n5} - r8c9{n9 n5} - c5n5{r8 .} ==> r6c9 ≠ 9

singles ==> r6c9 = 3, r6c6 = 8, r4c2 = 8

whip[2]: r6n9{c4 c1} - b7n9{r9c1 .} ==> r7c4 ≠ 9

whip[3]: r6n9{c4 c1} - r8n9{c1 c9} - r3n9{c9 .} ==> r5c5 ≠ 9

whip[3]: r9c5{n9 n4} - r5c5{n4 n5} - r6c4{n5 .} ==> r9c4 ≠ 9

whip[3]: r6n9{c1 c4} - r5n9{c4 c9} - b9n9{r9c9 .} ==> r7c1 ≠ 9

whip[4]: b4n9{r6c1 r4c3} - b6n9{r4c9 r5c9} - r8n9{c9 c5} - r3n9{c5 .} ==> r9c1 ≠ 9

whip[3]: b7n9{r7c2 r8c1} - b4n9{r5c1 r4c3} - b6n9{r4c9 .} ==> r7c9 ≠ 9

whip[4]: r9n9{c6 c9} - r8n9{c9 c1} - r5n9{c1 c4} - r6n9{c4 .} ==> r7c6 ≠ 9

whip[4]: b8n9{r9c6 r8c5} - r3n9{c5 c1} - r6n9{c1 c4} - r5n9{c4 .} ==> r9c9 ≠ 9

whip[1]: r9n9{c5 .} ==> r8c5 ≠ 9

whip[2]: r8n9{c9 c1} - b4n9{r5c1 .} ==> r4c9 ≠ 9
whip[3]: r6n9{c4 c1} - r3n9{c1 c9} - r8n9{c9 .} ==> r1c4 ≠ 9, r2c4 ≠ 9
whip[1]: c4n9{r6 .} ==> r5c6 ≠ 9
whip[3]: r6n9{c1 c4} - r5n9{c4 c9} - r8n9{c9 .} ==> r1c1 ≠ 9, r2c1 ≠ 9, r3c1 ≠ 9
whip[3]: r8n9{c9 c1} - r5n9{c1 c4} - r6n9{c4 .} ==> r3c9 ≠ 9
singles to the end

8.8.3.3. *Using only whips, W(P)>18*

Suppose now we wanted a solution with only whips. If a resolution path could be obtained with whips, some of them would have to be of length > 18, i.e. one has W(P) > 18. Actually, we did not try longer ones because of memory overflow problems and we did not insist because it did not seem interesting to go further.

8.8.3.4. *Using g-whips, gW(P)=4*

If we now use g-whips, we get gW(P) = 4, with a completely different resolution path (unrelated to the Quads in the first path):

***** SudoRules version 15b.1.12-gW *****
26 givens, 222 candidates and 1621 nrc-links
whip[2]: c8n5{r4 r7} - r8n5{c9 .} ==> r4c5 ≠ 5
whip[3]: r6c4{n9 n5} - r5c5{n5 n4} - r9c5{n4 .} ==> r4c5 ≠ 9

;;; after this point, the resolution path diverges completely with respect to the previous ones :
g-whip[3]: b6n9{r4c7 r456c9} - r3n9{c9 c5} - r8n9{c5 .} ==> r4c1 ≠ 9
g-whip[3]: b4n9{r4c3 r456c1} - r3n9{c1 c5} - r8n9{c5 .} ==> r4c9 ≠ 9
g-whip[3]: b7n9{r7c3 r789c1} - r3n9{c1 c9} - b9n9{r9c9 .} ==> r7c5 ≠ 9
g-whip[3]: b4n9{r6c1 r4c123} - b6n9{r4c7 r456c9} - b9n9{r9c9 .} ==> r7c1 ≠ 9
g-whip[3]: b7n9{r7c3 r789c1} - r5n9{c1 c456} - r6n9{c6 .} ==> r7c9 ≠ 9
whip[4]: b5n7{r4c4 r4c5} - b5n2{r4c5 r4c6} - b5n3{r4c6 r6c6} - b5n8{r6c6 .} ==> r4c4 ≠ 5, r4c4 ≠ 4, r4c4 ≠ 1, r4c4 ≠ 9
whip[4]: b5n7{r4c5 r4c4} - b5n2{r4c4 r4c6} - b5n3{r4c6 r6c6} - b5n8{r6c6 .} ==> r4c5 ≠ 4
whip[1] : r4n4{c9 .} ==> r5c9 ≠ 4
whip[4]: r6c4{n5 n9} - r5c6{n9 n1} - r5c4{n1 n4} - r5c5{n4 .} ==> r4c6 ≠ 5, r6c6 ≠ 5
whip[4]: r6c4{n9 n5} - r5c6{n5 n1} - r5c4{n1 n4} - r5c5{n4 .} ==> r4c6 ≠ 9, r6c6 ≠ 9
g-whip[3]: b9n9{r7c7 r789c9} - r6n9{c9 c1} - b7n9{r9c1 .} ==> r7c4 ≠ 9
g-whip[4]: b4n9{r5c1 r4c123} - b6n9{r4c7 r456c9} - r8n9{c9 c5} - r9n9{c6 .} ==> r3c1 ≠ 9
whip[3]: r3n9{c5 c9} - r8n9{c9 c1} - r6n9{c1 .} ==> r5c5 ≠ 9
whip[3]: r9c5{n9 n4} - r5c5{n4 n5} - r6c4{n5 .} ==> r9c4 ≠ 9
whip[4]: r9n7{c9 c1} - r3c1{n7 n8} - r3c9{n8 n9} - r2c8{n9 .} ==> r2c9 ≠ 7
whip[4]: r9n7{c9 c1} - r3c1{n7 n8} - r3c9{n8 n9} - r2c8{n9 .} ==> r1c9 ≠ 7
whip[4]: r2c8{n7 n9} - r3n9{c9 c5} - r3n7{c5 c9} - r9n7{c9 .} ==> r2c1 ≠ 7
whip[4]: r8c9{n5 n9} - r3n9{c9 c5} - r9c5{n9 n4} - r5c5{n4 .} ==> r8c5 ≠ 5
singles ==> r8c9 = 5, r4c8 = 5
whip[1]: c3n5{r1 .} ==> r1c1 ≠ 5, r2c1 ≠ 5
whip[3]: r8n9{c1 c5} - r9n9{c6 c9} - r3n9{c9 .} ==> r7c3 ≠ 9, r7c2 ≠ 9

whip[1]: b7n9{r9c1 .} ==> r1c1 ≠ 9, r2c1 ≠ 9, r5c1 ≠ 9, r6c1 ≠ 9
whip[1]: b4n9{r4c2 .} ==> r4c7 ≠ 9
whip[1]: b6n9{r6c9 .} ==> r9c9 ≠ 9
whip[1]: b9n9{r7c7 .} ==> r7c6 ≠ 9
whip[1]: b6n9{r6c9 .} ==> r1c9 ≠ 9, r2c9 ≠ 9, r3c9 ≠ 9
singles to the end

8.9. Subsets in n-Queens

Recalling that, in n-Queens, a label corresponds to a cell, we shall represent each transversal set in an S_p-subset pattern by p grey cells with the same shade of grey.

8.9.1. A Pair in 7-Queens with a transversal set not associated with a CSP variable

The instance of 7-Queens in Figure 8.9, with two queens already placed in r2c1 and r6c4 has a Pair for CSP variables Xr4 and Xr7, with transversal sets {r4c5, r7c2} and {r4c7, r7c7}. These sets are defined as the intersections of the two rows with respectively a diagonal and a column. The first thus provides an example of a transversal set not defined via a "transversal" CSP variable.

	c1	c2	c3	c4	c5	c6	c7
r1	o	o		o			B
r2	*	o	o	o	o	o	o
r3	o	o		o		A	o
r4	o	o	o	o	▒	o	▒
r5	o		o	o	o		C
r6	o	o	o	*	o	o	o
r7	o	▒	o	o	o	o	▒

Figure 8.9. A 7-Queens instance, with a Pair

***** Manual solution *****
whip[1]: r4{c5 .} ⇒ ¬r3c6 (A eliminated)
pair: {{Xr4, Xr7}, {{r4c5, r7c2}, {r4c7, r7c7}}} ⇒ ¬r1c7, ¬r5c7 (B and C eliminated)

Notice that A could have been eliminated by the Pair, because it is also linked to the first transversal set, but the whip[1] is applied before, because it is considered simpler. Both B and C are linked to the second transversal set.

Remember that the disjointness conditions of the definition bear on the candidates of the different CSP variables in the current resolution state and not on the transversal sets, let alone on the global transversal constraints (or transversal CSP variables) defining them, if any: here r2c7 is common to both constraints.

Finally, notice that, in conformance with the general theory, the Pair can be seen as a whip[2]:

whip[2]: $\Rightarrow$ r4{c7 c5} – r7{c2 .} $\Rightarrow$ ¬r1c7, ¬r5c7

8.9.2. A Pair in 10-Queens with transversal sets defined via transversal variables

Consider again the 10-Queens instance in Figure 5.7 (section 5.11.2), reproduced below as Figure 8.10. Suppose we do not see the second and the third long distance interaction whips. We can still eliminate B and C, based on Pairs in rows (CSP variables Xr3, Xr5), in which the transversal sets correspond to the intersections with columns ("transversal CSP variables" Xc1, Xc6).

	c1	c2	c3	c4	c5	c6	c7	c8	c9	c10
r1	o	o	o	o	o	o	*	o	o	o
r2	o	o	o	o	o	o	o	o	o	*
r3		o	o	o	o	+	o	o	o	o
r4	o	o	*	o	o	o	o	o	o	o
r5	+	o	o	o	o		o	o	o	o
r6	o	o	o	o	o	o	o	*	o	o
r7	o		o	+	o	o	o	o	o	o
r8	B	+	o	o		o	o	o	o	o
r9	o	o	o	o	o	o	o	o	*	o
r10	C	o	o	o	+	A	o	o	o	o

Figure 8.10. *A 10-Queens instance, with a Pair*

***** Manual solution *****

whip[1]: r3{c1 .} $\Rightarrow \neg$r10c6 (A eliminated)

pairs: {{Xr3, Xr5}, {c1{r3, r5}, c6{r3, r5}}} $\Rightarrow \neg$r8c1, $\neg$r10c1   (B, C eliminated)

single in r10: r10c5; single in r8: r8c2; single in r7: r7c4; single in r5: r5c1; single in r3: r3c6

Solution found in W_2.

8.9.3. Triplets in 9-Queens not subsumed by whip[3]

The instance of 9-Queens in Figure 8.11 has a complete Triplet (three candidates for the three CSP variables, i.e. all the optional candidates are present). The (unique) elimination (X) allowed by the Triplet cannot be replaced by a whip[3].

Here, the method is used to provide a simple proof that this instance has no solution.

***** Manual solution *****

triplets: {{Xr1, Xr3, Xr7}, {c1{r1, r3, r7}, c5{r1, r3, r7}, c7{r1, r3, r7}}} $\Rightarrow \neg$r6c1 (A eliminated)

whip[3]: r6{c2 c8} – r7{c7 c5} – r3{c5 .} $\Rightarrow \neg$r8c2 (B eliminated)

single in r8 $\Rightarrow$ r8c8

whip[1]: c1{r7 .} $\Rightarrow \neg$r7c5 (C eliminated)

single in r7 $\Rightarrow$ r7c1

This puzzle has no solution: no value for Xc2

	c1	c2	c3	c4	c5	c6	c7	c8	c9
r1		o	o	o		o		o	o
r2	o	o	*	o	o	o	o	o	o
r3		o	o	o		o		o	o
r4	o	o	o	o	o	o	o	o	*
r5	o	o	o	*	o	o	o	o	o
r6	A		o	o	o	o	o		o
r7	+	o	o	o	C	o		o	o
r8	o	B	o	o	o	o	o	+	o
r9	o	o	o	o	o	*	o	o	o

Figure 8.11. *A 9-Queens instance, with a complete Triplet*

09. Reversible Subset Chains, S_p-whips and S_p-braids

In this chapter, we define more complex types of chains than the whips, g-whips and corresponding braids introduced until now[7]. At least for the Sudoku CSP, this entails that we are dealing with exceptional instances, either because they cannot be solved by the previous patterns or because the new ones give them a smaller rating.

The main idea is that there are patterns that can be considered as elementary or "atomic" and there are ways to combine them into more complex ones. Until now, typical "atomic" patterns have been single candidates in chapter 5 and g-candidates in chapter 7. And the typical way of combining them has been to assemble them into chains, whips, g-whips, braids and g-braids via what we shall now call the "zt-ing principle": in the context of these chains, i.e. "modulo the target (z) and the previous right-linking candidates (t)", they appear as single candidates or as g-candidates.

We shall now show that this principle can be extended to the S_p-subset patterns of chapter 8, more precisely: given any Subset resolution theory S_p ($0{\leq}p{\leq}\infty$) for any CSP, one can define S_p-whips and S_p-braids as generalised whips or braids that accept patterns from this family of rules (i.e. $S_{p'}$-subsets for any p'$\leq$p), in addition to candidates and g-candidates, for their right-linking elements – whereas their left-linking elements remain mere candidates, as in the case of whips and g-whips. In a sense, allowing the inclusion of such patterns introduces a restricted kind of look-ahead with respect to the original non-anticipating (no look-ahead) whips and g-whips, because each $S_{p'}$-subset is inserted into the chain as a whole and it increases its length by p' (its size) instead of 1; but this form of look-ahead is strictly controlled by the p parameter and by the very specific type of pattern the $S_{p'}$-subsets are.

If we consider that, in the context of a whip or g-whip, the left-linking candidates have negative valence and the right-linking candidates or g-candidates have positive valence, then in the context of the new S_p-whips and S_p-braids, the right-linking S_p-subsets have positive valence, in the sense that, if the target was True in some resolution state RS, there would be some posterior resolution state in which they would appear as autonomous S_p-subsets.

---

[7] In the Sudoku context, we first introduced these extended whips and braids (with a different terminology) in the "Fully Supersymmetric Chains" thread of the late Sudoku Player's Forum (p. 14, October 17th, 2008).

In the next chapters, we shall see that one can go still further, but we think the intermediate step developed here is sufficiently interesting in its own. Moreover, it will be easier to justify certain choices we shall have to make later, after we have analysed the simpler case of S_p-whips (simpler mainly because, contrary to whips or braids, the S_p-subset patterns can be defined without any reference to their target).

Everything goes for S_p-whips as for g-whips (except that a few additional technicalities have to be faced). The main point to be noticed is that, when it comes to defining the concepts of S_p-links and S_p-compatibility, we always consider the S_p-labels underlying the S_p-subsets instead of the S_p-subsets themselves, in exactly the same way as we considered the full g-labels underlying the g-candidates when we defined g-links. The main reason for this choice is the same as that for g-links: we want all the notions related to linking and compatibility to be purely structural, i.e. we do not want them to depend on any particular resolution state; this will be essential for the confluence property of S_p-braid resolution theories (in section 9.4) and for the "T&E(S_p) vs S_p-braids" theorem (in section 9.5). But there are also important computational benefits in doing so (such as the possibility of pre-computing all the S_p-labels and S_p-links).

9.1. S_p-links; S_p-subsets modulo other Subsets; S_p-regular sequences

9.1.1. S_p-links, S_p-compatibility

Definition: a label l is *compatible with an S_p-label* S if l is not S_p-linked to S (i.e. if, for each transversal set TS of S, there is at least one label l' in TS such that l is not linked to l').

Definition: a label l is *compatible* with a set R of labels, g-labels and S-labels if l is compatible with each element of R (in the senses of "compatible" already defined separately for labels, g-labels and S_p-labels).

Definitions: a label l is *S_p-linked to an S_p-subset* S if l is S_p-linked to the S_p-label underlying S; a label l is compatible with an S_p-subset if l is not S_p-linked to it; a label l is *compatible* with a set R of candidates, g-candidates and Subsets if l is compatible with each element of R (in the senses of "compatible" already defined separately for candidates, g-candidates and S_p-subsets).

Notice that, in conformance with what we mentioned in the introduction to this chapter, according to the definition of "S_p-linked to an S_p-subset", it is not enough for label l to be linked to all the actual candidates of one of its transversal sets: it must be linked to all the labels of one of its transversal sets.

9.1.2. S_p-subsets modulo a set of labels, g-labels and S-labels

All our forthcoming definitions (Regular S_p-Chains, Reversible S_p-Chains, S_p-whips and S_p-braids) will be based on that of an S_p-subset modulo a set R of labels, g-labels and S-labels; in practice, R will be either the previous right-linking pattern or the set consisting of the target plus all the previous right-linking patterns (i.e. candidates, g-candidates and S_k-subsets).

Definition: in any resolution state of any CSP, given a set R of labels, g-labels and S-labels [or a set R of candidates, g-candidates and Subsets], a *Pair (or S_2-subset) modulo R* is an S_2-label {CSPVars, TransvSets}, where:

– CSPVars = {V_1, V_2},

– TransvSets is composed of the following transversal sets of labels:
    {<V_1, v_{11}>, <V_2, v_{21}>} for constraint c_1,
    {<V_1, v_{12}>, <V_2, v_{22}>} for constraint c_2,

such that:

– in RS, V_1 and V_2 are disjoint, i.e. they share no candidate;

– <V_1, v_{11}> ≠ <V_1, v_{12}> and <V_2, v_{22}> ≠ <V_2, v_{21}>;

– in RS, V_1 has the two mandatory candidates <V_1, v_{11}> and <V_1, v_{12}> compatible with R and no other candidate compatible with R;

– in RS, V_2 has the two mandatory candidates <V_2, v_{21}> and <V_2, v_{22}> compatible with R and no other candidate compatible with R.

Definition: in any resolution state of any CSP, given a set R of labels, g-labels and S-labels [or a set R of candidates, g-candidates and Subsets], a *Triplet (or S_3-subset) modulo R* is an S_3-label {CSPVars, TransvSets}, where:

– CSPVars = {V_1, V_2, V_3},

– TransvSets is composed of the following transversal sets of labels:

– {<V_1, v_{11}>, (<V_2, v_{21}>), <V_3, v_{31}>} for constraint c_1,

– {<V_1, v_{12}>, <V_2, v_{22}>, (<V_3, v_{32}>)} for constraint c_2,

– {(<V_1, v_{13}>), <V_2, v_{23}>, <V_3, v_{33}>} for constraint c_3,

such that:

– in RS, V_1, V_2 and V_3 are pairwise disjoint, i.e. no two of these variables share a candidate;

– <V_1, v_{11}> ≠ <V_1, v_{12}>, <V_2, v_{22}> ≠ <V_2, v_{23}> and <V_3, v_{33}> ≠ <V_3, v_{31}>;

– in RS, V_1 has the two mandatory candidates <V_1, v_{11}> and <V_1, v_{12}> compatible with R, one optional candidate <V_1, v_{13}> compatible with R (supposing this label exists) and no other candidate compatible with R;

– in RS, V_2 has the two mandatory candidates $<V_2, v_{22}>$ and $<V_2, v_{23}>$ compatible with R, one optional candidate $<V_2, v_{21}>$ compatible with R (supposing this label exists) and no other candidate compatible with R;

– in RS, V_3 has the two mandatory candidates $<V_3, v_{33}>$ and $<V_3, v_{31}>$ compatible with R, one optional candidate $<V_3, v_{32}>$ compatible with R (supposing this label exists) and no other candidate compatible with R.

We leave it to the reader to write the definitions of Subsets of larger sizes modulo R (S_p-subsets modulo R). The general idea is that, when one looks at some S_p-label S in RS "modulo R", i.e. when all the candidates in RS incompatible with R are "forgotten", what remains of S in RS satisfies the conditions of a non degenerated Subset of size p based on this S_p-label.

Definition: in all the above cases, *a target of the S_p-subset modulo R* is defined as a target of the S_p-subset itself (i.e. as a candidate S_p-linked to its underlying S_p-label). The idea is that, in any context (e.g. in a chain) in which all the elements in R have positive valence, the S_p-subset itself will have positive valence and any of its targets will have negative valence.

9.1.3. S_p-regular sequences

As in the case of chains built on mere candidates, it is convenient to introduce an auxiliary notion before we define Reversible S_p-chains, S_p-whips and S_p-braids.

Definition: let there be given an integer $1{\leq}p{\leq}\infty$, an integer $m{\geq}1$, a sequence (q_1, ..., q_m) of integers, with $1{\leq}q_k{\leq}p$ for all $1{\leq}k{\leq}m$, and let $n = \Sigma_{1{\leq}k{\leq}m}\ q_k$; let there also be given a sequence (W_1, ..., W_m) of different sets of CSP variables of respective cardinalities q_k and a sequence (V_1, ..., V_m) of CSP variables such that $V_k \in W_k$ for all $1{\leq}k{\leq}m$. We define *an S_p-regular sequence of length n associated with (W_1, ... W_m) and (V_1, ... V_m)* to be a sequence of length 2m [or 2m-1] (L_1, R_1, L_2, R_2, ... L_m, [R_m]), such that:

– $q_m{=}1$ and $W_m = \{V_m\}$;

– for $1{\leq}k{\leq}$ m, L_k is a candidate;

– for $1{\leq}k{\leq}$ m [or $1{\leq}k{<}m$], R_k is a candidate or a g-candidate if $q_k{=}1$ and it is a (non degenerated) Sq_k-subset if $q_k{>}1$;

– for each $1{\leq}k{\leq}m$ [or $1{\leq}k{<}m$], one has *"strong continuity", "strong g-continuity" or "strong Sq_k-continuity" from L_k to R_k*, namely:

  - if R_k is a candidate ($q_k{=}1$ and $W_k{=}\{V_k\}$), L_k and R_k have a representative with V_k: $<V_k, l_k>$ and $<V_k, r_k>$,

  - if R_k is a g-candidate ($q_k{=}1$ and $W_k{=}\{V_k\}$), L_k has a representative $<V_k, l_k>$ with V_k and R_k is a g-candidate $<V_k, r_k>$ for V_k (r_k being its set of values),

- if R_k is an Sq_k-subset ($q_k>1$), then W_k is its set of CSP variables and L_k has a representative $<V_k, l_k>$ with V_k.

The L_k are called the *left-linking candidates* of the sequence and the R_k the *right-linking objects (or elements or patterns)*.

Remarks:

– Notice the natural expression chosen for L_k to R_k continuity in case R_k is a Subset.

– The definition of Subsets implies a disjointness condition on the sets of candidates for the CSP variables inside each W_k, but the present definition puts no condition on the intersections of different W_k's. In particular, W_{k+i} may be a strict subset of W_k, if the right-linking elements in between give negative valence in W_{k+i} to some candidates that had no individual valence assigned in W_k. This is not considered as an inner loop of the sequence.

Exercise: after reading this chapter, comment on the condition $q_m=1$ and show that it entails no restriction in the sequel.

9.2. Reversible S_p-subset Chains (RS_pC)

Reversible S_p-Chains are an extension of bivalue chains in which right-linking candidates may be replaced by g-candidates or $S_{p'}$-subsets ($p'\leq p$). [One could introduce an intermediate notion, in which g-candidates are not allowed; but it does not seem to be very useful; we leave it as an exercise for the very motivated reader to prove that these "restricted" S_p-chains are reversible, to define the associated sequence of resolution theories and to prove that they have the confluence property.]

9.2.1. Definition of Reversible S_p-Chains

Definition: given an integer $1\leq p\leq\infty$ and a candidate Z (which will be a target), a *Reversible S_p-Chain* of length n ($n \geq 1$) built on Z, noted $RS_pC[n]$, is an S_p-regular sequence $(L_1, R_1, L_2, R_2, ... L_m, R_m)$ of length n associated with a sequence $(W_1, ... W_m)$ of sets of CSP variables and a sequence $(V_1, ... V_m)$ of CSP variables, such that:

– Z is neither equal to any candidate in $\{L_1, R_1, L_2, R_2, ... L_m, R_m\}$, nor a member of any g-candidate in this set, nor equal to any label in the Sq_k-label of R_k when R_k is an Sq_k-subset, for any $1\leq k<m$;

– Z is linked to L_1;

– for each $1 < k \leq m$, L_k is linked or g-linked or Sq_{k-1}-linked to R_{k-1}; this is the natural way of defining *"continuity" from R_{k-1} to L_k*;

– R_1 is a candidate or a g-candidate or an Sq_1-subset modulo Z: R_1 is the only candidate or g-candidate or is the unique Sq_1-subset composed of all the candidates C for the CSP variables in W_1 such that C is compatible with Z;

– for any $1 < k \leq m$, R_k is a candidate or a g-candidate or an Sq_k-subset modulo R_{k-1}: R_k is the only candidate or g-candidate or is the unique Sq_k-subset (if $k \neq m$) composed of all the candidates C for the CSP variables in W_k such that C is compatible with R_{k-1};

– Z is not a label for V_m;

– Z is linked or g-linked to R_m.

Theorem 9.1 (Reversible S_p-chain rule for a general CSP): in any resolution state of any CSP, if Z is a target of a Reversible S_p-chain, then it can be eliminated (formally, this rule concludes ¬candidate(Z)).

Proof: if Z was True, then L_1 would be eliminated by ECP and R_1 would be asserted by S (if it is a candidate) or it would be a g-candidate or an Sq_1-subset; in any case, L_2 would be eliminated by ECP or W_1 or Sq_1. After iteration: R_m would be asserted by S or it would be a g-candidate – which would contradict Z being True.

9.2.2. Reversibility of Reversible S_p-Chains in the general CSP

The following theorem justifies the name we have given these chains. Notice that it does in no way depend on the fact that the transversal sets defining the Subsets would be defined by "transversal" CSP variables.

Theorem 9.2: a Reversible S_p-Chain is reversible.

Proof: the main point of the proof is the construction of the reversed chain (a generalisation of the construction in section 7.2).

This construction can be followed in part using Figure 9.1. It gives a symbolic representation of the end of a Reversible S_2-chain and the start of the associated reversed chain. Horizontal solid lines represent CSP variables (both chains use the same global set of CSP variables); vertical dotted lines represent transversal sets: on horizontal lines, candidates can only exist at the intersections with dotted lines (here "horizontal" and "vertical" are in no way related to an underlying grid on which the CSP would have to be defined). Octagons are symbolic containers for the candidates in the right-linking S_2-subsets (solid lines for the initial chain, dotted lines for the reversed chain); they also show how CSP variables are grouped in each chain to define their respective Subsets.

Given a Reversible S_p-chain (L_1, R_1, L_2, R_2, ... L_m, R_m) of length n built on Z and associated with the sequence (W_1, ... W_m) of sets of CSP variables and the sequence (V_1, ... V_m) of CSP variables, let us define a reversed S_p-chain of same length, with the same target Z and associated with a sequence $(W'_1, ..., W'_m)$ of sets of CSP variables and a sequence $(V'_1, ..., V'_m)$ of CSP variables that are closely related, but not identical, to the reversed sequences of $(W_1, ..., W_m)$ and $(V_1, ..., V_m)$ respectively, and with a sequence of sizes $(q'_1, ..., q'_m)$ such that its first m-1 elements are those of $(q_1, ..., q_{m-1})$ in reversed order and $q'_m=1$. Let $L'_1 = R_m$.

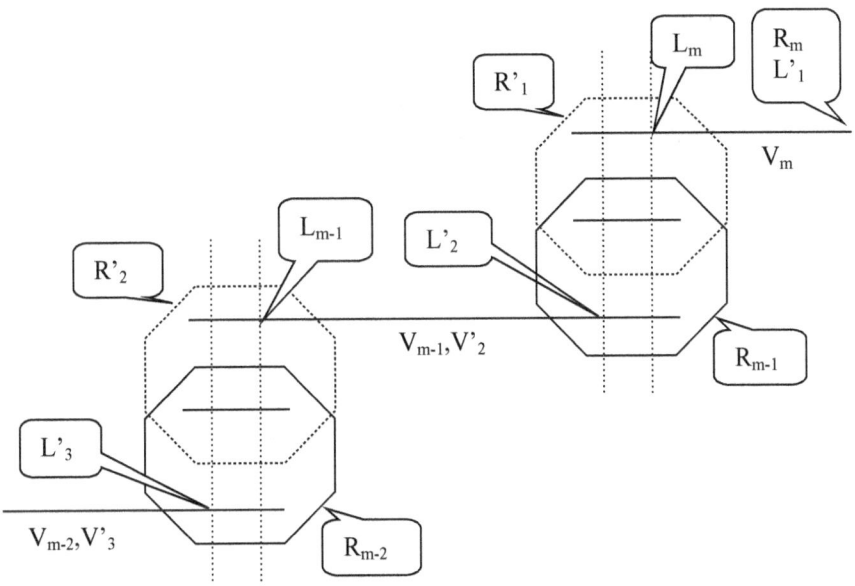

Figure 9.1. *A symbolic representation of the end of a Reversible S_2-chain and the start of the associated reversed chain.*

We can now define W'_1, V'_1, R'_1 and L'_2, depending on what R_{m-1} is:

– if $q_{m-1}=1$ and R_{m-1} is a candidate or a g-candidate and it is linked to only one candidate for V_m (which implies that this candidate can only be L_m), then let $W'_1 = \{V_m\}$, $V'_1 = V_m$, $q'_1=1$ and $R'_1 = L_m$ (R'_1 is a candidate); let $L'_2 = R_{m-1}$ if R_{m-1} is a candidate and $L'_2 =$ any candidate in R_{m-1} if R_{m-1} is a g-candidate;

– if $q_{m-1}=1$ and R_{m-1} is a candidate or a g-candidate and it is linked to several candidates for V_m (which implies that these candidates can only be elements of a g-label for V_m, say g), let $W'_1 = \{V_m\}$, $V'_1 = V_m$, $q'_1=1$ and let R'_1 be the subset of g consisting of these candidates (R'_1 is a thus g-candidate); as before, let $L'_2 = R_{m-1}$ if R_{m-1} is a candidate and $L'_2 =$ any candidate in R_{m-1} if R_{m-1} is a g-candidate;

– if $q_{m-1} > 1$, then R_{m-1} is an Sq_{m-1}-subset; let $W'_1 = W_{m-1} \cup \{V_m\} - \{V_{m-1}\}$; let $V'_1 = V_m$; and let R'_1 be the set of all the candidates for variables in W'_1. Because R_m is the only candidate for V_m modulo R_{m-1}, all the candidates for V_m other than $L'_1 = R_m$ can only be in the transversal sets of R_{m-1}. Thus, forgetting L'_1, R'_1 together with the same transversal sets as R_{m-1} is an Sq_{m-1}-subset and it has all the candidates for V_{m-1} in R_{m-1} as targets (and we take any of these as L'_2). As a result, all the other candidates for V_{m-1} (i.e. all those that are compatible with R'_1) can only be in the transversal sets of R_{m-2}.

We are now in a situation in which L'_2 is defined and the above construction can be iterated, using L'_2 instead of L'_1, R_{m-2} instead of R_{m-1}, W_{m-1} instead of W_m and V_{m-1} instead of V_m (once L'_1 was defined, the fact that $q_m = 1$, i.e. that R_m was a candidate or a g-candidate played no role in the above construction).

All this can be iterated until we can define the final $W'_m = \{V'_m\}$ with $V'_m = V_1$; L_1 or the g-candidate consisting of L_1 and the other candidates for V_1 linked to Z can be taken as R'_m. qed.

Notice that, in this construction: even though $q_m = 1$, one can have $q'_1 \neq 1$; and even if $q_1 \neq 1$, one always has $q'_m = 1$, as in the definition of a Reversible S_p-chain.

Exercise: check that this reversed chain does satisfy all the conditions in the definition of a Reversible S_p-chain.

9.2.3. RS_pC resolution theories and the RS_pC ratings

As is now usual when we have new rules, we can define a new increasing family of resolution theories. Here, we can do it for each p.

Definition: for each p, $1 \leq p \leq \infty$, one can define an increasing sequence $(RS_pC_n,$ $n \geq 0)$ of resolution theories:

– $RS_pC_0 = BRT(CSP)$,
– $RS_pC_1 = RS_pC_0 \cup \{$rules for Reversible S_p Chains of length 1$\} = W_1$,
– $RS_pC_2 = RS_pC_1 \cup S_2$ (if $p \geq 2$) $\cup \{$rules for Reversible S_p Chains of length 2$\}$,
– ....
– $RS_pC_n = RS_pC_{n-1} \cup S_n$ (if $p \geq n$) $\cup \{$rules for Reversible S_p Chains of length n$\}$,
– $RS_pC_\infty = \cup_{n \geq 0} RS_pC_n$.

For p=1, $S_1W_n = gW_n$. For p=∞, i.e. for Reversible S_p Chains built on Subsets of *a priori* unrestricted size, we also write RSC_n instead of $RS_\infty C_n$.

Definition: for any $1 \leq p \leq \infty$, the **RS_pC-rating** of an instance P, noted $RS_pC(P)$, is the smallest $n \leq \infty$ such that P can be solved within RS_pC_n.

Theorem 9.3: all the RS_pC_n resolution theories (for $1{\leq}p{\leq}\infty$ and $n \geq 0$) are stable for confluence; therefore, they have the confluence property.

Proof: we leave it as an exercise for the reader (using reversibility to propagate the consequences of value assertions and candidate deletions, it can be obtained via a drastic simplification of the proof for the S_p-braids case, theorem 9.8).

9.2.4. S_p-subsets versus Reversible S_p-chains

9.2.4.1. Targets of S_p-subsets are targets of $RS_{p-1}C[p]$

Theorem 9.4: a target of an S_p-subset is always a target of a Reversible S_{p-1} Chain of length p.

Proof: Proof: almost obvious. After renumbering the CSP variables, one can always suppose that Z is S_p-linked to transversal set TS_1 and that V_1 has a candidate $L_1 = <V_1, l_1>$ to which Z is linked. Let $L_p = <V_p, l_p>$ be a candidate for V_p not in TS_1 (there must be one if the S_p-subset is not degenerated). Let S2 be the S_{p-1}-subset: $\{\{V_1,..., V_{p-1}\}, \{TS_2, ..., TS_p\}\}$. Then the desired chain is:
$RS_{p-1}C[p]$: $\{L_1 \ S2\} - V_p\{l_p \ .\}$.

9.2.4.2. Type-2 targets of S_p-subsets are targets of Reversible S_{p-2} Chains

An S_p-subset that has transversal sets with non-void intersections allows more eliminations than the "standard" ones defined in chapter 8. (This can happen only for p>2.)

Definition: a type-2 target of an S_p-subset is a candidate belonging to (at least) two of its transversal sets.

Theorem 9.5: a type-2 target of an S_p-subset can be eliminated.

Proof: suppose the type-2 target Z is a candidate for variable V_1 and belongs to transversal sets TS_1 and TS_2. If Z was True, then all the other candidates in TS_1 or TS_2 or in a g-candidate in TS_1 or TS_2 would be eliminated by ECP. This would leave only p-2 possibilities for the remaining p-1 CSP variables – which is contradictory, in exactly the same way as in the case of a normal target.

Notice however that this is a very unusual kind of elimination. Until now, for all the rules we have met, the target did not belong to the pattern. The following theorem shows that this "cannibalistic" abnormality can be palliated. It also justifies that we did not consider type-2 targets of S_p-subsets in chapter 8: these abnormal targets can always be eliminated by a simpler pattern. An illustration of this theorem will appear in section 10.3 for the more general case of gS_p-subsets.

Theorem 9.6: A type-2 target of an S_p-subset is always the (normal) target of a shorter Reversible S_{p-2} Chain of length p-1.

Proof: in a resolution state RS, let Z be a type-2 target of an S_p-subset with CSP variables $V_1, \dots V_p$ and transversal sets $TS_1, \dots TS_p$. One can always suppose that V_1 is the CSP variable for which Z is a candidate (there can be only one in RS) and that TS_1 and TS_2 are the two transversal sets to which Z belongs.

Firstly, each of the CSP variables $V_2, V_3, \dots V_p$ must have at least one candidate belonging neither to TS_1 nor to TS_2 (if it has several, choose one arbitrarily and name it $<V_2, c_2>, \dots <V_p, c_p>$, respectively). Otherwise, the initial S_p-subset would be degenerated; more precisely, Z could be eliminated by a whip[1] (or even by ECP after a Single) associated with (any of) the CSP variable(s) that has no such candidate.

Secondly, in TS_1 or TS_2, there must be at least one candidate for at least one of the CSP variables $V_2, \dots V_p$. Otherwise, the initial S_p-subset would be degenerated; more precisely, it would contain, among others, the S_{p-2}-subset $\{\{V_3, \dots, V_p\}, \{TS_3, \dots, TS_p\}\}$; this would allow to eliminate all the candidates for V_1 and V_2 that are not in TS_1 or TS_2; Z could then be eliminated by a whip[1] associated with V_2; and V_1 would have no candidate left. One can always suppose that there exists such a candidate L_2 for V_2, i.e. $L_2 = <V_2, l_2>$.

Modulo Z, we therefore have an S_{p-2} subset S2 with CSP variables $V_2, \dots V_{p-1}$ and transversal sets $TS_3, \dots TS_p$. Then, Z is a (normal) target of the following Reversible S_{p-2} Chain of length p-1:

$$RS_{p-2}C[p-1]: V_2\{l_2 \ S2\} - V_p\{c_p .\} \Rightarrow \neg candidate(Z).$$

9.2.5. Reversible Subset Chains in Sudoku: grouped ALS chains and AICs

Non Sudoku experts can skip this sub-section or see the classical definitions of ALS chains (Almost Locked Set chains) and AICs (Alternating Inference Chains / Nice Loops) in the over-abundant Sudoku literature, e.g. at www.sudopedia.org. Our main purpose here is to notice that the above Reversible Subset Chains, defined for any CSP, correspond in Sudoku to these well-known patterns (though the above presentation provides a very unusual perspective of them).

In Sudoku, if one considers only the X_{rc} CSP variables, Reversible Subset Chains correspond to the classical grouped ALS-chains ("grouped" because we allow g-candidates as right-linking patterns). The only difference is, we never mention "Almost Locked Sets" (ALSs) or "Restricted Commons", we deal only with Subsets ("Locked Sets") modulo something.

If one uses all the X_{rc}, X_{rn}, X_{cn} and X_{bn} CSP variables, Reversible Subset Chains correspond to the grouped AICs (Alternating Inference Chains).

[Historical note: what an AIC is has never been very clear in the Sudoku literature. (In what it differs from "Nice Loops", apart from being written in a different notation has never been very clear either; it seems to be more a matter of competition between different people than anything else). On the one hand, the definition of AICs is so vague that, transposed into our vocabulary, almost anything could be used as a right-linking pattern.

On the other hand, i.e. on the concrete side of things, the fact that "Fish" (our Super-Hidden Subsets) could be included in AICs has been mentioned only long after we introduced the more general S_p-whips and S_p-braids (in a different terminology); as the definition of the latter was fully supersymmetric and included all types of Subsets from the start, there was no need to make a special mention of Fish Subsets; in particular, all our classification results with Subsets in *HLS*, or those with S_p-braids mentioned in section 9.6, included Fish.

From an epistemological point of view, it is interesting to explore the reasons of this late recognition. In our opinion, there are four:
  – the various notions involved lacked being formalised;
  – in particular, there was an incomplete view of all the logical symmetries;
  – the notions of an Almost Locked Set and of a Restricted Common, at the basis of ALS chains, are much more complicated than the notion of a Locked Set modulo something; they are difficult to deal with; in particular, their correct transposition to AICs, i.e. their extension to the rn, cn and bn spaces, seems to be difficult to do without having a complete logical formalisation; they also lead to the introduction of several levels of almosting: AALSs, AAALSs (all of which are taken care of by the more general zt-ing principle);
  – there was a strong insistence on chains having to be reversible (without any definition of this property); even for chains effectively reversible, this blocked any view of them, such as the one exposed here, that would have allowed to shortcut the notion of a Restricted Common.]

9.3. S_p-whips and S_p-braids

S_p-whips and S_p-braids are an extension of g-whips and g-braids in which $S_{p'}$-subsets ($p' \leq p$) may appear as right-linking patterns. They can also be seen as extensions of the Reversible S_p-chains: starting from the same S_p-subset bricks, the "almosting-principle" used to assemble Reversible S_p-chains (a principle that only allows to "forget" candidates linked to the previous right-linking pattern) has to be replaced by the much more powerful "zt-ing principle" (a principle that allows to "forget" candidates linked to any of the previous right-linking patterns or to the target). In this replacement, reversibility is lost, but the most important property, non-anticipativeness, is preserved.

9.3.1. Definition of S_p-whips

Definition: given an integer $1 \leq p \leq \infty$ and a candidate Z (which will be the target), an S_p-whip of length n (n ≥ 1) built on Z is an S_p-regular sequence $(L_1, R_1, L_2, R_2, \ldots L_m)$ [notice that there is no R_m] of length n, associated with a sequence $(W_1, \ldots W_m)$ of sets of CSP variables and a sequence $(V_1, \ldots V_m)$ of CSP variables (with $W_m = \{V_m\}$), such that:

– Z is neither equal to any candidate in $\{L_1, R_1, L_2, R_2, \ldots L_m\}$ nor a member of any g-candidate in this set nor equal to any element in the Sq_k-label of R_k when R_k is an Sq_k-subset, for any $1 \leq k \leq m$;

– L_1 is linked to Z;

– for each $1 < k \leq m$, L_k is linked or g-linked or Sq_{k-1}-linked to R_{k-1}; this is the natural way of defining "continuity" from R_{k-1} to L_k;

– for any $1 \leq k < m$, R_k is a candidate or a g-candidate or an Sq_k-subset modulo Z and all the previous right-linking patterns: either R_k is the only candidate or g-candidate compatible with Z and with all the R_i with $1 \leq i < k$, or R_k is the unique Sq_k-subset composed of all the candidates C for some of the CSP variables in W_k such that C is compatible with Z and with all the R_i with $1 \leq i < k$;

– Z is not a label for V_m;

– V_m has no candidate compatible with the target and with all the previous right-linking objects (but V_m has more than one candidate).

Theorem 9.7 (S_p-whip rule for a general CSP): in any resolution state of any CSP, if Z is a target of an S_p-whip, then it can be eliminated (formally, this rule concludes ¬candidate(Z)).

Proof: the proof is an easy adaptation of that for g-whips. If Z was True, all the z-candidates would be eliminated by ECP and, iterating upwards from k=2: R_{k-1} would be asserted by S or it would be a g-candidate or an Sq_{k-1}-subset; R_{k-1} to L_k continuity ensures that L_k would be eliminated by ECP, W_1 or Sq_{k-1}; and the t-candidates would be eliminated by these rules. When m-1 is reached, R_{m-1} would be asserted by S or it would be a whip[1] (a g-candidate) or a Subset with target L_m; finally, there would be no value left for V_m (because Z itself is not a label for V_m).

9.3.2. Definition of S_p-braids

Definition: given an integer $1 \leq p \leq \infty$ and a candidate Z (which will be the target), an S_p-braid of length n (n ≥ 1) built on Z is an S_p-regular sequence $(L_1, R_1, L_2, R_2, \ldots L_m)$ [notice that there is no R_m] of length n, associated with a sequence $(W_1, \ldots W_m)$ of sets of CSP variables and a sequence $(V_1, \ldots V_m)$ of CSP variables (with $W_m = \{V_m\}$), such that:

– Z is neither equal to any candidate in $\{L_1, R_1, L_2, R_2, \ldots L_m\}$ nor a member of any g-candidate in this set nor equal to any element in the Sq_k-label of R_k when R_k is an Sq_k-subset, for any $1 \leq k \leq m$;

– L_1 is linked to Z;

– for each $1 < k \leq m$, L_k is linked or g-linked or S-linked to Z or to some of the R_i, i<k; this is the only difference with S_p-whips;

– for any $1 \leq k < m$, R_k is a candidate or a g-candidate or an Sq_k-subset modulo Z and all the previous right-linking patterns: either R_k is the only candidate or g-candidate compatible with Z and with all the R_i with $1 \leq i < k$, or R_k is the unique Sq_k-subset composed of all the candidates C for some of the CSP variables in W_k such that C is compatible with Z and with all the R_i with $1 \leq i < k$;

– Z is not a label for V_m;

– V_m has no candidate compatible with the target and with all the previous right-linking objects (but V_m has more than one candidate).

Theorem 9.8 (S_p-braid rule for a general CSP): in any resolution state of any CSP, if Z is a target of an S_p-braid, then it can be eliminated (formally, this rule concludes ¬candidate(Z)).

Proof: almost the same as in the S_p-whips case. The Z or R_i (i<k) to L_k condition replacing the R_{k-1} to L_k continuity condition allows the same intermediate conclusion for L_k.

Definition: in any of the above defined Reversible Subset Chains, S_p-whips or S_p-braids, a candidate other than L_k for any of the CSP variables ("global" variable V_k or inner variables $V_{k,i}$ if R_k is an inner Subset), is called a t- [respectively a z-] candidate if it is incompatible with a previous right-linking pattern [resp. with the target]. Notice that a candidate can be z- and t- at the same time and that the t- and z- candidates are not considered as being part of the pattern.

9.3.3. S_p-whip and S_p-braid resolution theories; S_pW and S_pB ratings

In exactly the same way as in the cases of whips, g-whips, braids and g-braids, one can now, for each p, define an increasing sequence of resolution theories. They now have two parameters, one (n) for the total length of the chain and one (p) for the maximum size of included Subsets. By convention, p=1 means no Subset, only candidates and g-candidates.

Definition: for each $1 \leq p \leq \infty$, one can define an increasing sequence $(S_pW_n, n \geq 0)$ of resolution theories (similar definitions can be given for S_p-braids, merely by replacing everywhere "whip" by "braid" and "W" by "B"):

– $S_pW_0 = BRT(CSP)$,

- $S_pW_1 = S_pW_0 \cup$ {rules for S_p-whips of length 1} = W_1,
- $S_pW_2 = S_pW_1 \cup S_2$ (if p≥2) $\cup$ {rules for S_p-whips of length 2},
- ....
- $S_pW_n = S_pW_{n-1} \cup S_n$ (if p≥n) $\cup$ {rules for S_p-whips of length n},
- $S_pW_\infty = \cup_{n\geq0} S_pW_n$.

For p=1, $S_1W_n = gW_n$. For p=∞, i.e. for S-whips built on Subsets of *a priori* unrestricted size (but, in practice, p < n), we also write SW_n instead of $S_\infty W_n$.

Definition: for any 1≤p≤∞, the *S_pW-rating* of an instance P, noted $S_pW(P)$, is the smallest n ≤ ∞ such that P can be solved within S_pW_n. By convention, $S_pW(P)$ = ∞ means that P cannot be solved by S_p-whips of any length; $SW(P)$ = ∞ means that P cannot be solved by S-whips of any length including subsets of any size.

Definition: similarly, for any 1≤p≤∞, the *S_pB-rating* of an instance P, noted $S_pB(P)$, is the smallest n ≤ ∞ such that P can be solved within S_pB_n. By convention, $S_pB(P)$ = ∞ means that P cannot be solved by S_p-braids of any length.

For any 1≤p≤∞, the S_pW and S_pB ratings are defined in a purely logical way, independent of any implementation; the S_pW and S_pB ratings of an instance are intrinsic properties of this instance; moreover, as will be shown in the next section, for any fixed p (1≤p≤∞), the S_pB rating is based on an increasing sequence of theories (S_pB_n, n≥0) with the confluence property and it can therefore be computed with a simplest first strategy based on the global length of the S_p-braids involved.

For any puzzle P, one has obviously $W(P) \geq gW(P) = S_1W(P) \geq S_2W(P) \geq S_pW(P) \geq S_{p+1}W(P) \geq$ ... and similar inequalities for the $S_pB(P)$.

Beware of not confusing the definitions in this section with those in section 8.6.3. In the latter case, whips and Subsets of same size are merely put together in the same set of rules; in the present section, whips and subsets are fused into more complex structures. The respective notations can be remembered with the following mnemonic (and a similar one for braids): the "+" sign (and the repetition of size n) in W+S (and in W_n+S_n) indicate(s) the juxtaposition of two different things; the absence of a space between W and S in WS and in WS_n indicates their fusion into new patterns.

Notice that consistent definitions of length for S_p-whips or S_p-braids and of the associated S_pW and S_pB ratings are highly constrained:

- by the fact that they are generalisations of the RS_pC chains;
- by the subsumption theorems of section 8.7 and their obvious generalisations to Subsets of any size: in "many" cases of inclusion of such Subsets in an S_p-whip or S_p-braid, it will be possible to replace them by equivalent g-whips or g-braids and to transform the original S_p-whip or S_p-braid into an equivalent g-whip or g-braid. It

seems natural to impose that, in such cases, the two visions of the "same" pattern lead to the same length (especially as length is taken as the measure of complexity of instances).

Finally, the confluence property of all the S_pB_n resolution theories for each p, $1 \le p \le \infty$, (see section 9.4) allows to superimpose on S_pB_n a "simplest first" strategy compatible with the S_pB rating.

9.3.4. Accepting type-2 targets of S_p-subsets in RS_pC, S_pW and S_pB?

Theorem 9.6 alone does not guarantee that type-2 targets of S_p-subsets, if allowed to be used as left-linking candidates in the definitions of Reversible S_p Chains, S_p-whips or S_p-braids, could not lead to (slightly) more general patterns than those in our current definitions. The following, if true, would provide a negative answer and it would complete the justification for not accepting type-2 targets in S_p-subsets: "for any Reversible S_p Chains, S_p-whip or S_p-braid, according to an extended definition that would allow using type-2 targets of $S_{p'}$-subsets as left-linking candidates, there is an equivalent standard (i.e. satisfying the definitions of this chapter) Reversible S_p Chains, S_p-whip or S_p-braid, respectively, of same or shorter length and with the same target". But, although we have no counter-example, this does not seem to be true in general.

In order to understand why it may not be true, consider the following tentative proof, using the notations of the definitions. If the situation occurs several times in the chain, the same kind of actions as defined below could be repeated.

If left-linking candidate L_{k+1} is a type-2 target of Subset S_k, then consider CSP variable V_{k+1} (which cannot be the unique CSP variable of S_k for which L_{k+1} is a candidate) modulo S_k (in the RS_pC case, setting $S_0 = Z$) or modulo Z and all the previous right-linking objects (in the S_p-whip and S_p-braid cases):
  – either it has another candidate, say L'_k, linked to all the candidates in some of the transversal sets of S_k to which L_{k+1} does not belong; then, one can replace L_{k+1} with L'_{k+1} in the original chain;
  – or it has no such candidate but it has a candidate linked to Z or to a previous right-linking object; then, in the RS_pC and S_p-braid cases but *a priori* not in the S_p-whip case, one can replace L_{k+1} with this candidate in the original chain and, in addition, excise S_p from it in the RS_pC case;
  – or it has no such candidate and no candidate linked to Z or to a previous right-linking object; then no possibility seems to be available.

We think that the cases in which this construction does not work and there are no alternative resolution paths are extremely rare and that allowing type-2 targets of $S_{p'}$-subsets inside RS_pC-chains, S_p-whips or S_p-braids is not worth until experimental results show the contrary. Until then, we shall stick to our original definitions. Moreover, see section 10.2.3.6.

Remark: the main interest of the above tentative proof may be that it can be transposed to other situations; e.g. it can explain why, given an $S_{p'}$-subset S allowing the elimination of a target L that could be replaced by a whip elimination (according to the subsumption theorems), if the same S appears inside an S_p-whip or S_p-braid modulo the target and the previous right-linking patterns of this S_p-whip or S_p-braid and if L is used in this S_p-whip or S_p-braid as the next left-linking candidate, S can nevertheless not always be replaced by a sub-whip inside this global S_p-whip or S_p-braid. In particular, given that W_2 subsumes S_2, this gives an idea why S_2W_n is not equal to gW_n for n>2 (or why $S_2W_5 \not\subset gW_5$, as shown by the example in section 9.7.1; or why S_2B is not equal to gB in table 9.1).

9.4. The confluence property of the S_pB_n resolution theories

Theorem 9.9: each of the S_pB_n resolution theories (for $1 \leq p \leq \infty$, $0 \leq n \leq \infty$) is stable for confluence; therefore, it has the confluence property.

Proof: in order to keep the same notations as in the proof for the g-braid resolution theories (section 7.6), we prove the result for S_rB_n, r and n fixed. The proof follows the same general lines as that for g-braids in section 7.5. We keep the same numbering of the various cases to be considered. However, some new sub-cases appear and some cases have to be split into three, in order to take into account the different kinds of right-linking patterns. Marks now extend from case f to case d.

We must show that, if an elimination of a candidate Z could have been done in a resolution state RS_1 by an S_r-braid B of length n' $\leq$ n and with target Z, it will always still be possible, starting from any further state RS_2 obtained from RS_1 by consistency preserving assertions and eliminations, if we use a sequence of rules from S_rB_n. Let B be: $\{L_1 R_1\} - \{L_2 R_2\} - .... - \{L_p R_p\} - \{L_{p+1} R_{p+1}\} - ... - \{L_m .\}$, with target Z, where the R_k's are now candidates or g-candidates or Subsets from S_r modulo Z and the previous R_i's.

Consider first the state RS_3 obtained from RS_2 by applying repeatedly the rules in S_r until quiescence. As S_r has the confluence property (by theorem 8.4), this state is uniquely defined.

If, in RS_3, target Z has been eliminated, there remains nothing to prove. If target Z has been asserted, then the instance of the CSP is contradictory; if not yet detected in RS_3, this contradiction can be detected by CD in a state posterior to RS_3, reached by a series of applications of rules from S_r, following the S_r-braid structure of B.

Otherwise, we must consider all the elementary events related to B that can have happened between RS_1 and RS_3 as well as those we must provoke in posterior resolution states RS. For this, we start from B' = what remains of B in RS_3 and we let RS = RS_3. At this point, B' may not be an S_r-braid in RS. We progressively update RS and B' by repeating the following procedure, for p = 1 to p = m, until it produces a new (possibly shorter) S_r-braid B' with target Z in resolution state RS – a situation that is bound to happen. (As in the g-braids case, and because we have included W_1 in S_r, we have to consider a state RS posterior to RS_3). Return from this procedure as soon as B' is a g-braid in RS. All the references below are to the current RS and B'.

a) If, in RS, any candidate that had negative valence in B – i.e. the left-linking candidate, or any t- or z- candidate, of CSP variable V_p, or any t- or z- candidate of R_p in case R_p is an inner Subset – has been asserted (this can only be between RS_1 and RS_3), then all the candidates linked to it have been eliminated by relevant rules from S_r in RS_3, in particular: Z and/or all the candidate(s) R_k (k<p) to which it is linked, and/or all the elements of the g-candidate(s) R_k (k<p) to which it is g-linked, and/or all the candidates of the CSP variable in W_k to which it belongs and/or all the candidates in the transversal set(s) of the R_k's (k<p) to which it is S-linked (by the definition of an S_r-braid); if Z is among them, there remains nothing to prove; otherwise, the procedure has already either been successfully terminated by case f1 or f2α or dealt with by case d2 of the first previous such k.

b) If, in RS, left-linking candidate L_p has been eliminated (but not asserted), it can no longer be used as a left-linking candidate in an S_r-braid. Suppose that either CSP variable V_p still has a z- or a t- candidate C_p, or R_p is an inner braid and there is another CSP variable V_p' in its W_p such that V_p' still has a z- or a t- candidate C_p; then replace L_p by C_p and (in the latter case) V_p by V_p'. Now, up to C_p, B' is a partial S_r-braid in RS with target Z. Notice that, even if L_p was linked or g-linked or S_r-linked to R_{p-1} (e.g. if B was an S_r-whip) this may not be the case for C_p; therefore trying to prove a similar theorem for S_r-whips would fail here.

c) If, in RS, any t- or z- candidate of V_p or of the inner Subset S_p has been eliminated (but not asserted), this has not changed the basic structure of B (at stage p). Continue with the same B'.

d) Consider now assertions occurring in right-linking objects. There are two cases instead of one for g-braids.

d1) If, in RS, right-linking candidate R_p or a candidate R_p' in right-linking g-candidate R_p has been asserted (p can therefore not be the last index of B'), R_p can no longer be used as an element of an S_r-braid, because it is no longer a candidate or a g-candidate. As in the proof for g-braids, and only because of this d1 case, we cannot be sure that this assertion occurred in RS_3. We must palliate this. First eliminate by ECP or W_1 any left-linking or t- candidate for any CSP variable of B' after p, including those in the inner Subsets, that is incompatible with R_p, i.e. linked or g-linked to it, if it is still present in RS. Now, considering the S_r-braid structure of B upwards from p, more eliminations and assertions can been done by rules from S_r. (Notice that, as in the g-braids case, we are not trying to do more eliminations or assertions than needed to get a g-braid in RS; in particular, we continue to consider R_p, not R_p'; in any case, it will be excised from B'; but, most of all, we do not have to find the shortest possible S_r-braid!)

Let q be the smallest number strictly greater than p such that CSP variable V_q or some CSP variable V_q' in W_q still has a left-linking, t- or z- candidate C_q that is not linked, g-linked or S-linked to any of the R_i for $p \leq i < q$ (C_q is therefore linked, g-linked or S-linked to Z or to some R_i with $i < p$). (For index q, there is thus a V_q' in W_q and a candidate C_q for V_q' such that C_q is linked, g-linked or S-linked to Z or to some R_i with $i < p$.)

Apply the following rules from S_r (if they have not yet been applied between RS_2 and RS) for each of the CSP variables V_u (and all the $V_{u,i}$ in W_u if R_u is an inner Subset) with index (or first index) u increasing from p+1 to q-1 included:
- eliminate its left-linking candidate (L_u) by ECP or W_1 or some $S_{r'}$ ($r' \leq r$);
- at this stage, CSP variable V_u has no left-linking candidate and there remains no t- or z- candidate in W_u if R_u is an inner Subset;
- if R_u is a candidate, assert it by S and eliminate by ECP all the candidates for CSP variables after u, including those in the inner Subsets, that are incompatible with R_u in the current RS;
- if R_u is a g-candidate, it cannot be asserted; eliminate by W_1 all the candidates for CSP variables after u, including those in the inner Subsets, that are incompatible with R_u in the current RS;
- if R_u is an Sq_u-subset, it cannot be asserted by Sq_u; eliminate by Sq_u all the candidates for CSP variables after u, including those in the inner Subsets, that are incompatible with R_u in the current RS.

In the new RS thus obtained, excise from B' the part related to CSP variables and inner Subsets p to q-1 (included); if L_q has been eliminated in the passage from RS_2 to RS, replace it by C_q (and, if necessary, replace V_q by V_q'); for each integer $s \geq p$, decrease by q-p the index of CSP variable V_s, of its candidates and inner right-linking pattern (g-candidate or S_r-subset) and of the set W_s, in the new B'. In RS, B' is now, up to p (the ex q), a partial S_r-braid in S_rB_n with target Z.

d2) If, in RS, a candidate C_p in a right-linking Sq_p-subset R_p has been asserted or eliminated or marked in a previous step, R_p can no longer be used as such as a right-linking Subset of an S_r-braid, because it may no longer be a (conditional) Sq_p-subset. Moreover, there may be several such candidates in R_p; consider them all at once. Notice that candidates can only have been asserted as values in the transition

from RS_1 to RS_3 (the candidates asserted in case d1 are all excised from B') and that all the candidates for their CSP variables or in their transversal sets have also been eliminated in this transition. Delete from R_p the CSP variables and the transversal sets corresponding to these asserted candidates (as we do not have type-2 targets, there is the same number of each). Call R_p' what remains of R_p and replace R_p by R_p' in B'. A few more questions must be dealt with:
- is there still a candidate for one of the CSP variables of R_p' that could play the role of a left-linking candidate for R_p'? If not, R_p' has already become an autonomous Subset in RS_3; excise it from B', together with a whole part of B' after it, along the same lines as in case d1;
- is R_p' still linked to the next part of B'? If not, excise it from B', together with a whole part of B' after it, as in the previous case;
- R_p' may be degenerated (modulo Z and the previous R_k's); this can easily be fixed by replacing R_p' with the corresponding Reversible Subset Chain (modulo Z and the previous R_k's);
- R_p' or the Reversible Subset Chain (modulo Z and the previous R_k's) replacing it may have more targets than R_p; if any of these is a right-linking candidate or an element of a right-linking g-candidate or of an S_r-subset of B' for an index after p, then mark it so that the information can be used in cases d2, f1, f2 or f3 of later steps.

In RS, B' is now, up to p (the ex q), a partial S_r-braid in S_rB_n with target Z.

e) If, in RS, a left-linking candidate L_p has been eliminated (but not asserted) and CSP variable V_p has no t- or z- candidate in RS_2 (complementary to case b), we now have to consider three cases instead of the two we had for g-braids.

e1) If R_p is a candidate, then V_p has only one possible value, namely R_p; if R_p has not yet been asserted by S somewhere between RS_2 and RS, do it now; this case is now reducible to case d1 (because the assertion of R_p also entails the elimination of L_p); go back to case d1 for the same value of p (in order to prevent an infinite loop, mark this case as already dealt with for the current step).

e2) If R_p is a g-candidate, then R_p cannot be asserted by S; however, it can still be used, for any CSP variable after p, to eliminate by W_1 any of its t-candidates that is g-linked to R_p. Let q be the smallest number strictly greater than p such that, in RS, CSP variable V_q still has a left-linking, t- or z- candidate C_q that is not linked or g-linked or S-linked to any of the R_i for $p \leq i < q$. Replace RS by the state obtained after all the assertions and eliminations similar to those in case d1 above have been done. Then, in RS, excise the part of B' related to CSP variables p to q-1 (included), replace L_q by C_q (if L_q has been eliminated in the passage from RS_2 to RS) and re-number the posterior elements of B', as in case d1. In RS, B' is now, up to p (the ex q), a partial S_r-braid in S_rB_n with target Z.

e3) If R_p is an Sq_p-subset, then R_p is no longer linked via L_p to a previous right-linking element of the braid. If none of the CSP variables V_p' in W_p has a z- or t-candidate C_p that can be linked, g-linked or S-linked to Z or to a previous R_i, (situation complementary to case b), it means that the elimination of L_p has turned R_p into an unconditional Sq_p-subset. Let q be the smallest number strictly greater than p such that, in RS, CSP variable V_q has a left-linking, t- or z- candidate C_q that is not linked or g-linked or S-linked to any of the R_i for $p \leq i < q$. Replace RS by the state obtained after all the assertions and eliminations similar to those in case d1 above have been done. Then, in RS, excise the part of B' related to CSP variables p to q-1 (included), replace L_q by C_q (if L_q has been eliminated in the passage from RS_2 to RS) and re-number the posterior elements of B', as in case d1. In RS, B' is now, up to p (the ex q), a partial S_r-braid in S_rB_n with target Z.

f) Finally, consider eliminations occurring in a right-linking object R_p. This implies that p cannot be the last index of B'. There are three cases.

f1) If, in RS, right-linking candidate R_p of B has been eliminated (but not asserted) or marked, then replace B' by its initial part:
$\{L_1 R_1\} - \{L_2 R_2\} - .... - \{L_p .\}$. At this stage, B' is in RS a (possibly shorter) S_r-braid with target Z. Return B' and stop.

f2) If, in RS, a candidate in right-linking g-candidate R_p has been eliminated (but not asserted) or marked, then:

f2α) either there remains no unmarked candidate of R_p in RS; then replace B' by its initial part: $\{L_1 R_1\} - \{L_2 R_2\} - .... - \{L_p .\}$; at this stage, B' is in RS a (possibly shorter) S_r-braid with target Z; return B' and stop;

f2β) or the remaining unmarked candidates of R_p in RS still make a g-candidate and B' does not have to be changed;

f2γ) or there remains only one unmarked candidate C_p of R_p; replace R_p by C_p in B'. We must also prepare the next steps by putting marks. Any t-candidate of B that was g-linked to R_p, if it is still present in RS, can still be considered as a t-candidate in B', where it is now linked to C_p instead of g-linked to R_p; this does not raise any problem. However, this substitution may entail that candidates that were not t-candidates in B become t-candidates in B'; if they are left-linking candidates of B', this is not a problem either; but if any of them is a right-linking candidate or an element of a right-linking g-candidate or of an S_r-subset of B', then mark it so that the same procedure (i.e. f1, f2 or f3) can be applied to it in a later step.

f3) If, in RS, a candidate C_p in right-linking Sq_p-subset R_p has been eliminated (but not asserted) or marked, this has been dealt with in case d2.

Notice that, as was the case for braids and g-braids, this proof works only because the notions of being linked, g-linked or S-linked do not depend on the resolution state.

9.5. The "T&E(S_p) vs S_p-braids" theorem, $1 \leq p \leq \infty$

Any resolution theory T stable for confluence has the confluence property and the procedure T&E(T) can therefore be defined (see section 5.6.1). Taking $T = S_p$, it is obvious that any elimination done by an S_p-braid can be done by T&E(S_p). As was the case for braids and for g-braids, the converse is true:

Theorem 9.10: for any $1 \leq p \leq \infty$, any elimination done by T&E(S_p) can be done by an S_p-braid.

The proof is very similar to the g-braids case.

Proof: Let RS be a resolution state and let Z be a candidate eliminated by T&E(S_p, Z, RS) using some auxiliary resolution state RS'. Following the steps of resolution theory S_p in RS', we progressively build an S_p-braid in RS with target Z. First, remember that S_p contains only five types of rules: ECP (which eliminates candidates), W_1 (whips of length 1, which eliminates candidates), $S_{p'}$ (which eliminates targets of $S_{p'}$-subsets, $p' \leq p$), S (which asserts a value for a CSP variable) and CD (which detects a contradiction on a CSP variable).

Consider the sequence $(P_1, P_2, ..., P_k, ...P_m)$ of rule applications in RS' based on rules from S_p different from ECP and suppose that P_m is the first occurrence of CD (there must be at least one occurrence of CD if Z is eliminated by T&E(S_p, Z, RS)). We first define the R_k, V_k, W_k and q_k sequences, for $k < m$:
- if P_k is of type S, then it asserts a value R_k for some CSP variable V_k; let $W_k = \{V_k\}$ and $q_k = 1$;
- if P_k is of type whip[1]: $\{M_k .\} \Rightarrow \neg candidate(C_k)$ for some CSP variable V_k, then define R_k as the g-candidate of V_k that contains M_k and is g-linked to C_k; (notice that C_k will not necessarily be L_{k+1}); let $W_k = \{V_k\}$ and $q_k = 1$;
- if P_k is of type $S_{p'}$, then define R_k as the non degenerated $S_{p'}$-subset used by the condition part of P_k, as it appears at the time when P_k is applied; let W_k be the set of CSP variables of R_k and $q_k = p'$; in this case, V_k will be defined later.

We shall build an S_p-braid[n] in RS with target Z, with the R_k's as its sequence of right-linking candidates or g-candidates or S_{q_k}-subsets, with the W_k's as its sequence of sets of CSP variables, with the q_k's as its sequence of sizes and with $n = \Sigma_{1 \leq k \leq m} q_k$ (setting $q_m = 1$). We only have to define properly the L_k's, q_k's and V_k's with $V_k \in W_k$. We do this by recursion, successively for $k = 1$ to $k = m$. As the proofs for $k = 1$ and for the passage from k to $k+1$ are almost identical, we skip the case k = 1. Suppose we have done it until k and consider the set W_{k+1} of CSP variables.

Whatever rule P_{k+1} is (S or whip[1] or $S_{p'}$), the fact that it can be applied means that, apart from R_{k+1} (if it is a candidate) or the candidates contained in R_{k+1} (if it is a g-candidate or an $S_{p'}$-subset), all the other candidates for all the CSP variables in W_{k+1} that were still present in RS (and there must be at least one, say L_{k+1}, for some CSP variable $V_{k+1} \in W_{k+1}$) have been eliminated in RS' by the assertion of Z and the previous rule applications. But these previous eliminations can only result from being linked or g-linked or S-linked to Z or to some R_i, i≤k. The data L_{k+1}, R_{k+1} and $V_{k+1} \in W_{k+1}$ therefore define a legitimate extension for our partial S_p-braid.

End of the procedure: at step m, a contradiction is obtained by CD for a CSP variable V_m. It means that all the candidates for V_m that were still candidates for V_m in RS (and there must be at least one, say L_m) have been eliminated in RS' by the assertion of Z and the previous rule applications. But these previous eliminations can only result from being linked or g-linked or S-linked to Z or to some R_i, i<m. L_m is thus the last left-linking candidate of the S_p-braid we were looking for in RS and we can take $W_m=\{V_m\}$. qed.

Remarks:

– here again, this proof works only because the existence of a link, g-link or S_p-link between a candidate and a pattern does not depend on the resolution state;

– as in the previous cases of braids and g-braids, it is very unlikely that following the T&E(S_p) procedure to produce an S_p-braid, as in the construction in this proof, would produce the shortest available one in resolution state RS (and this intuition is confirmed by experience).

9.6. The scope of S_p-braids (in Sudoku)

The "T&E(S_p) vs S_p-braids" theorem can be used to estimate with simple calculations the scope of S_p-braids (which is also an upper boundary for the scope of S_p-whips) for any p, without having to find effectively the resolution paths.

Several times in this book, we mentioned that all the Sudoku puzzles in a set of random collections of about 10,000,000 minimal puzzles generated according to different methods (several independent implementations of the bottom-up, top-down and controlled-bias algorithms) could be solved by T&E or (equivalently) by braids (indeed, we also mentioned that they can all be solved by whips).

As a result, Sudoku puzzles that are not in the scope of braids are extremely rare and if one wants to compare the scopes of several types of more complex S_p-braids, one can only do this on a collection of exceptionally hard puzzles. The natural choice for this is gsf's highly non random, manually selected collection of 8152 puzzles; it contains puzzles of varying levels of difficulty, with a very strong bias for the hardest ones (and a tendency for repetition of similar puzzles); its top level part has long been considered as containing the hardest known puzzles.

Resolution theory → ↓ slice of puzzles	B_∞	gB_∞	S_2B_∞	S_3B_∞	S_4B_∞	$S_{4Fin}B_\infty$	$+x_2y_2$
1-500	0	187	336	414	443	466	489
500-1000	0	178	335	415	460	480	497
1001-1500	0	163	382	451	486	494	500
1501-2000	0	168	397	476	490	496	499
2001-2500	0	135	367	434	474	489	497
2501-3000	0	116	334	443	479	493	499
3001-3500	1	120	335	424	473	486	498
3501-4000	0	113	325	426	472	493	499
4001-4500	1	104	298	395	448	471	497
4501-5000	0	231	399	450	482	494	499
5001-5500	47	348	487	500			
5501-6000	434	490	500				
6001-6500	487	500					
6501-7000	494	500					
7001-8152	1152						
Total solved	2616	4505	6647	7480	7859	8014	8126
Total unsolved	5536	3647	1505	672	293	136	26

Table 9.1. Number of puzzles solved by S$_p$-braids, for each slice of 500 puzzles in gsf's list. Missing cells in a row are intended to make it easier to see when a slice is completely solved.

In Table 9.1[8], the first column defines the sets of puzzles under consideration: gsf's list is decomposed into slices of 500 puzzles each (but the last ones), starting from the top (i.e. from the hardest in his classification). The next columns show how many puzzles of each slice can be solved using the S_p-braids mentioned in the first row.

This table shows that almost all the hardest puzzles, known at the time when gsf's list was published, can be solved with braids (of unspecified total length) built on (Naked, Hidden and Super-Hidden) Subsets, on Finned Fish (a variant of Fish with additional candidates linked to the target) and on x2y2-belts[9].

However, "Eleven" recently reported he generated more than four million minimal puzzles that cannot be solved by T&E(S_4). He used a kind of genetic programming algorithm, filtered by T&E(S_4) and with evaluation function the widely used SER rating[10] of puzzles. The question of a resolution theory T such that T&E(T), or associated T-braids, could solve all the known Sudoku puzzles is thus more open (although these are extremely exceptional instances: four millions in approximately 2.5×10^{25} non essentially equivalent minimal puzzles, as estimated in section 6.3.2, is not much). More on this in section 11.3.

9.7. Examples

As examples of Reversible Subset Chains abound in the Sudoku web forums, we shall not give any here. They can be found under the names of Alternating Inference Chains (AICs) or Nice Loops, as explained in section 9.2.4. Still more examples of a very special case, ALS chains (chains of Almost Locked Sets), can also be found; they are AICs (in the broad sense we have given them here) restricted to rc-space.

9.7.1. $S_2W_5 \not\subset gW_5$: an S_2-whip[5] not subsumed by a g-whip[5] (+ $S_2W_5 \not\subset gW_{13}$)

The puzzle P in Figure 9.2 provides an example of an S_2-whip[5] that is not equivalent to a whip or a g-whip of same length. This puzzle has moderate complexity (though it is on the high side of the fuzzy boundary of puzzles solvable by humans): $W(P) = gW(P) = 13$.

---

[8] We first published these results on the late Sudoku Player's Forum, "Abominable T&E and Lovely Braids" thread, p. 3, October 2008.

[9] See our website for the definition of x_2y_2 belts; they are our formal interpretation of a pattern known as a "hidden-pairs loop" or "SK-loop". This extremely symmetric pattern originated in the famous EasterMonster puzzle created by "jpf" and was discovered by Steven Kurzhals; EasterMonster has long been considered as the hardest puzzle and it has given rise to many variants; it is over-represented in gsf's list.

[10] See the first note of chapter 6.

Figure 9.2 — left grid (puzzle P):

7		8			3			
		2	1					
5								
	4				8	2	6	
3			8					
		1				9	3	
	9	6						4
			7		5			

Figure 9.2 — right grid (solution):

7	2	8	9	4	6	3	1	5
9	3	4	2	5	1	6	7	8
5	1	6	7	3	8	2	4	9
1	4	7	5	9	3	8	2	6
3	6	9	4	8	2	1	5	7
8	5	2	1	6	7	4	9	3
2	9	3	6	1	5	7	8	4
4	8	1	3	7	9	5	6	2
6	7	5	8	2	4	9	3	1

Figure 9.2. *A puzzle P with W(P) = 13*

Figure 9.3 — candidate grid (resolution state RS_1):

	c1	c2	c3	c4	c5	c6	c7	c8	c9
r1	7	n1 n2 n6	8	n4 n5 n9	n4 n5 n6 n9	n4 n5 n6 n9	3	n1 n4 n5 n6	n1 n2 n5 n9
r2	n4 n6 n9	n3 n6	n3 n4 n6 n9	2	n3 n4 n5 n6 n9	1	n4 n6 n7 n9	n4 n5 n6 n7 n8	n5 n7 n8
r3	5	n1 n2 n3 n6	n1 n2 n3 n4 n6 n9	n7 n8	n3 n4 n6 n9	n7 n8	n1 n2 n4 n6	n1 n4 n6	n1 n2 n9
r4	n1 n9	4	n1 n5 n7 n9	n3 n5 n7 n9	n5 n7 n9	n3 n5 n7 n9	8	2	6
r5	3	n2 n6 n7	n2 n6 n7 n9	n4 n7 n9	8	n2 n4 n6 n9	n1 n4 n7	n1 n4 n5 n7	n1 n5 n7
r6	n2 n6 n8	n2 n5 n6 n7 n8	n2 n5 n6 n7	1	n2 n4 n5 n6 n7	n2 n4 n5 n6 n7	n4 n7	9	3
r7	n1 n2 n8	9	n1 n2 n3 n5 n7	6	n1 n2 n5	n2 n3 n5 n8	n1 n2 n7	n1 n3 n7 n8	4
r8	n1 n2 n4 n6 n8	n1 n2 n3 n6 n8	n1 n2 n3 n4 n6	n3 n4 n8 n9	7	n2 n3 n4 n8 n9	5	n1 n3 n8	n1 n2 n6 n8 n9
r9	n1 n2 n4 n6 n8	n1 n5 n6 n7 n8	n1 n2 n3 n4 n5 n6 n7	n3 n4 n5 n8 n9	n1 n2 n4 n5 n9	n2 n3 n4 n5 n8 n9	n1 n2 n4 n5 n8 n9	n1 n3 n6 n7 n9	n1 n2 n6 n7 n8 n9

Figure 9.3. *Resolution state RS_1 of puzzle P in Figure 9.2*

The first (easy) steps of the resolution paths with whips or g-whips are identical.

***** SudoRules version 15b.1.12-W *****
20 givens, 267 candidates and 2000 nrc-links
whip[1]: r4n1{c1 .} ==> r5c3 ≠ 1, r5c2 ≠ 1

whip[1]: r2n8{c8 .} ==> r3c9 ≠ 8, r3c8 ≠ 8
whip[1]: r2n7{c7 .} ==> r3c9 ≠ 7, r3c8 ≠ 7, r3c7 ≠ 7
whip[1]: b6n5{r5c9 .} ==> r5c2 ≠ 5, r5c3 ≠ 5, r5c4 ≠ 5, r5c6 ≠ 5
whip[2]: b2n7{r3c4 r3c6} - b2n8{r3c6 .} ==> r3c4 ≠ 4, r3c4 ≠ 3
whip[2]: b2n8{r3c4 r3c6} - b2n7{r3c6 .} ==> r3c4 ≠ 9
whip[2]: b2n7{r3c6 r3c4} - b2n8{r3c4 .} ==> r3c6 ≠ 6, r3c6 ≠ 4, r3c6 ≠ 3
whip[1]: b2n3{r2c5 .} ==> r9c5 ≠ 3, r7c5 ≠ 3, r4c5 ≠ 3
whip[2]: b2n8{r3c6 r3c4} - b2n7{r3c4 .} ==> r3c6 ≠ 9
whip[3]: c1n2{r9 r6} - b4n8{r6c1 r6c2} - c2n5{r6 .} ==> r9c2 ≠ 2
whip[4]: b3n8{r2c9 r2c8} - r2n5{c8 c5} - r4c5{n5 n9} - c1n9{r4 .} ==> r2c9 ≠ 9
whip[5]: r4n7{c6 c3} - b4n1{r4c3 r4c1} - b4n9{r4c1 r5c3} - r5n6{c3 c2} - r5n2{c2 .} ==> r5c6 ≠ 7

The resolution state RS$_1$ reached at this point is displayed in Figure 9.3.

After RS$_1$, both resolution paths with whips or g-whips continue with a whip[6] and a whip[8]:

whip[6]: c2n7{r5 r9} - c9n7{r9 r2} - b3n8{r2c9 r2c8} - r2n5{c8 c5} - r4c5{n5 n9} - r5n9{c6 .} ==> r5c3 ≠ 7
whip[8]: c1n2{r7 r6} - b4n8{r6c1 r6c2} - c2n5{r6 r9} - c2n7{r9 r5} - b6n7{r5c7 r6c7} - r7c7{n7 n1} - b8n1{r7c5 r9c5} - c5n2{r9 .} ==> r7c3 ≠ 2

After these two whips, the two resolution paths diverge (one has either a whip[12] or a g-whip[8]), but they finally both give a rating of 13: W(P)=gW(P)=13. As they have nothing noticeable, we skip them.

What is interesting in the context of this chapter is that, in state RS$_1$, there appears a shorter pattern than those provided by whips or g-whips, a S$_2$-whip[5] (Although n7 appears on row r2 in column c7, i.e. outside the two cells of the hidden pair, n7r2c7 is a z-candidate and can be "forgotten": we have a hidden pair modulo the target.):

S$_2$-whip[5]: c1n9{r2 r4} - r4c5{n9 n5} - r2{c5n5 HP:[c8 c9][n7 n8]} –r2n7{c8 .} => r2c7 ≠ 9

Indeed, the two full resolution paths with whips and g-whips show more: as the elimination r2c7≠9 appears only after a whip[13], it shows that the above S$_2$-whip[5] cannot be replaced by a g-whip, even longer, with length less than 13.

10. g-Subsets, Reversible g-Subset Chains, gS_p-whips and gS_p-braids

This chapter extends the definitions and results of chapters 8 and 9 by allowing the basic elements of Subsets (the "intersections" between the CSP variables and the transversal sets) to be g-candidates instead of candidates. While gS_p-subsets are an extension of S_p-subsets in which g-transversal sets of candidates or g-candidates replace transversal sets of candidates, gS_p-whips (respectively gS_p-braids) are an extension of S_p-whips (resp. S_p-braids) in which gS_p-subsets replace S_p-subsets. The situation is similar to that in chapter 7, when we extended all the definitions and results from whips (resp. braids) to g-whips (resp. g-braids). For this reason and the following additional ones, we shall give precise definitions and theorems (at the risk of some apparent redundancy) but we shall be rather sketchy for their proofs:

– the first two parts of this chapter strictly parallel chapters 8 and 9 respectively;

– it seems that exploiting all the possibilities of these new g-Subsets is rather difficult in practice; in Sudoku, g-Subsets appear either as "Franken Fish" or as "Mutant Fish" (see section 10.1.6); as far as we know, these exotic Fish patterns have never before been considered as g-Subsets, i.e. as the "grouped" version of Subsets (for the reason that Subsets themselves have never been considered in the full generality allowed by the CSP point of view developed in chapter 8);

– our personal opinion is that, most of the time, it is often easier to find and understand a solution with whips or g-whips, when it exists (see the subsumption results in section 10.1.5), than with such patterns; but we acknowledge that some Sudoku Fishermen may have a different opinion; the main advantage of a g-Subset is that, like a Subset, it often allows several eliminations at once (see section 10.3);

– although this is not a problem for their general theory, finding explicitly all the non degenerated subcases of gS_p-subsets is very difficult for $p>3$;

– as for the gS_p-whips and gS_p-braids obtained by allowing these new subsets as right-linking objects, even if their theories can easily be developed in a strict parallel to those of S_p-whips and S_p-braids (as shown in section 10.2), they seem to be rather complex structures; in Sudoku, they include the "Fishy Cycles" (which are already "almost" subsumed by the simpler S_p-whips and S_p-braids).

For a better understanding of the concepts involved, it may be a good idea to read the detailed example in section 10.3 in parallel with the first two sections.

10.1. g-Subsets

10.1.1. g-transversality, gS$_p$-labels and gS$_p$-links

In the same way as, in chapters 7 and 8, we had to introduce a distinction between g-labels or S$_p$-labels (defined as maximal sets of labels) and g-candidates or S$_p$-subsets (that did not have to be maximal), we must now introduce a distinction between gS$_p$-labels, which can only refer to CSP variables and transversal sets of labels and g-labels (which can be considered as a kind of maximality condition on gS$_p$-labels), and gS$_p$-subsets, in which considerations about mandatory and optional candidates or g-candidates will appear.

10.1.1.1. Set of labels and g-labels g-transversal to a set of disjoint CSP variables

Definition: for p>1, given a set of p different CSP variables $\{V_1, V_2, ..., V_p\}$, we say that a set S of at most p different labels and g-labels is *g-transversal* with respect to $\{V_1, V_2, ..., V_p\}$ for constraint c if:

− none of the labels in S or contained in a g-label in S has a representative for two of these CSP variables;

− all the labels in S or contained in a g-label in S are pairwise linked by some constraint;

− all the labels in S are pairwise linked by constraint c;

− each g-label in S contains a "distinguished" label that is linked by constraint c to all the labels in S and to all the other distinguished labels of all the g-labels of S;

− S is maximal, in the sense that no label or g-label pertaining to one of these CSP variables could be added to it without contradicting the first two conditions.

Remarks:

− as in the definition of transversal sets, the first condition will always be true for pairwise strongly disjoint CSP variables, but, for the same reasons as before, we do not take this as a necessary condition;

− conditions 2, 3 and 4 together express that constraint c plays a role for the whole transversal set; forgetting the idea of such a global constraint and adopting only condition 2 would not change the general theory developed in this chapter (but the totality of the second remark in section 8.1.1 after the definition of a transversal set applies here also);

− conversely, one could imagine replacing conditions 2, 3 and 4 by the stronger one: all the labels in S or contained in a g-label in S are pairwise linked by constraint c; in Sudoku, it is obvious that this would not change anything (because all the g-labels involve blocks); but for the general CSP, this may be too restrictive.

10.1.1.2. gS$_p$-labels and gS$_p$-links

Definitions: for any integer p>1, a *gS$_p$-label* is a couple of data: {CSPVars, TransvSets}, where CSPVars is a set of p different CSP variables and TransvSets is a set of p different g-transversal sets of labels and g-labels for these variables (each one for a well defined constraint). A gS-label is a gS$_p$-label for some p >1.

Definition: a label l is *gS$_p$-linked* or simply *gS-linked* to a gS$_p$-label S = {CSPVars, TransvSets} if there is some k, 1≤k≤p, such that:

– l is linked or g-linked to all the labels and g-labels in the k-th element TransvSets$_k$ of TransvSets,

– l is linked by the constraint c$_k$ of TransvSets$_k$ to all the labels and all the distinguished labels contained in all the g-labels in TransvSets$_k$.

In these conditions, l is also called *a potential target of the gS$_p$-label.*

Definition: Two sets of labels or g-labels are said to be "strongly transitively disjoint" if no label appearing in one of them (even as an element of a g-label) can appear in the other (even as an element of a g-label). This is much stronger than saying that these two sets are disjoint (in the usual set-theoretic sense of "disjoint"); if these sets are only disjoint, two g-labels, one in each of the sets, can be different but share a label (in Sudoku, <Xr1n1, r1n1c123> and <Xc1n1, c1n1r123> share the label with representative <r1c1, n1>); or a label in one set can be contained in a g-label in the other (e.g. <r1n1, c1> in <Xc1n1, c1n1r123>).

Definition: In a resolution state RS, two sets of candidates or g-candidates are said to be "transitively disjoint" if no candidate effectively present in one of them (even as an element of a g-candidate) is effectively present in the other (even as a candidate in a g-candidate); again this is much stronger than saying that these two sets are disjoint. However, this is weaker than saying that the sets obtained by considering the labels and g-labels underlying the candidates and g-candidates in these two sets are strongly transitively disjoint.

Miscellaneous remarks about gS$_p$-labels:

– with the above definition of a gS$_p$-label, a label and a g-label are not gS$_p$-labels (due to the condition p>1); for labels, this is a mere matter of convention, but this choice is more convenient for the sequel;

– different transversal sets in a gS$_p$-label are not required to be pairwise transitively disjoint, let alone pairwise disjoint; such conditions will appear only in the definitions of g-Subsets and only with respect to candidates (not labels);

– a gS$_p$-label corresponds to the maximal extent of a possible gS$_p$-subset (as defined below).

Notation: in the definition of g-Subsets, as in the case of Subsets, we shall need a means of specifying that, in some g-transversal sets, some labels or g-labels must exist while others may exist or not. We shall write this as e.g. $\{<V_1, v_1>, <V_2, v_2>, ..., (<V_k, v_k>), ...\}$. This should be understood as follows: a label or g-label not surrounded with parentheses must exist; a "label" or "g-label" surrounded with parentheses may exist or not; if it exists, then it is named $<V_k, v_k>$.

10.1.2. g-Pairs

Definition: in any resolution state RS of any CSP, a *g-Pair* (or *gS_2-subset*) is a gS_2-label {CSPVars, TransvSets}, where:

– CSPVars = $\{V_1, V_2\}$,

– TransvSets is composed of the following g-transversal sets of labels and g-labels:

$\{<V_1, v_{11}>, <V_2, v_{21}>\}$ for constraint c_1,

$\{<V_1, v_{12}>, <V_2, v_{22}>\}$ for constraint c_2,

such that:

– in RS, V_1 and V_2 are disjoint, i.e. they share no candidate;

– in RS, $\{<V_1, v_{11}>\}$ and $\{<V_1, v_{12}>\}$ are transitively disjoint; $\{<V_2, v_{22}>\}$ and $\{<V_2, v_{21}>\}$ are transitively disjoint;

– in RS, V_1 has the two mandatory candidates or g-candidates $<V_1, v_{11}>$ and $<V_1, v_{12}>$ and no other candidate or g-candidate;

– in RS, V_2 has the two mandatory candidates or g-candidates $<V_2, v_{21}>$ and $<V_2, v_{22}>$ and no other candidate or g-candidate.

A *target of a g-Pair* is a candidate gS_2-linked to the underlying gS_2-label.

Theorem 10.1 (gS_2 rule): in any CSP, a target of a g-Pair can be eliminated.

Proof: as the two g-transversal sets play similar roles, we can suppose that Z is linked or g-linked to $<V_1, v_{11}>$ and $<V_2, v_{21}>$. If Z was True, these candidates or all the candidates these g-candidates contain would be eliminated by ECP. As V_1 and V_2 have only two candidates or g-candidates each, their other candidate or g-candidate ($<V_1, v_{12}>$, respectively $<V_2, v_{22}>$) would be or would contain their real value, which is impossible, as both are linked or g-linked. Here again, the proof works only because V_1 and V_2 share no candidate in RS (and in no posterior resolution state).

10.1.3. g-Triplets

There may be several formulations of g-Triplets. Here again, as in the case of ordinary Triplets, we adopt one that is neither too restrictive (the presence of some of the candidates or g-candidates potentially involved is not mandatory) nor too comprehensive (i.e., by making mandatory the presence of some of the candidates or g-candidates involved, it does not allow degenerated cases).

Definition: in any resolution state RS of any CSP, a *g-Triplet* (or *gS$_3$-subset*) is a gS$_3$-label {CSPVars, TransvSets}, where:

– CSPVars = {V$_1$, V$_2$, V$_3$},

– TransvSets is composed of the following g-transversal sets of labels and g-labels:

   – {<V$_1$, v$_{11}$>, (<V$_2$, v$_{21}$>), <V$_3$, v$_{31}$>} for constraint c$_1$,

   – {<V$_1$, v$_{12}$>, <V$_2$, v$_{22}$>, (<V$_3$, v$_{32}$>)} for constraint c$_2$,

   – {(<V$_1$, v$_{13}$>), <V$_2$, v$_{23}$>, <V$_3$, v$_{33}$>} for constraint c$_3$,

such that:

– in RS, V$_1$, V$_2$ and V$_3$ are pairwise disjoint;

– in RS, {<V$_1$, v$_{11}$>} and {<V$_1$, v$_{12}$>} are transitively disjoint; {<V$_2$, v$_{22}$>} and {<V$_2$, v$_{23}$>} are transitively disjoint; {<V$_3$, v$_{33}$>} and {<V$_3$, v$_{31}$>} are transitively disjoint;

– in RS, V$_1$ has the two mandatory candidates or g-candidates <V$_1$, v$_{11}$> and <V$_1$, v$_{12}$>, one optional candidate or g-candidate <V$_1$, v$_{13}$> (supposing this label or g-label exists) and no other candidate or g-candidate;

– in RS, V$_2$ has the two mandatory candidates or g-candidates <V$_2$, v$_{22}$> and <V$_2$, v$_{23}$>, one optional candidate or g-candidate <V$_2$, v$_{21}$> (supposing this label or g-label exists) and no other candidate or g-candidate;

– in RS, V$_3$ has the two mandatory candidates or g-candidates <V$_3$, v$_{33}$> and <V$_3$, v$_{31}$>, one optional candidate or g-candidate <V$_3$, v$_{32}$> (supposing this label or g-label exists) and no other candidate or g-candidate.

A *target of a g-Triplet* is defined as a candidate gS$_3$-linked to the underlying gS$_3$-label.

Theorem 10.2 (gS$_3$ rule): in any CSP, a target of a g-Triplet can be eliminated.

Proof: as the three g-transversal sets play similar roles, we can suppose that Z is gS$_3$-linked to the first, i.e. linked or g-linked to <V$_1$, v$_{11}$>, <V$_2$, v$_{21}$> and <V$_3$, v$_{31}$> if it exists. If Z was True, these candidates or all the candidates these g-candidates contain (if they are present) would be eliminated by ECP. Each of V$_1$, V$_2$ and V$_3$ would have at most two candidates or g-candidates left. Any choice for V$_1$ would reduce to at most one the number of possibilities (in terms of candidates and g-candidates) for each of V$_2$ and V$_3$ (due to the pairwise contradictions between members of each g-transversal sets). Finally, the unique choice for V$_2$ (still in terms of candidates and g-candidates), if any, would in turn reduce to zero the number of possibilities for V$_3$.

10.1.4. g-Quads

Finding the proper formulation for g-Quads, guaranteeing that it covers no degenerated case, is less obvious than for g-Triplets. Borrowing to the Quads case, we consider two types of g-Quads: Cyclic and Special, and we choose to write the Special g-Quad in such a way that it is does not cover any case already covered by the Cyclic g-Quad.

Definition: in any resolution state RS of any CSP, a *Cyclic g-Quad* (or *Cyclic gS_4-subset*) is a gS_4-label {CSPVars, TransvSets}, where:

– CSPVars = $\{V_1, V_2, V_3, V_4\}$,

– TransvSets is composed of the following g-transversal sets of labels and g-labels:

– $\{<V_1, v_{11}>, (<V_2, v_{21}>), (<V_3, v_{31}>), <V_4, v_{41}>\}$ for constraint c_1,

– $\{<V_1, v_{12}>, <V_2, v_{22}>, (<V_3, v_{32}>), (<V_4, v_{42}>)\}$ for constraint c_2,

– $\{(<V_1, v_{13}>), <V_2, v_{23}>, <V_3, v_{33}>, (<V_4, v_{43}>)\}$ for constraint c_3,

– $\{(<V_1, v_{14}>), (<V_2, v_{24}>), <V_3, v_{34}>, <V_4, v_{44}>\}$ for constraint c_4,

such that:

– in RS, V_1, V_2, V_3 and V_4 are pairwise disjoint, i.e. no two of these variables share a candidate;

– in RS, $\{<V_1, v_{11}>\}$ and $\{<V_1, v_{12}>\}$ are transitively disjoint; $\{<V_2, v_{22}>\}$ and $\{<V_2, v_{23}>\}$ are transitively disjoint; $\{<V_3, v_{33}>\}$ and $\{<V_3, v_{34}>\}$ are transitively disjoint; $\{<V_4, v_{44}>\}$ and $\{<V_4, v_{41}>\}$ are transitively disjoint;

– in RS, V_1 has the two mandatory candidates or g-candidates $<V_1, v_{11}>$ and $<V_1, v_{12}>$, two optional candidates or g-candidates $<V_1, v_{13}>$ and $<V_1, v_{14}>$ (supposing any of these labels exists) and no other candidate or g-candidate;

– in RS, V_2 has the two mandatory candidates or g-candidates $<V_2, v_{22}>$ and $<V_2, v_{23}>$, two optional candidates or g-candidates $<V_2, v_{24}>$ and $<V_2, v_{21}>$ (supposing any of these labels exists) and no other candidate or g-candidate;

– in RS, V_3 has the two mandatory candidates or g-candidates $<V_3, v_{33}>$ and $<V_3, v_{34}>$, two optional candidates or g-candidates $<V_3, v_{31}>$ and $<V_3, v_{32}>$ (supposing any of these labels exists) and no other candidate or g-candidate;

– in RS, V_4 has the two mandatory candidates or g-candidates $<V_4, v_{44}>$ and $<V_4, v_{41}>$, two optional candidates or g-candidates $<V_4, v_{42}>$ and $<V_4, v_{43}>$ (supposing any of these labels exists) and no other candidate or g-candidate.

Definition: in any resolution state RS of any CSP, a *Special g-Quad* (or *Special gS_4-subset*) is a gS_4-label {CSPVars, TransvSets}, where:

– CSPVars = $\{V_1, V_2, V_3, V_4\}$,

– TransvSets is composed of the following transversal sets of labels and g-labels:

$\{<V_1, v_{11}>, <V_2, v_{21}>, <V_3, v_{31}>, (<V_4, v_{41}>\})$ for constraint c_1,
$\{<V_1, v_{12}>, (<V_2, v_{22}>), (<V_3, v_{32}>), <V_4, v_{42}>\}$ for constraint c_2,
$\{(<V_1, v_{13}>), <V_2, v_{23}>, (<V_3, v_{33}>), <V_4, v_{43}>\}$ for constraint c_3,
$\{(<V_1, v_{14}>), (<V_2, v_{24}>), (<V_3, v_{34}>), <V_4, v_{44}>\}$ for constraint c_4,

such that:

– in RS, V_1, V_2, V_3 and V_4 are pairwise disjoint, i.e. no two of these variables share a candidate;

– in RS, $<V_1, v_{11}>$ and $<V_1, v_{12}>$ are transitively disjoint; $<V_2, v_{21}>$ and $<V_2, v_{23}>$ are transitively disjoint; $<V_3, v_{31}>$ and $<V_3, v_{34}>$ are transitively disjoint; moreover, $<V_4, v_{42}>$, $<V_4, v_{43}>$ and $<V_4, v_{44}>$ are pairwise transitively disjoint;

– in RS, V_1 has the two mandatory candidates or g-candidates $<V_1, v_{11}>$ and $<V_1, v_{12}>$ and no other candidate or g-candidate;

– in RS, V_2 has the two mandatory candidates or g-candidates $<V_2, v_{21}>$ and $<V_2, v_{23}>$ and $<V_1, v_{12}>$ and no other candidate or g-candidate;

– in RS, V_3 has the two mandatory candidates or g-candidates $<V_3, v_{31}>$ and $<V_3, v_{34}>$ and $<V_1, v_{12}>$ and no other candidate or g-candidate;

– in RS, V_4 has the three mandatory candidates or g-candidates $<V_4, v_{42}>$, $<V_4, v_{43}>$ and $<V_4, v_{44}>$ and no other candidate or g-candidate.

In both cases, a *target of a g-Quad* is defined as a candidate gS_4-linked to the underlying gS_4-label.

Theorem 10.3 (gS_4 rule): in any CSP, a target of a Cyclic or Special g-Quad can be eliminated.

Proof for the cyclic case: as the four g-transversal sets play similar roles, we can suppose that Z is linked or g-linked to $<V_1, v_{11}>$, $<V_2, v_{21}>$, $<V_3, v_{31}>$ if it exists and $<V_4, v_{41}>$ if it exists. If Z was True, these candidates or all the candidates these g-candidates contain (if they are present) would be eliminated by ECP. Each of V_1, V_2, V_3 and V_4 would have at most three candidates or g-candidates left. Any choice for V_1 would reduce to at most two the number of possibilities (in terms of candidates and g-candidates) for V_2, V_3 and V_4. Any further choice among the remaining candidates or g-candidates for V_2 would reduce to at most one the number of possibilities (still in terms of candidates and g-candidates) for V_3 and V_4. Finally the unique choice left for V_3 (still in terms of candidates and g-candidates), if any, would reduce to zero the number of possibilities for V_4.

Proof for the special case: there are four subcases (the last two of which are similar to the second):
- suppose Z is linked or g-linked to both $<V_1, v_{11}>$, $<V_2, v_{21}>$, $<V_3, v_{31}>$ and $<V_4, v_{41}>$ if it exists. If Z was True, these candidates or all the candidates these g-

candidates contain (if they are present) would be eliminated by ECP. Each of V_1, V_2, V_3, would have only one candidate or g-candidate left; choosing these candidates or any candidate in these g-candidates as their respective values would reduce to zero the number of possibilities for V_4.

- suppose Z is linked to both $<V_1, v_{12}>$, $<V_2, v_{22}>$ if it exists, $<V_3, v_{32}>$ and $<V_4, v_{42}>$ if it exists. If Z was True, $<V_1, v_{12}>$ and $<V_4, v_{42}>$ or all the candidates they contain would be eliminated by ECP; $<V_1, v_{11}>$ would then be True, which would eliminate $<V_2, v_{21}>$ and $<V_3, v_{31}>$ or all the candidates they contain. Then $<V_2, v_{23}>$ and $<V_3, v_{34}>$ would be True. This would leave no possibility for V_4.

If we wanted to introduce larger g-Subsets, it would get harder and harder to write separate formulæ guaranteeing non-degeneracy of each subcase. We leave this as an exercise for the reader. Contrary to Subsets, in the 9×9 Sudoku case, there can be g-Subsets of size larger than 4: see the example of a Franken Squirmbag (size 5) in section 10.3.

10.1.5. gS$_p$-subset theories and confluence

All of section 8.6 can be transposed and extended from S_p-subsets to gS_p-subsets: definition of the gS_p resolution theories, proof of their stability for confluence, definition of the $gW_p + gS_p$ and $gB_p + gS_p$ resolution theories (in which g-whips[p] or g-braids[p] are added to gS_p-subsets) and proof of their stability for confluence.

10.1.6. Subsumption results for g-Subsets

10.1.6.1. g-Pairs

Theorem 10.4: $gS_2 \subseteq gW_2$ (g-whips of length 2 subsume all the g-Pairs).

Proof: keeping the notations of theorem 10.1 and considering a target Z of the g-Pair that is gS_2-linked to the first g-transversal set, i.e. linked or g-linked to both $<V_1, v_{11}>$ and $<V_2, v_{21}>$, the following g-whip[2] eliminates Z:
g-whip[2]: $V_1\{v^*_{11} \ v_{12}\} - V_2\{v^*_{22} \ .\} \Rightarrow \neg candidate(Z)$,
where v^*_{11} [respectively v^*_{22}] is v_{11} [resp. v_{22}] if $<V_1, v_{11}>$ [resp. $<V_2, v_{22}>$] is a candidate and it is any element chosen in v_{11} [resp. v_{22}] if $<V_1, v_{11}>$ [resp. $<V_2, v_{22}>$] is a g-candidate. In case $<V_1, v_{11}>$ is a g-candidate, the candidates in $<V_1, v_{11}>$ other than $<V_1, v^*_{11}>$ are z-candidates in the whip[2]; in case $<V_2, v_{22}>$ is a g-candidate, the candidates in $<V_2, v_{22}>$ other than $<V_2, v^*_{22}>$ are t-candidates in the whip[2].

The converse of the above theorem is false: $gW_2 \not\subseteq gS_2$; indeed, $W_2 \not\subseteq gS_2$. For a deep understanding of both whips and g-Subsets, this is as interesting as the theorem itself. Using the example in section 8.8.1, we have concluded in section 8.7.1 that $W_2 \not\subseteq S_2$. But the same example can now be used to show that $W_2 \not\subseteq gS_2$.

The three whips[2] defined in section 8.8.1 can be considered no more as g-Pairs than as Pairs.

It is nevertheless instructive to understand how a tentative proof of the inclusion gW$_2$ ⊆ gS$_2$ would fail. We would have to proceed as follows. Let
g-whip[2]: $V_1\{v_{11}\ v_{12}\} - V_2\{v_{22}\ .\} \Rightarrow \neg$candidate(Z), be a g-whip[2] with target Z, associated with CSP variables (V_1, V_2). Consider all the possible candidates or g-candidates for each of these variables.

For V_1, they can only be:

– $<V_1, v'_{11}>$ = the candidate or g-candidate consisting of $<V_1, v_{11}>$ and all the candidates for V_1 linked to Z;

– and $<V_1, v'_{12}>$ = $<V_1, v_{12}>$ (a candidate or a g-candidate, with no element linked to Z).

For V_2, they can only be:

– $<V_2, v'_{22}>$ = the candidate or g-candidate consisting of $<V_2, v_{22}>$ and all the candidates for V_2 linked to $<V_1, v_{12}>$;

– and $<V_2, v'_{21}>$ = the candidate or g-candidate consisting of all the candidates for V_2 linked to Z but not to $<V_1, v_{12}>$.

We have thus built a g-transversal set $\{<V_1, v'_{12}>, <V_2, v'_{22}>\}$, but $\{<V_1, v'_{11}>, <V_2, v'_{21}>\}$ may not be a g-transversal set: the target Z is linked to these two candidates or g-candidates that may not be linked together by any constraint. This is exactly the situation with the three whips[2] in the example of section 8.8.1.

10.1.6.2. g-Triplets

Theorem 10.5: gW$_3$ subsumes "almost all" the g-Triplets.

Proof: keeping the notations of theorem 10.3 and considering a target Z of the g-Triplet that is gS-linked to the first g-transversal set (the three of them play similar roles), the following g-whip eliminates Z in any CSP:
g-whip[3]: $V_1\{v^*_{11}\ v_{12}\} - V_2\{v^*_{22}\ v_{23}\} - V_3\{v^*_{33}\ .\} \Rightarrow \neg$candidate(Z),
provided that $<V_1, v_{13}>$ is not a candidate or a g-candidate for V_1.
Here, v^*_{11} [respectively v^*_{22}, v^*_{33}] is v_{11} [resp. v_{22}, v_{33}] if $<V_1, v_{11}>$ [resp. $<V_2, v_{22}>$, $<V_3, v_{33}>$] is a candidate and it is any element chosen in v_{11} [resp. v_{22}, v_{33}] if $<V_1, v_{11}>$ [resp. $<V_2, v_{22}>$, $<V_3, v_{33}>$] is a g-candidate.
The optional candidates and the elements of the optional g-candidates of the g-Triplet appear in the g-whip as z- or t- candidates.

Considering that, in the above situation, the three CSP variables play symmetrical roles, there is only one case of a g-Triplet elimination that cannot be replaced by a g-whip[3] elimination. It occurs when the optional candidates or g-candidates for variables V_1, V_2 and V_3 in the g-transversal set to which the target is gS-linked correspond to existing labels or g-labels and are all effectively present in the resolution state.

10.1.6.3. g-Quads

Theorem 10.6: gW_4 subsumes "almost all" the Cyclic g-Quads.

Keeping the notations of theorem 10.5, the following g-whip eliminates a target Z of the Cyclic g-Quad in any CSP:

g-whip[4]: $V_1\{v^*_{11} \quad v_{12}\} \quad - \quad V_2\{v^*_{22} \quad v_{23}\} \quad - \quad V_3\{v^*_{33} \quad v_{34}\} \quad - \quad V_4\{v^*_{41} \quad .\} \quad \Rightarrow$ ¬candidate(Z),

provided that $<V_1, v_{13}>$ and $<V_1, v_{14}>$ are not candidates for V_1 and $<V_2, v_{23}>$ is not a candidate for V_2,

with the v^*_{xy} defined as before.

The optional candidates and the elements of the optional g-candidates of the g-Quad appear in the g-whip as z- or t- candidates.

Theorem 10.7: gB_4 subsumes all the Special g-Quads.

Keeping the notations of theorem 10.5, let Z be a target of the Special g-Quad:

- if Z is gS_4-linked to the first g-transversal set, the following g-braid eliminates Z:

g-braid[4]: $V_1\{v^*_{11} \quad v_{12}\} \quad - \quad V_2\{v^*_{21} \quad v_{23}\} \quad - \quad V_3\{v^*_{31} \quad v_{34}\} \quad - \quad V_4\{v^*_{44} \quad .\} \quad \Rightarrow$ ¬candidate(Z),

in which the first three left-linking candidates are linked to Z;

- if Z is gS_4-linked to another g-transversal set, say the second, the following g-whip eliminates Z:

g-whip[4]: $V_1\{v^*_{12} \quad v_{11}\} \quad - \quad V_2\{v^*_{21} \quad v_{23}\} \quad - \quad V_4\{v^*_{43} \quad v_{44}\} \quad - \quad V_3\{v^*_{34} \quad .\} \quad \Rightarrow$ ¬candidate(Z),

in which candidate $<V_4, v_{42}>$ appears as a z-candidate for the third CSP variable.

10.1.7. g-Subsets in Sudoku

Although the concept of a g-Subset has never been considered as such in Sudoku, the point we want to make here is that g-Subsets have been in existence for a long time, under other names: they appear as the "Franken Fish" and "Mutant Fish" patterns.

The difference between the two kinds depends on the specific geometry of Sudoku and is of little interest for the general theory developed here. Let us therefore mention it quickly, transposed into the vocabulary of this book. For a given number n°, a standard Fish in rows [respectively in columns] (of size p) uses only p different Xrn° [resp. Xcn°] CSP variables and p different transversal sets defined by Xcn° [resp. Xrn°] constraints. A Franken Fish in rows (of size p) is defined as an extended Fish (Super-Hidden Subset) pattern of size p in which either some of the p CSP variables are of type Xbn° instead of Xrn° [resp. Xcn°] *or* some of the p transversal sets are defined by Xbn° constraints instead of Xcn° [resp. Xrn°] constraints. In a Mutant Fish, blocks may appear in both CSP variables (i.e. there may be Xbn° CSP variables) and in constraints defining the transversal sets, which makes them much more complex than Franken Fish.

For more details and for examples, see sudopedia.org. For a (maybe not exhaustive) review of the various possibilities, we direct the reader to the specialised forums, where he will find that there is a handful of people who consecrate their time to studying and naming them (together with their "finned", "sushi", "sashimi" and other extensions). There is also a recent free java Sudoku solver, specialised in Fish: Hodoku – as far as we know, the only solver implementing (almost) all the known possibilities. See also the detailed example in section 10.3 below.

10.2. Reversible gS$_p$ Chains, gS$_p$-whips and gS$_p$-braids

When we try to apply the zt-ing principle to g-Subsets, everything goes for gS$_p$-whips as for S$_p$-whips. Here again, when it comes to defining the concepts of gS$_p$-links and gS$_p$-compatibility, we always consider the gS$_p$-labels underlying the gS$_p$-subsets instead of the gS$_p$-subsets themselves, in exactly the same way as we considered the full S$_p$-labels underlying the S$_p$-subsets when we defined S$_p$-links. The main reason for this choice is the same as in the S$_p$-links case: we want all the notions related to linking and compatibility to be purely structural (see chapter 9 for more detail).

10.2.1. gS$_p$-links; gS$_p$-subsets modulo other g-Subsets; gS$_p$-regular sequences

10.2.1.1. gS$_p$-links, gS$_p$-compatibility

Definition: a label l is *compatible with a gS$_p$-label* S if l is not gS$_p$-linked to S (i.e. if, for each g-transversal set TS of S, there is at least one label or g-label l' in TS such that l is not linked or g-linked to l').

Definition: a label l is *compatible* with a set R of labels, g-labels, S-labels and gS-labels if l is compatible with each element of R (in the senses of "compatible" already defined separately for labels, g-labels, S$_p$-labels and gS$_p$-labels).

Definitions: a label l is *gS$_p$-linked to a gS$_p$-subset* S if l is gS$_p$-linked to the gS$_p$-label underlying S; a label l is compatible with a gS$_p$-subset if l is not gS$_p$-linked to it; a label l is *compatible* with a set R of candidates, g-candidates, Subsets and g-Subsets if l is compatible with each element of R (in the senses of "compatible" already defined separately for candidates, g-candidates, S$_p$-subsets and gS$_p$-subsets).

Notice that, in conformance with what we mentioned in the introduction to section 10.2, according to the definition of "gS$_p$-linked to a gS$_p$-subset", it is not enough for label l to be linked or g-linked to all the actual candidates and g-candidates of one of its transversal sets: it must be linked or g-linked to all the labels and g-labels of one of its transversal sets.

10.2.1.2. gS_p-subsets modulo a set of labels, g-labels, S-labels and gS-labels

All our forthcoming definitions (Regular gS_p-Chains, gS_p-whips and gS_p-braids) will be based on that of a gS_p-subset modulo a set R of labels, g-labels, S-labels and gS-labels; in practice, R will be either the previous right-linking pattern or the set consisting of the target plus all the previous right-linking patterns (i.e. candidates, g-candidates, S_k-subsets and gS_k-subsets).

Definition: in any resolution state of any CSP, given a set R of labels, g-labels, S-labels and gS-labels [or a set R of candidates, g-candidates, Subsets and g-Subsets], a *g-Pair (or gS_2-subset) modulo R* is a gS_2-label {CSPVars, TransvSets}, where:

– CSPVars = $\{V_1, V_2\}$,

– TransvSets is composed of the following transversal sets of labels and g-labels:

– $\{<V_1, v_{11}>, <V_2, v_{21}>\}$ for constraint c_1,

– $\{<V_1, v_{12}>, <V_2, v_{22}>\}$ for constraint c_2,
such that:

– in RS, V_1 and V_2 are disjoint, i.e. they share no candidate;

– in RS, $\{<V_1, v_{11}>\}$ and $\{<V_1, v_{12}>\}$ are transitively disjoint; in RS, $\{<V_2, v_{22}>\}$ and $\{<V_2, v_{21}>\}$ are transitively disjoint;

– in RS, V_1 has the two mandatory candidates or g-candidates $<V_1, v_{11}>$ and $<V_1, v_{12}>$ compatible with R and no other candidate or g-candidate compatible with R;

– in RS, V_2 has the two mandatory candidates or g-candidates $<V_2, v_{21}>$ and $<V_2, v_{22}>$ compatible with R and no other candidate or g-candidate compatible with R.

Definition: in any resolution state of any CSP, given a set R of labels, g-labels, S-labels and gS-labels [or a set R of candidates, g-candidates, Subsets and g-Subsets], a *g-Triplet (or gS_3-subset) modulo R* is a gS_3-label {CSPVars, TransvSets}, where:

– CSPVars = $\{V_1, V_2, V_3\}$,

– TransvSets is composed of the following transversal sets of labels and g-labels:

– $\{<V_1, v_{11}>, (<V_2, v_{21}>), <V_3, v_{31}>\}$ for constraint c_1,

  $- \{<V_1, v_{12}>, <V_2, v_{22}>, (<V_3, v_{32}>)\}$ for constraint c_2,

  $- \{(<V_1, v_{13}>), <V_2, v_{23}>, <V_3, v_{33}>\}$ for constraint c_3,
such that:

  $-$ in RS, V_1, V_2 and V_3 are pairwise disjoint;

  $-$ in RS, $\{<V_1, v_{11}>\}$ and $\{<V_1, v_{12}>\}$ are transitively disjoint; $\{<V_2, v_{22}>\}$ and $\{<V_2, v_{23}>\}$ are transitively disjoint; $\{<V_3, v_{33}>\}$ and $\{<V_3, v_{31}>\}$ are transitively disjoint;

  $-$ in RS, V_1 has the two mandatory candidates or g-candidates $<V_1, v_{11}>$ and $<V_1, v_{12}>$ compatible with R, one optional candidate or g-candidate $<V_1, v_{13}>$ compatible with R (supposing this label or g-label exists), and no other candidate or g-candidate compatible with R;

  $-$ in RS, V_2 has the two mandatory candidates or g-candidates $<V_2, v_{22}>$ and $<V_2, v_{23}>$ compatible with R, one optional candidate or g-candidate $<V_2, v_{21}>$ compatible with R (supposing this label or g-label exists), and no other candidate or g-candidate compatible with R;

  $-$ in RS, V_3 has the two mandatory candidates or g-candidates $<V_3, v_{33}>$ and $<V_3, v_{31}>$ compatible with R, one optional candidate or g-candidate $<V_3, v_{32}>$ compatible with R (supposing this label or g-label exists), and no other candidate or g-candidate compatible with R.

We leave it to the reader to write the definitions of g-Subsets of larger sizes modulo R (gS$_p$-subsets modulo R). The general idea is that, when one looks at some gS$_p$-label S in RS "modulo R", i.e. when all the candidates and g-candidates in RS incompatible with R are "forgotten", what remains of S in RS satisfies the conditions of a non degenerated g-Subset of size p based on this gS$_p$-label.

Definition: in all the above cases, *a target of the gS$_p$-subset modulo R* is defined as a target of the gS$_p$-subset itself (i.e. as a candidate gS$_p$-linked to its underlying gS$_p$-label).

The idea is that, in any context (e.g. in a chain) in which the elements in R have positive valence, the gS$_p$-subset itself will have positive valence and any of its targets will have negative valence.

10.2.1.3. gS$_p$-regular sequences

As in the previous chapter, it is convenient to introduce an auxiliary notion before we define Reversible gS$_p$-chains, gS$_p$-whips and gS$_p$-braids.

Definition: let there be given an integer $1 \leq p \leq \infty$, an integer $m \geq 1$, a sequence $(q_1, ..., q_m)$ of integers, with $1 \leq q_k \leq p$ for all $1 \leq k \leq m$, and let $n = \sum_{1 \leq k \leq m} q_k$; let there also be given a sequence $(W_1, ..., W_m)$ of different sets of CSP variables of respective cardinalities q_k and a sequence $(V_1, ..., V_m)$ of CSP variables such that $V_k \in W_k$ for all $1 \leq k \leq m$. We define *a gS$_p$-regular sequence of length n associated with (W$_1$, ... W$_m$) and (V$_1$, ... V$_m$)* to be a sequence of length 2m [or 2m-1] (L$_1$, R$_1$, L$_2$, R$_2$, ... L$_m$, [R$_m$]), such that:

    – q_m=1 and $W_m = \{V_m\}$,

    – for $1 \leq k \leq$ m, L_k is a candidate;

    – for $1 \leq k \leq$ m [or $1 \leq k < m$], R_k is a candidate or a g-candidate if q_k=1 and it is a (non degenerated) Sq$_k$-subset or gSq$_k$-subset if q_k>1;

    – for each $1 \leq k \leq$ m [or $1 \leq k < m$], one has *"strong continuity", "strong g-continuity", "strong Sq$_k$-continuity" or "strong gSq$_k$-continuity"* from L_k to R_k:

      - if R_k is a candidate (q_k=1 and $W_k=\{V_k\}$), L_k and R_k have a representative with V_k: <V_k, l_k> and <V_k, r_k>,

      - if R_k is a g-candidate (q_k=1 and $W_k=\{V_k\}$), L_k has a representative <V_k, l_k> with V_k and R_k is a g-candidate <V_k, r_k> for Vk (r_k being its set of values),

      - if R_k is an Sq$_k$-subset or a gSq$_k$-subset (q_k>1), then W_k is its set of CSP variables and L_k has a representative with V_k.

The L_k are called the *left-linking candidates* of the sequence and the R_k the *right-linking objects (or elements or patterns)*. Notice that the natural expression of L_k to R_k continuity in case R_k is a g-Subset is the same as if it is a Subset.

Notice also that the definition of a g-Subset implies a disjointness condition on the sets of candidates for the CSP variables inside each W_k, but for gS$_p$-regular sequence there is no condition on the intersections of different W_k's. In particular, W_{k+i} may be a strict subset of W_k, if the right-linking elements in between give negative valence in W_{k+i} to some candidates or g-candidates that had no individual valence assigned in W_k. This is not considered as an inner loop of the sequence.

10.2.2. Reversible g-Subset Chains

Reversible gS$_p$-Chains are an extension of Reversible S$_p$-Chains in which right-linking S$_{p'}$-subsets may be replaced by gS$_{p'}$-subsets (p'≤p).

10.2.2.1. Definition of Reversible gS$_p$-Chains

Definition: given an integer $1 \leq p \leq \infty$ and a candidate Z (which will be a target), a *Reversible gS$_p$-Chain* of length n (n ≥ 1) is a gS$_p$-regular sequence (L$_1$, R$_1$, L$_2$, R$_2$, ... L$_m$, R$_m$) of length n associated with a sequence (W$_1$, ... W$_m$) of sets of CSP variables and a sequence (V$_1$, ... V$_m$) of CSP variables, such that:

    – Z is neither equal to any candidate in {L$_1$, R$_1$, L$_2$, R$_2$, ... L$_m$, R$_m$} nor a member of any g-candidate in this set nor equal to any label in the Sq$_k$-label or gSq$_k$-label of R_k when R_k is an Sq$_k$-subset or a gSq$_k$-subset, for any $1 \leq k < m$;

    – Z is linked to L$_1$;

– for each $1 < k \leq m$, L_k is linked or g-linked or Sq$_{k-1}$-linked or gSq$_{k-1}$-linked to R_{k-1}; this is the natural way of defining *"continuity" from R_{k-1} to L_k*;

– R_1 is a candidate or a g-candidate or an Sq$_1$-subset or a gSq$_1$-subset modulo Z: R_1 is the only candidate or g-candidate or is the unique Sq$_1$-subset or is the unique gSq$_1$-subset composed of all the candidates C for the CSP variables in W_1 such that C is compatible with Z;

– for any $1 < k \leq m$, R_k is a candidate or a g-candidate or an Sq$_k$-subset or a gSq$_k$-subset modulo R_{k-1}: R_k is the only candidate or g-candidate or is the unique Sq$_k$-subset (if $k \neq m$) or is the unique gSq$_k$-subset (if $k \neq m$) composed of all the candidates C for the CSP variables in W_k such that C is compatible with R_{k-1};

– Z is not a label for V_m;

– Z is linked to L_1 and to R_m.

Theorem 10.8 (Reversible gS$_p$-chain rule for a general CSP): in any resolution state of any CSP, if Z is a target of a Reversible gS$_p$-chain, then it can be eliminated (formally, this rule concludes $\neg candidate(Z)$).

Proof: if Z was True, then L_1 would be eliminated by ECP and R_1 would be asserted by S (if it is a candidate) or it would be a g-candidate or an Sq$_1$-subset or a gSq$_1$-subset; in any case, L_2 would be eliminated by ECP. By induction, we arrive at: R_m would be asserted by S or it would be a g-candidate – which would contradict Z being True.

10.2.2.2. Reversibility of Reversible gS$_p$-Chains in the general CSP

The following theorem justifies the name we have given these chains.

Theorem 10.9: a Reversible gS$_p$-Chain is reversible.

Proof: the main point of the proof is the construction of the reversed chain. As it is a simple transposition of the proof for Reversible S$_p$-Chains in section 9.2.2, we leave it as an exercise for the reader. Figure 9.1 can still be used as a partial visual support for the proof, but now the intersections between horizontal lines (CSP variables) and vertical lines (transversal sets) must be interpreted as candidates or g-candidates instead of only candidates.

10.2.2.3. Targets of gS$_p$-subsets versus targets of Reversible gS$_p$-subset Chains

The results in section 9.2.4 can be extended with only slight changes.

Theorem 10.10: a target Z of a gS$_p$ subset is always a target of a Reversible gS$_{p-1}$ Chain of length p.

Proof: almost obvious. One can always suppose that Z is gS$_p$-linked to TS$_1$ and that V_1 has a candidate or a g-candidate to which Z is linked or g-linked. Let $L_1 =$

$<V_1, l_1>$ be this candidate or any candidate in this g-candidate. Let $L_p = <V_p, l_p>$ be a candidate for V_p not in TS_1 (there must be one if the gS_p-subset is not degenerated). Let S2 be the gS_{p-1}-subset: $\{\{V_1, ..., V_{p-1}\}, \{TS_2, ..., TS_p\}\}$. Then the desired chain is $RgS_{p-1}C[p]$: $\{L_1 \ S2\} - V_p\{l_p .\}$.

A gS_p-subset that has g-transversal sets with "transitively non-void" intersections allows more eliminations than the "standard" ones defined in section 10.1. (This can happen only for $p>2$.)

Definition: a type-2 target of a gS_p-subset is a candidate belonging, either as an element or as a member of a g-label, to (at least) two of its g-transversal sets.

Theorem 10.11: a type-2 target of a gS_p-subset can be eliminated.

Proof: suppose a type-2 target Z is a candidate for variable V_1 and belongs to g-transversal sets TS_1 and TS_2. If Z was True, then all the candidates in TS_1 or TS_2 or in a label in TS_1 or TS_2 would be eliminated by ECP. This would leave only $p-2$ possibilities (in terms of candidates or g-candidates) for the remaining $p-1$ CSP variables – which is contradictory, in the same way as if it was a normal target.

As in the case of S_p-subsets, this is a very unusual kind of elimination. The following theorem shows that, here also, this abnormality can be palliated. It also justifies that we did not consider type-2 targets of gS_p-subsets in section 10.1: these abnormal targets can always be eliminated by a simpler pattern. An illustration of the following theorem will appear in section 10.3 (for a "Franken Squirmbag").

Theorem 10.12: A type-2 target of a gS_p-subset is always the (normal) target of a shorter Reversible gS_{p-2} Chain of length p-1.

Proof: in a resolution state RS, let Z be a type-2 target of a gS_p-subset with CSP variables $V_1, ... V_p$ and g-transversal sets $TS_1, ... TS_p$. One can always suppose that V_1 is the CSP variable for which Z is a candidate (there can be only one in RS) and that TS_1 and TS_2 are the two g-transversal sets to which Z belongs.

Fisrtly, each of the CSP variables $V_2, V_3, ... V_p$ must have at least one candidate or g-candidate of the gS_p-subset that is not in TS_1 or TS_2 (if it has several, choose one arbitrarily and name it $<V_2, c_2>, ... <V_p, c_p>$, respectively). Otherwise, the gS_p-subset would be degenerated; more precisely, Z could be eliminated by a whip[1] (or even by ECP after a Single) associated with (any of) the CSP variable(s) that has no such candidate.

Secondly, in TS_1 or TS_2, there must be at least one candidate for at least one of the CSP variables $V_2, ... V_p$. Otherwise, the initial gS_p-subset would be degenerated; more precisely, it would contain, among others, the gS_{p-2}-subset $\{\{V_3, ..., V_p\}, \{TS_3, ..., TS_p\}\}$; this would allow to eliminate all the candidates for V_1 and V_2 that are not in TS_1 or TS_2; Z could then be eliminated by a whip[1] associated with V_2;

and V_1 would have no candidate left. One can always suppose that there exists such a candidate L_2 for V_2, i.e. $L_2 = <V_2, l_2>$.

Modulo Z, we therefore have a gS$_{p-2}$ subset S2 with CSP variables $V_2, \ldots V_{p-1}$ and g-transversal sets TS$_3$, … TS$_p$. Let c_p* be c_p if it is a candidate or any element in c_p if it is a g-candidate. Then, Z is a (normal) target of the following Reversible gS$_{p-2}$ Chain of length p-1: RgS$_{p-2}$C[p-1]: $V_2\{l_2 \, S2\} - V_p\{c_p .\} \Rightarrow \neg candidate(Z)$.

10.2.3. gS$_p$-whips and gS$_p$-braids

gS$_p$-whips and gS$_p$-braids are an extension of g-whips and g-braids in which gS$_{p'}$-subsets (p'≤p) may appear as right-linking patterns. They can also be seen as extensions of the Reversible gS$_p$-chains by application of the zt-ing principle.

10.2.3.1. Definition of gS$_p$-whips

Definition: given an integer $1 \leq p \leq \infty$ and a candidate Z (which will be the target), a *gS$_p$-whip* of length n (n ≥ 1) built on Z is a gS$_p$-regular sequence ($L_1, R_1, L_2, R_2,$ …. L_m) [notice that there is no R_m] of length n, associated with a sequence ($W_1, \ldots W_m$) of sets of CSP variables and a sequence ($V_1, \ldots V_m$) of CSP variables (with $W_m = \{V_m\}$), such that:

– Z is neither equal to any candidate in $\{L_1, R_1, L_2, R_2, \ldots. L_m\}$ nor a member of any g-candidate in this set nor equal to any label in the Sq$_k$-label or gSq$_k$-label of R_k when R_k is an Sq$_k$-subset or a gSq$_k$-subset, for any $1 \leq k < m$;

– L_1 is linked to Z;

– for each $1 < k \leq m$, L_k is linked or g-linked or S-linked or gS-linked to R_{k-1}; this is a form of "continuity" from R_{k-1} to L_k;

– for any $1 \leq k < m$, R_k is a candidate or a g-candidate or an Sq$_k$-subset or a gSq$_k$-subset modulo Z and all the previous right-linking patterns: either R_k is the only candidate or g-candidate compatible with Z and with all the R_i with $1 \leq i < k$, or R_k is the unique Sq$_k$-subset or gSq$_k$-subset composed of all the candidates C for some of the CSP variables in W_k such that C is compatible with Z and with all the R_i with $1 \leq i < k$;

– Z is not a label for V_m;

– V_m has no candidate compatible with the target and with all the previous right-linking objects (but V_n has more than one candidate).

Theorem 10.13 (gS$_p$-whip rule for a general CSP): in any resolution state of any CSP, if Z is a target of a gS$_p$-whip, then it can be eliminated (formally, this rule concludes $\neg candidate(Z)$).

Proof: the proof is an easy adaptation of that for the S$_p$-whips. Supposing Z was True and iterating upwards: R_{k-1} would be asserted by S or it would be a g-candidate or an Sq_{k-1}-subset or a gSq_{k-1}-subset; due to R_{k-1} to L_k continuity, L_k would be eliminated by ECP; as usual the z- and t- candidates would be progressively eliminated. When m-1 is reached, R_{m-1} would have positive valence and there would be no possible value left for V_m (because Z itself is not a label for V_m).

10.2.3.2. Definition of gS_p-braids

Definition: given an integer $1\leq p\leq\infty$ and a candidate Z (which will be the target), a *gS_p-braid* of length n (n $\geq$ 1) built on Z is a gS_p-regular sequence (L_1, R_1, L_2, R_2, ... L_m) [notice that there is no R_m] of length n, associated with a sequence (W_1, ... W_m) of sets of CSP variables and a sequence (V_1, ... V_m) of CSP variables (with W_m = {V_m}), such that:

– Z is neither equal to any candidate in {L_1, R_1, L_2, R_2, ... L_m} nor a member of any g-candidate in this set nor equal to any label in the Sq_k-label or gSq_k-label of R_k when R_k is an Sq_k-subset or a gSq_k-subset, for any $1\leq k<m$;

– L_1 is linked to Z;

– for each $1 < k \leq m$, L_k is linked or g-linked or S-linked or gS-linked to Z or to some of the R_i, i<k; this is the only difference with gS_p-whips;

– for any $1 \leq k < m$, R_k is a candidate or a g-candidate or an Sq_k-subset or a gSq_k-subset modulo Z and all the previous right-linking patterns: either R_k is the only candidate or g-candidate compatible with Z and with all the R_i with $1\leq i< k$, or R_k is the unique Sq_k-subset or gSq_k-subset composed of all the candidates C for some of the CSP variables in W_k such that C is compatible with Z and with all the R_i with $1\leq i< k$;

– Z is not a label for V_m;

– V_m has no candidate compatible with the target and with all the previous right-linking objects (but V_n has more than one candidate).

Theorem 10.14 (gS_p-braid rule for a general CSP): in any resolution state of any CSP, if Z is a target of a gS_p-braid, then it can be eliminated (formally, this rule concludes ¬candidate(Z)).

Proof: almost the same as the proof for gS_p-whips. The condition replacing R_{k-1} to L_k continuity still allows the elimination of L_k by ECP.

10.2.3.3. gS_p-whip and gS_p-braid resolution theories; gS_pW and gS_pB ratings

In the same way as for the S_p-whips or S_p-braids cases, one can define increasing sequences of resolution theories, with two parameters, one (n) for the total length of the chain and one (p) for the maximum size of included Subsets or g-Subsets. By convention, p=1 means no Subset or g-Subset, only candidates and g-candidates.

Definition: for each p, $1 \leq p \leq \infty$, one can define an increasing sequence (gS$_p$W$_n$, $n \geq 0$) of resolution theories (similar definitions can be given for gS$_p$-braids, merely by replacing everywhere "gS$_p$-whip" by "gS$_p$-braid" and "gS$_p$W" by "gS$_p$B"):

- gS$_p$W$_0$ = BRT(CSP),
- gS$_p$W$_1$ = gS$_p$W$_0$ $\cup$ {rules for gS$_p$-whips of length 1} = W$_1$,
- gS$_p$W$_2$ = gS$_p$W$_1$ $\cup$ gS$_2$ (if p$\geq$2) $\cup$ {rules for gS$_p$-whips of length 2},
- ....
- gS$_p$W$_n$ = gS$_p$W$_{n-1}$ $\cup$ gS$_n$ (if p$\geq$n) $\cup$ {rules for gS$_p$-whips of length n},
- gS$_p$W$_\infty$ = $\cup_{n \geq 0}$ gS$_p$W$_n$.

For p=∞, i.e. for gS-whips built on g-Subsets of *a priori* unrestricted size, we also write gSW$_n$ instead of gS$_\infty$W$_n$.

Definitions: for any $1 \leq p \leq \infty$, the **gS$_p$W-rating** of an instance P, noted gS$_p$W(P), is the smallest $n \leq \infty$ such that P can be solved within gS$_p$W$_n$. Similarly, for any $1 \leq p \leq \infty$, the **gS$_p$B-rating** of an instance P, noted gS$_p$B(P), is the smallest $n \leq \infty$ such that P can be solved within gS$_p$B$_n$.

Obviously, setting p=1, gS$_1$W$_n$ = gW$_n$ and gS$_1$W(P) = gW(P) for any instance. Similarly, gS$_1$B$_n$ = gB$_n$ and gS$_1$B(P) = gB(P) for any instance

For any $1 \leq p \leq \infty$, the gS$_p$W and gS$_p$B ratings are defined in a purely logical way, independent of any implementation; the gS$_p$W and gS$_p$B ratings of an instance are intrinsic properties of this instance; moreover, for any fixed p ($1 \leq p \leq \infty$), the gS$_p$B rating is based on an increasing sequence of theories (gS$_p$B$_n$, $n \geq 0$) with the confluence property.

For any puzzle P, one has obviously gW(P) = gS$_1$W(P) $\geq$ gS$_2$W(P) $\geq$ gS$_p$W(P) $\geq$ gS$_{p+1}$W(P) $\geq$ ... and similar inequalities for the gS$_p$B(P).

10.2.3.4. The confluence property of all the gS$_p$B$_n$ resolution theories

Theorem 10.15: in any CSP, each of the gS$_p$B$_n$ resolution theories ($1 \leq p \leq \infty$, $0 \leq n \leq \infty$) is stable for confluence; therefore, it has the confluence property.

Proof: as it is a simple adaptation of the proof for the S$_p$B$_n$ resolution theories, we leave it as an exercise for the reader. We could even allow type-2 targets.

As usual, the confluence property of all the gS$_p$B$_n$ resolution theories for each p, $1 \leq p \leq \infty$, allows to superimpose on gS$_p$B$_n$ a "simplest first" strategy compatible with the gS$_p$B rating.

10.2.3.5. The "T&E(gS$_p$) vs gS$_p$-braids" theorem, $1 \leq p \leq \infty$

Any resolution theory T stable for confluence has the confluence property and the procedure T&E(T) can therefore be defined (see section 5.6.1). Taking T = gS$_p$, it is obvious that any elimination done by a gS$_p$-braid can be done by T&E(gS$_p$). As was the case for braids, for g-braids and for S$_p$-braids, the converse is true:

Theorem 10.16: for any $1 \leq p \leq \infty$, any elimination done by T&E(gS$_p$) can be done by a gS$_p$-braid.

As the proof closely follows that for S$_p$-braids, we leave it to the reader.

10.2.3.6. Type-2 targets of gSp-subsets used as left-linking candidates

Theorem 10.8 allows to replace any elimination of a candidate Z as a type-2 target for a gS$_p$[n] by the elimination of Z as a normal target for a RgS$_{p-2}$S[n-1] Reversible Subset Chain. But, as was the case for S$_p$-subsets in section 9.3.4 and for similar reasons, this is not enough to guarantee that type-2 targets of gS$_p$-subsets, if allowed to be used as left-linking candidates in the definitions of Reversible gS$_p$ Chains, gS$_p$-whips or gS$_p$-braids, could not lead to (slightly) more general patterns than those in our current definitions, due to the (probably rare) cases similar to those evoked in section 9.3.4. However, in the present case, one can prove the following:

Theorem 10.17: for any $1 \leq p \leq \infty$, for any $n > 2$, if a RgS$_p$C[n] (respectively a gS$_p$W[n], a gS$_p$B[n]) has a left-linking candidate L_k that is a type-2 target of an inner gS$_p$-subset, then it can be seen as a normal (i.e. with no inner type-2 targets) RgS$_q$C[n] (resp. gS$_q$W[n], gS$_q$B[n]), for some $q > p$, i.e. with larger inner g-Subsets.

Proof: almost obvious. Every time a left-linking candidate appears as a type-2 target, it suffices to merge its g-Subset with the next pattern in the sequence. Notice that this would not work for a "non-g" version of this theorem, because, even in this case, the next pattern could be a g-candidate.

Unfortunately, g-Subsets obtained by this (rather artificial) method tend to be very close to degeneracy.

10.3. A detailed example

We shall use the puzzle in Figure 10.1 (from the Hodoku examples) for several purposes:

– it will provide an example of a gS$_5$-subset and illustrate that, in conformance with our definition, the g-transversal sets do not have to meet all its CSP variables;

– it will illustrate the application of theorem 10.12 to the type-2 targets of a gS$_5$-subset;

– it will provide an example of a Reversible gS$_2$-subset Chain;

– it will illustrate alternative solutions using either gS$_5$-subsets and Reversible Chains or g-whips[5].

				6	1		4	7
				4	8		5	2
	1	5			2			6
	7	4		5			1	8
	2	8		7	5			4
	5	6		3	4		2	1

8	4	2	5	9	7	1	6	3
5	9	3	2	6	1	8	4	7
7	6	1	3	4	8	9	5	2
6	8	9	4	1	3	2	7	5
3	1	5	7	8	2	4	9	6
2	7	4	6	5	9	3	1	8
4	3	7	1	2	6	5	8	9
1	2	8	9	7	5	6	3	4
9	5	6	8	3	4	7	2	1

Figure 10.1. *A puzzle P with W(P) = 4, gW(P) = 5, gSW(P) = 5*

10.3.1. Solution using only gS$_p$-subsets and Reversible gS$_p$ Chains

Let us first see what is obtained if we use the Hodoku software mentioned in section 10.1.7, when only basic rules plus xy-chains, Subsets, Finned-Fish, Franken Fish, Mutant Fish and Kraken Fish (a kind of Fish Chains to be discussed below) are activated. We keep Hodoku's self-explaining notation. In the first three patterns, the Finned Swordfish, the various "f" indicate the fins. "Finned Swordfish" is a classical variant of Swordfish with additional candidates linked to the target; in our view, it is merely a "Swordfish modulo the target"; the eliminations allowed here by the three instances of this pattern can also be done by g-whips[3] (see section 10.3.2).

***** Hodoku 2.0.1 *****
Finned Swordfish: 3 c239 r147 fr2c2 fr2c3 fr3c2 fr3c3 => r1c1≠3
Finned Swordfish: 9 r239 c147 fr2c2 fr2c3 fr3c2 fr3c3 => r1c1≠9
Finned Swordfish: 9 c569 r147 fr5c5 fr6c6 => r4c4≠9
;;; Resolution state RS$_1$

Now, Hodoku reaches a resolution state RS$_1$ (displayed in Figure 10.2) with a "Franken Squirmbag in columns" for Number 9: in the five Columns c2, c3, c5, c6 and c9 (in light grey), Number 9 appears only in Rows r1, r4 and r7 and in Blocks b1 and b5 (in dark grey).

Franken Squirmbag: 9 c23569 r147b15 => r1c23478,r2347c1,r4c5678,r567c4,r7c78≠9

In the approach of this chapter, this is a gS$_5$-subset: the five CSP variables are X_{c2n9}, X_{c3n9}, X_{c5n9}, X_{c6n9} and X_{c9n9} (symbolised by light grey columns); the five g-transversal sets are defined by CSP variables (considered as constraints) X_{r1n9}, X_{r4n9}, X_{r7n9}, X_{b1n9} and X_{b5n9} (symbolised by dark grey rows and blocks). The targets of the

gS_5-subset are all the candidates (the fourteen ones in bold underlined characters in Figure 10.1) S_5-linked to one of the transversal sets, i.e. all the Numbers 9 in r1, r4, r7, b1 or b5 but not in any of c2, c3, c5, c6 or c9.

	c1	c2	c3	c4	c5	c6	c7	c8	c9	
r1	n1 n2 n4 n5 n6 n7 n8	n3 n4 n6 n8 n9	n1 n2 n3 n7 n9	n2 n3 n5 n7 n9	n2 n9	n3 n7 n9	n1 n3 n6 n8 n9	n3 n6 n8 n9	n3 n9	r1
r2	n2 n3 n5 n8 n9	n3 n8 n9	n2 n3 n9	n2 n3 n5 n9	n6	n1	n3 n8 n9	n4	n7	r2
r3	n1 n3 n6 n7 n9	n3 n6 n9	n1 n3 n7 n9	n3 n7 n9	n4	n8	n1 n3 n6 n9	n5	n2	r3
r4	n2 n3 n6 n8 n9	n3 n6 n8 n9	n2 n3 n9	n1 n3 n4 n6 n7 n8	n1 n8 n9	n3 n6 n7 n9	n2 n3 n4 n5 n7 n9	n3 n7 n9	n3 n5 n9	r4
r5	n3 n8 n9	n1	n5	n3 n4 n7 n8 n9	n8 n9	n2	n3 n4 n7 n9	n3 n7 n9	n6	r5
r6	n2 n3 n6 n9	n7	n4	n3 n6 n9	n5	n3 n6 n9	n2 n3 n9	n1	n8	r6
r7	n1 n3 n4 n7 n9	n3 n4 n9	n1 n3 n7 n9	n1 n2 n6 n8 n9	n1 n2 n6 n8 n9	n9	n3 n5 n6 n7 n8 n9	n3 n6 n7 n8 n9	n3 n5 n9	r7
r8	n1 n3 n9	n2	n8	n1 n6 n9	n7	n5	n3 n6 n9	n3 n6 n9	n4	r8
r9	n7 n9	n5	n6	n8 n9	n3	n4	n7 n8 n9	n2	n1	r9
	c1	c2	c3	c4	c5	c6	c7	c8	c9	

Figure 10.2. *Resolution state RS_1 of P, in which there appears a Franken Squirmbag*

At first sight, this Franken Squirmbag leads to a very impressive result: eighteen eliminations done by a single pattern. Notice that, contrary to whips that generally eliminate only one candidate at a time (though associated whips obtained by permutations can often eliminate more candidates), a Subset often eliminates several candidates; but eighteen is really exceptional.

However, a closer look shows some difference in the eliminations done by Hodoku for its Franken Squirmbag and those done by our gS_5-subset: in addition to the fourteen candidates of the latter, the former eliminates the following four candidates (in underlined but not bold characters): n9r1c2, n9r1c3, n9r4c5 and n9r4c6. These are examples of the type-2 targets evoked in section 10.2.2.3. We shall take advantage of them to illustrate how, according to theorem 10.8, they could be eliminated by shorter Reversible g-Subset chains with smaller g-Subsets. Here, we have p=5, so we find $RgS_3C[4]$:

RgS$_3$C[4]: n9{r1c5 gS$_3$:{c5 c6 c9}{r4 r7 b5}} – c3n9{r4 .} ==> r1c2 ≠ 9
RgS$_3$C[4]: n9{r1c5 gS$_3$:{c5 c6 c9}{r4 r7 b5}} – c2n9{r4 .} ==> r1c3 ≠ 9
RgS$_3$C[4]: n9{r4c5 gS$_3$:{c2 c3 c9}{r1 r7 b1}} – c6n9{r1 .} ==> r4c5 ≠ 9
RgS$_3$C[4]: n9{r4c5 gS$_3$:{c2 c3 c9}{r1 r7 b1}} – c6n9{r1 .} ==> r4c6 ≠ 9

The end of the Hodoku resolution path has nothing noticeable:

XYZ-Wing: 7/6/3 in r4c68,r6c4 => r4c4≠3
XYZ-Wing: 9/3/6 in r6c46,r7c6 => r4c6≠6
Naked Pair: 3,7 in r4c68 => r4c12379≠3, r4c47≠7
Locked Candidates Type 1 (Pointing): 3 in b4 => r2378c1≠3
Locked Candidates Type 1 (Pointing): 3 in b7 => r7c789≠3
Hidden Single: r1c9=3
Locked Candidates Type 1 (Pointing): 3 in b2 => r56c4≠3
Singles: r6c4=6, r7c6=6
Locked Candidates Type 1 (Pointing): 9 in b3 => r5689c7≠9
Locked Candidates Type 2 (Claiming): 9 in r1 => r23c4≠9
Locked Candidates Type 2 (Claiming): 9 in c4 => r7c5≠9
Naked Pair: 7,8 in r7c8,r9c7 => r7c7≠7, r7c7≠8
Singles: r7c7=5, r7c9=9, r4c9=5, r5c8=9, r5c5=8, r5c1=3, r4c5=1, r4c4=4, r5c4=7, r5c7=4, r3c4=3, r4c6=3, r6c6=9, r1c6=7, r6c1=2, r6c7=3, r4c3=9, r4c7=2, r4c8=7, r7c5=2, r1c5=9, r7c8=8, r1c8=6, r8c8=3, r7c4=1, r8c4=9, r9c4=8, r8c1=1, r8c7=6, r9c7=7, r9c1=9
Naked Triple: 4,5,8 in r1c12,r2c1 => r2c2≠8
XY-Chain: 5 5- r1c4 -2- r1c3 -1- r3c3 -7- r3c1 -6- r4c1 -8- r2c1 -5 => r1c1,r2c4<>5
Singles to the end

10.3.2. Solution using only g-whips

The resolution path with whips has nothing noticeable; it gives W(P) = 9. We shall skip it. But the path with g-whips gives gW(P) = 5. The "SQ" comment at the end of a line indicates that the elimination is one available with the Franken Squirmbag; "SQ2" indicates that it is a type-2 target.

As can be seen, most of the Squirmbag eliminations can be done by shorter g-whips or even shorter whips. The "<<<<<" comment indicates a whip[4] elimination not available to the Franken Squirmbag but that can be done before it.

***** SudoRules version 15b.1.12-gW *****
28 givens, 201 candidates and 1425 nrc-links
g-whip[3]: b2n9{r1c5 r123c4} - r9n9{c4 c7} - b3n9{r1c7 .} ==> r1c1 ≠ 9
g-whip[3]: b4n9{r4c3 r456c1} - r9n9{c1 c7} - r8n9{c7 .} ==> r4c4 ≠ 9
g-whip[3]: b4n3{r5c1 r4c123} - c9n3{r4 r7} - r8n3{c8 .} ==> r1c1 ≠ 3
;;; Resolution state RS$_1$
g-whip[3]: b8n9{r7c6 r789c4} - r2n9{c4 c123} - r3n9{c3 .} ==> r7c7 ≠ 9 ; SQ
whip[4]: b2n3{r3c4 r1c6} - c9n3{r1 r7} - b7n3{r7c3 r8c1} - b4n3{r4c1 .} ==> r4c4 ≠ 3 ; <<<<<
g-whip[4]: b4n9{r6c1 r4c123} - c9n9{r4 r1} - c6n9{r1 r6} - c5n9{r4 .} ==> r7c1 ≠ 9 ; SQ
g-whip[4]: b7n9{r8c1 r7c123} - c9n9{r7 r1} - c6n9{r1 r6} - c5n9{r4 .} ==> r4c1 ≠ 9 ; SQ

g-whip[4]: b2n9{r1c5 r123c4} - b8n9{r8c4 r7c456} - c2n9{r7 r4} - c9n9{r4 .} ==> r1c3 ≠ 9 ; SQ2
g-whip[4]: b2n9{r1c5 r123c4} - b8n9{r8c4 r7c456} - c3n9{r7 r4} - c9n9{r4 .} ==> r1c2 ≠ 9 ; SQ2
g-whip[4]: b4n9{r4c3 r456c1} - b7n9{r8c1 r7c123} - c6n9{r7 r1} - c9n9{r1 .} ==> r4c5 ≠ 9 ; SQ2
g-whip[4]: b4n9{r4c3 r456c1} - b7n9{r8c1 r7c123} - c5n9{r7 r1} - c9n9{r1 .} ==> r4c6 ≠ 9 ; SQ2
g-whip[4]: b3n9{r3c7 r1c789} - b2n9{r1c5 r123c4} - r9n9{c4 c1} - b4n9{r5c1 .} ==> r4c7 ≠ 9 ; SQ
g-whip[4]: b2n9{r3c4 r1c456} - b3n9{r1c8 r123c7} - r6n9{c7 c1} - r9n9{c1 .} ==> r5c4 ≠ 9 ; SQ
g-whip[4]: b7n9{r7c3 r789c1} - b4n9{r5c1 r4c123} - c9n9{r4 r1} - b2n9{r1c4 .} ==> r7c4 ≠ 9 ; SQ
g-whip[4]: b8n9{r8c4 r7c456} - c9n9{r7 r4} - c3n9{r4 r123} - c2n9{r3 .} ==> r1c4 ≠ 9 ; SQ
g-whip[5]: b2n9{r1c6 r123c4} - r9n9{c4 c1} - r8n9{c1 c8} - c9n9{r7 r4} - b4n9{r4c2 .} ==> r1c7 ≠ 9
; SQ
g-whip[5]: b4n9{r4c3 r456c1} - b7n9{r8c1 r7c123} - c9n9{r7 r1} - c6n9{r1 r6} - c5n9{r5 .} ==>
r4c8 ≠ 9 ; SQ
whip[3]: r5n4{c7 c4} - r5n7{c4 c8} - r4c8{n7 .} ==> r5c7 ≠ 3
whip[4]: r4c8{n3 n7} - r5n7{c8 c4} - r5n3{c4 c1} - b7n3{r8c1 .} ==> r7c8 ≠ 3
whip[5]: b5n4{r4c4 r5c4} - b5n7{r5c4 r4c6} - r4c8{n7 n3} - r5n3{c8 c1} - b4n8{r5c1 .} ==>
r4c4 ≠ 8
whip[5]: c6n7{r1 r4} - r4c8{n7 n3} - c9n3{r4 r7} - r8n3{c8 c1} - b4n3{r4c1 .} ==> r1c6 ≠ 3
whip[1]: c6n3{r6 .} ==> r6c4 ≠ 3, r5c4 ≠ 3
whip[4]: r6c4{n9 n6} - b8n6{r8c4 r7c6} - c6n9{r7 r1} - b3n9{r1c9 .} ==> r6c7 ≠ 9
whip[5]: r4c5{n1 n8} - r5c5{n8 n9} - b6n9{r5c7 r4c9} - b6n5{r4c9 r4c7} - r4n4{c7 .} ==> r4c4 ≠ 1
hidden-single-in-a-block ==> r4c5 = 1
whip[1]: r4n8{c1 .} ==> r5c1 ≠ 8
whip[4]: b7n3{r7c3 r8c1} - r5c1{n3 n9} - b6n9{r5c8 r4c9} - r1c9{n9 .} ==> r7c9 ≠ 3
whip[5]: b1n4{r1c2 r1c1} - b1n5{r1c1 r2c1} - c1n8{r2 r4} - c1n2{r4 r6} - b4n6{r6c1 .} ==> r1c2 ≠ 6
whip[5]: b1n5{r2c1 r1c1} - c1n8{r1 r4} - c1n2{r4 r6} - r6c7{n2 n3} - r5n3{c8 .} ==> r2c1 ≠ 3
whip[5]: b6n4{r5c7 r4c7} - c7n5{r4 r7} - r7c9{n5 n9} - c8n9{r8 r1} - c5n9{r1 .} ==> r5c7 ≠ 9
g-whip[3]: b7n9{r7c3 r789c1} - b4n9{r5c1 r4c123} - b6n9{r4c9 .} ==> r7c8 ≠ 9
whip[5]: b6n9{r4c9 r5c8} - c5n9{r5 r7} - c6n9{r7 r6} - b5n3{r6c6 r4c6} - c9n3{r4 .} ==> r1c9 ≠ 9
naked-single ==> r1c9 = 3
whip[3]: c9n9{r7 r4} - r5n9{c8 c1} - b7n9{r9c1 .} ==> r7c5 ≠ 9
whip[3]: r6c4{n9 n6} - b8n6{r8c4 r7c6} - b8n9{r7c6 .} ==> r3c4 ≠ 9, r2c4 ≠ 9
whip[1]: b2n9{r1c5 .} ==> r1c8 ≠ 9
whip[1]: b3n9{r3c7 .} ==> r9c7 ≠ 9, r8c7 ≠ 9
whip[3]: r9n9{c4 c1} - r5n9{c1 c8} - r8n9{c8 .} ==> r6c4 ≠ 9 ; SQ
singles ==> r6c4 = 6, r7c6 = 6
whip[2]: r9c7{n7 n8} - r7c8{n8 .} ==> r7c7 ≠ 7
whip[2]: r9c7{n8 n7} - r7c8{n7 .} ==> r7c7 ≠ 8
whip[2]: r4c8{n3 n7} - r4c6{n7 .} ==> r4c1 ≠ 3, r4c2 ≠ 3, r4c3 ≠ 3
whip[1]: b4n3{r6c1 .} ==> r3c1 ≠ 3, r7c1 ≠ 3, r8c1 ≠ 3
whip[1]: r8n3{c8 .} ==> r7c7 ≠ 3
singles ==> r7c7 = 5, r7c9 = 9, r4c9 = 5, r5c8 = 9, r5c1 = 3, r5c5 = 8, r7c5 = 2, r1c5 = 9, r1c6 = 7,
r3c4 = 3, r4c6 = 3, r6c6 = 9, r6c1 = 2, r4c3 = 9, r6c7 = 3, r8c7 = 6, r8c8 = 3, r4c8 = 7, r5c7 = 4, r4c7 =
2, r5c4 = 7, r7c8 = 8, r9c7 = 7, r9c1 = 9, r8c1 = 1, r8c4 = 9, r9c4 = 8, r1c8 = 6, r7c4 = 1, r4c4 = 4
whip[3]: r2c1{n8 n5} - r1c1{n5 n4} - r1c2{n4 .} ==> r2c2 ≠ 8
whip[4]: r2c7{n8 n9} - r2c2{n9 n3} - r7c2{n3 n4} - r1c2{n4 .} ==> r2c1 ≠ 8
singles to the end

11. W_p-whips and B_p-braids

In chapters 7, 9 and 10, we have extended the possibilities for right-linking elements of whips and braids from candidates to respectively g-candidates, Subsets and g-Subsets – whereas we always kept left-linking elements restricted to mere candidates. In the present chapter, we shall show that whips and braids themselves can be used as right-linking patterns. For each $1 \leq p \leq \infty$, we shall define two increasing sequences of resolution theories ($W_p W_n$ and $B_p B_n$, $0 \leq n \leq \infty$) and we shall associate with them two new ratings, $W_p W$ and $B_p B$.

We shall prove two main results for B_p-braids, similar to those proven for all our previous generalised braid theories: the confluence property of all the $B_p B_n$ resolution theories (providing the $B_p B$ ratings with all the good properties of the previous similar ratings) and a "T&E(B_p) vs B_p-braids" theorem.

We shall also prove that there is a close relationship between B-braids and an iterated (depth 2) Trial-and-Error procedure, the "T&E(2) vs B-braids" theorem. As very fast programs can easily be written for T&E(2), this theorem provides an easy way of checking if an instance of a CSP can be solved by B-braids, without actually finding its resolution path and its BB rating. A practical consequence for Sudoku is that, as all the known minimal puzzles can be solved by T&E(2), they all have a finite BB rating.

11.1. W_p-labels and B_p-labels; W_p-whips and B_p-braids

11.1.1. W_p-labels and B_p-labels; W_p-links and B_p-links

When one wants to allow a pattern P as a right-linking object of a whip or a braid, the first step is to explicit the P-label underlying its definition, independently of any resolution state. The following definition of a W-label extracts from the definition of a whip its structural part: only the part, but all the part, that does not depend on the resolution state, i.e. that can be expressed with labels and links, without referring to actual candidates.

Definition: for any $n \geq 1$, a W_n-label is a structured list $(Z, (V_1, L_1, R_1), ..., (V_{n-1}, L_{n-1}, R_{n-1}), (V_n, L_n))$, such that:
– for any $1 \leq k \leq n$, V_k is a CSP variable;

– Z, all the L_k's and all the R_k's are labels;

– in the sequence of labels $(L_1, R_1, ..., L_{n-1}, R_{n-1}, L_n)$, any two consecutive elements are different;

– Z does not belong to $\{L_1, R_1, L_2, R_2, ... L_n\}$;

– L_1 is linked to Z;

– right-to-left continuity: for any $1 < k \leq n$, L_k is linked to R_{k-1};

– strong left-to-right continuity: for any $1 \leq k < n$, L_k and R_k are labels for V_k;

– L_n is a label for V_n;

– Z is not a label for V_n.

Definition: a B_n-label is a structured sequence as above, with the right-to-left continuity condition replaced by:

– for any $1 < k \leq n$, L_k is linked to Z or to a previous R_i.

Definitions: a label l is W_n-linked [respectively B_n-linked] to a W_n-label [resp. a B_n-label] $(Z, (V_1, L_1, R_1), ..., (V_{n-1}, L_{n-1}, R_{n-1}), (V_n, L_n))$ if l equal to Z. The index n in "W_n-linked" or "B_n-linked" may be dropped, as there can be no ambiguity. A label l is compatible with the above W_n-label [resp. B_n-label] if it is not W_n-linked [resp. B_n-linked] to it.

One can now give an alternative equivalent definition of a whip [or a braid], in which the structural and non-structural conditions are completely separated:

Definition: in a resolution state RS, given a candidate Z (which will be the target), a whip [respectively a braid] of length n $(n \geq 1)$ built on Z is a W_n-label [resp. a B_n-label] $(Z, (V_1, L_1, R_1), ..., (V_{n-1}, L_{n-1}, R_{n-1}), (V_n, L_n))$, such that:

– all the L_k's and all the R_k's are candidates (not only labels);

– for any $1 \leq k < n$, R_k is the only candidate for V_k compatible with Z and with all the previous right-linking candidates (i.e. with Z and with all the R_i, $1 \leq i < k$);

– V_n has no candidate compatible with Z and with all the previous right-linking candidates (but V_n has more than one candidate – a non-degeneracy condition).

11.1.2. Equivalence of whips or braids

Until now, we have been very strict on the targets of whips [or braids]: a whip [or a braid] has only one target, specified in its definition. But, sometimes there is another whip [braid] with an underlying W_n-label [B_n-label] strongly equivalent (definition below) to that of the first whip [braid] and allowing to eliminate its own target. This entailed no problem until now, because the second whip [braid] could be written after the first and it did not change the W [B] rating of an instance. But if a whip [braid] is to be inserted into another one as a right-linking pattern, then it should not be counted several times if it serves to justify several t-candidates. The following definitions palliate this problem. Notice that, in all our previous resolution paths, we have implicitly used them, every time two eliminations appear in the same line.

Definition (structural): two W_n-labels [B_n-labels] are strongly equivalent if they differ only by their targets. This is obviously an equivalence relation.

Definition (non-structural): in a resolution state RS, two whips [braids] of length n are strongly equivalent if their underlying W_n-labels [B_n-labels] are strongly equivalent. This is an equivalence relation.

Remarks about strongly equivalent whips [braids]:

– the definition entails that the two whips [braids] have the same t-candidates;

– it also supposes that, for each CSP variable of the common label, every candidate that is not a left-linking, t- or right- linking candidate must be a z-candidate for both whips [braids] simultaneously (i.e. it is linked to their two different targets);

– due to the second remark, there is no simple way to replace this definition by a purely structural one; but if it is satisfied in a resolution state RS, then it will be satisfied in any posterior resolution state; we say that it is *persistent*, which, for some purposes, is almost as good as being structural;

– having no z-candidates, as in t-whips [t-braids], gives rise to strongly equivalent whips [braids]; but this is not a necessary condition;

– a W_n-label [B_n-label] can be interpreted as a potential whip [braid], waiting for the elimination of some candidates from its CSP variables before it becomes an actual one.

Definition (non-structural): an *extended target of a whip W in a resolution state RS* is a target of any whip strongly equivalent to W in RS.

Remarks:

– there is an obvious correspondence between a W_1-label and the set consisting of a g-label and its targets (seen as whip[1] targets); and a label l is g-linked to a g-label g if and only if l is an extended target of the W_1-label corresponding to g;

– if Z' is an extended target of a whip W in a resolution state RS, then it remains one in any posterior resolution state in which W (or a strongly equivalent whip) is still present and Z' is still a candidate; being an extended target is a persistent property;

– however, a candidate Z' that is not an extended target for W in RS may become one in a posterior resolution state RS', in case all the z-candidates of W in RS that are not t-candidates of W and that are not linked to Z', have been eliminated along the path between RS and RS'.

Definition (non-structural): a candidate C is compatible with a whip W in a resolution state RS if it is not an extended target of W in RS. Transposing the above remarks: if C is incompatible with a whip W in RS, then it remains incompatible with W in any posterior resolution state in which W is still present and C is still a candidate; but if C is compatible with W in RS, it may become incompatible with W in a posterior resolution state. It is therefore necessary to be always clear about the resolution state under consideration. Said otherwise, incompatibility with a whip is persistent, compatibility is not.

11.1.3. Definition of W_p-whips, W_p-braids and B_p-braids

Special care must be taken with the definition of whips accepting inner whips as right-linking patterns:

– global variables of the global whip and inner variables of each of its inner sub-whips must not be confused;

– similarly, global and inner left-to-right linking conditions must not be confused;

– for a proper definition of the global size, the conditions must be written in a form that does not allow degeneracy of the inner whips; fortunately, this is much easier to do than for inner Subsets: one only has to make sure that contradictions in the inner whips can only occur on their last CSP variables (i.e. not before);

– it must not be forgotten that, as is always the case for all the inner patterns of generalised whips, inner whips will appear as "reversed" whips (modulo the target and the previous right-linking objects), in the sense that their targets will have to appear as the next left-linking candidate.

Definition: in any resolution state RS of any CSP, for any $n \geq 1$ and $1 \leq p < n$, a W_p-whip[n] is a structured list $(Z, (V_1, L_1, R_1, q_1), \ldots, (V_{m-1}, L_{m-1}, R_{m-1}, q_{m-1}), (V_m, L_m, q_m))$, with $m \leq n$, that satisfies the following structural and non-structural conditions:

structural conditions (that could be considered as defining a "W_p-regular sequence of length n"):

– $1 \leq q_k \leq p$ for all $1 \leq k \leq m$, $q_m = 1$ and $n = \sum_{1 \leq k \leq m} q_k$;

– for any $1 \leq k \leq m$, V_k is a CSP variable;

– for each $1 \leq k \leq m$, L_k is a label for V_k;

– for each $1 \leq k < m$, R_k is a label or a W_1-label if $q_k = 1$ and it is a W_{q_k}-label if $q_k > 1$;

– L_1 is linked to Z;

– right-to-left continuity: for any $1 \leq k \leq m$, L_k is linked or W_{k-1}-linked to R_{k-1};

– for any $1 \leq k < m$, the following "strong continuity or strong W-continuity from L_k to R_k", implying *"continuity or W-continuity from L_k to R_k"*, is satisfied:

- if $q_k=1$ and R_k is a label, then R_k (as well as L_k) is a label for V_k;
- if $q_k \geq 1$ and R_k is a Wq_k-label, then V_k is one of its CSP variables (not necessarily the last one);
− Z is not a label for V_m;

non-structural conditions:
− Z and all the L_k's are candidates (not only labels);
− for any $1 \leq k < m$: if $q_k=1$ and R_k is a label, then, R_k is the only candidate for V_k compatible *in RS* with Z and with all the previous right-linking patterns R_i; if $q_k \geq 1$ and R_k is a Wq_k-label $(Z_k, (V_{k,1}, L_{k,1}, R_{k,1}), ..., (V_{k,i}, L_{k,i}, R_{k,i}), ..., (V_{k,qk-1}, L_{k,qk-1}, R_{k,qk-1}), (V_{k,qk}, L_{k,qk}))$, then:
  - for each $i < q_k$: $L_{k,i}$ and $R_{k,i}$ are candidates (not only labels) for CSP variable $V_{k,i}$ of R_k;
  - for each $i < q_k$: $R_{k,i}$ is the only candidate for $V_{k,i}$ compatible *in RS* with Z, with the previous right-linking patterns R_i ($i<k$) of the global W_p-whip[n] being defined, with the previous right-linking candidates $R_{k,i'}$ ($i'<i$) inside R_k, *and with* Z_k;
  - $L_{k,qk}$ is a candidate for $V_{k,qk}$ (not only a label); $V_{k,qk}$ has no candidate compatible *in RS* with Z, with the previous right-linking patterns R_i ($i<k$) of the global W_p-whip[n] being defined, with the previous right-linking candidates $R_{k,i}$ ($i<q_k$) inside R_k *and with* Z_k (but $V_{k,qk}$ has more than one candidate compatible *in RS* with Z and with the previous right-linking objects R_i ($i<k$) of the global W_p-whip − this is the non-degeneracy condition of the inner R_k whip);
− V_m has no candidate compatible *in RS* with the target and with all the previous right-linking objects of the global W_p-whip (but V_m has more than one candidate − the usual non-degeneracy condition of the global W_p-whip).

Remark: for all n, a W_1-whip[n] is the same thing as a g-whip[n].

Definition: for any $n \geq 1$ and $1 \leq p < n$, a *W_p-braid[n]* is a structured list as above, with the structural right-to-left continuity condition of a W_p-whip[n] replaced by:
− for any $1 \leq k \leq m$, L_k is linked or W-linked to Z or to a previous R_i.

Definition: in the previous definition, if the inner Wq_k-labels are replaced by Bq_k-labels, one obtains B_p-braids[n].

Definitions: in any of the above defined W_p-whips or B_p-braids, a candidate other than L_k for any of the "global" CSP variables V_k is called a global t- [respectively global z-] candidate if it is incompatible with a previous right-linking pattern [resp. with the target Z]; a candidate for a "local" or "inner" CSP variable $V_{k,i}$ of an inner braid R_k is called a local (or inner) t- [respectively local (or inner) z-]

candidate if it is incompatible with a previous local right-linking candidate $R_{k,j}$, $j<i$ [resp. with the local target Z_k of R_k].

Notice that a candidate can be at the same time global and local, z- and t-. Notice also that, as in all our previous definitions, the (global or local) z- and t- candidates are not considered as being part of the W_p-whip or B_p-braid patterns.

Remarks:

– in the above definitions, as in any of the previously defined types of generalised whips or braids, left-linking elements of the global W_p-whip [B_p-braid] are mere candidates (and not more general patterns);

– as shown by the fact that inner whips or braids are "reversed" (see Figure 11.2 or the proof of the W_p-whip elimination theorem), the acceptance of whips[p] or braids[p] as right-linking patterns amounts to accepting some form of look-ahead of size p (a form different from that accepted in S_p-whips or S_p-braids, globally less restricted);

– in the same way as all the types of braids we have met before, B_p-braids[n] are interesting for the confluence property of the B_pB_n theories and for the "T&E(B_p) vs B_pB" theorem (see proofs below); and W_p-whips are interesting as a simpler (and hopefully good) approximation of B_p-braids;

– one could also define B_p-whips[n], but there does not seem to be any good reason for imposing an "outer" continuity condition if the inner bricks do not enjoy their own inner continuity.

11.1.4. Graphico-symbolic representations

The symbolic representations in Figures 11.1 and 11.2 may help understand how a partial W_2-whip[3] differs from an ordinary partial whip[3]. In these Figures:

– black horizontal lines represent CSP variables; they are supposed to have candidates only at their extremities or at their meeting points with arrows;

– dark grey straight oblique arrows represent links from Z to L_1 or from R_k to L_{k+1} and also, in the second Figure, inner links from $R_{i,k}$ to $L_{i,k+1}$;

– light grey arrows represent links to z- or t- candidates in the global whip and (in the second Figure) in an inner whip (the straight ones represent links to candidates in the same g-label as the next left-linking candidate);

– the straight double-sided dark grey arrow in the second Figure represents the double role of L_2 as a target of the inner whip (descending arrow) and as the next left-linking candidate (ascending arrow);

– the orientations of arrows represent the way links are used in the proof of the whip or W_2-whip rule; by themselves, links are not orientated; but these orientations also illustrate the idea that inner whips correspond to some form of look-ahead.

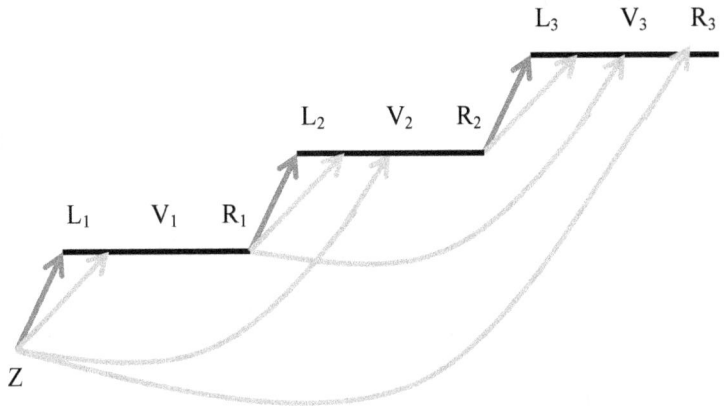

Figure 11.1. *A graphico-symbolic representation of a partial whip[3]*

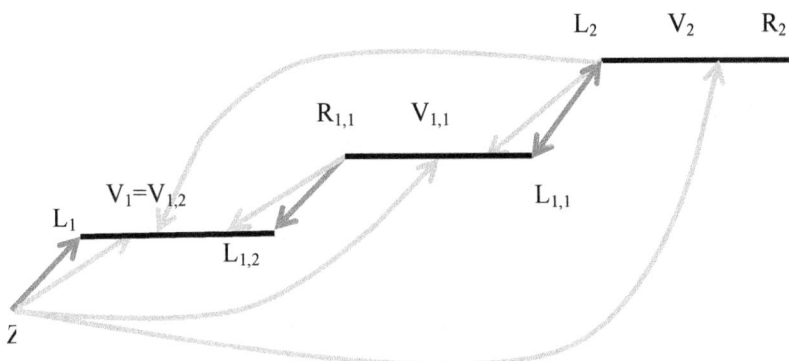

Figure 11.2. *A graphico-symbolic representation of a partial W$_2$-whip[3]. One can see an inner whip[2] modulo Z, with target L$_2$: (L$_2$, (V$_{1,1}$, L$_{1,1}$, R$_{1,1}$) (V$_{1,2}$, L$_{1,2}$)). Here, V$_1$ is taken equal to the last CSP variable (V$_1$=V$_{1,2}$) of the inner whip R$_2$, but it is not a necessary condition; if V$_{1,2}$ had no candidate linked to Z, but V$_{1,1}$ had one, one would take V$_1$=V$_{1,1}$.*

11.1.5. Elimination theorems

Theorem 11.1 (W_p-whip elimination theorem): given a W_p-whip, one can eliminate its target.

Proof: obvious. The main point was having the correct definitions. If Z was True, then L_1 and all the candidates linked to Z (the z-candidates) would be eliminated by ECP; if R_1 is a label, then it would be asserted by S; if R_1 is a Wq_1-label, then, after these first series of eliminations, it would be a whip[q_1] with target L_2 and L_2 would be eliminated by rule Wq_1. We can iterate until we reach L_m would be eliminated by ECP or by rule Wq_{m-1}. (As usual, the t-candidates of the global whip would be progressively eliminated by ECP or some Wq_k). The last condition implies that V_m would have no possible value.

Theorem 11.2 (B_p-braid elimination theorem): given a B_p-braid, one can eliminate its target.

Proof: almost the same as the proof for W_p-whips (with any reference to Wq_k replaced by one to Bq_k), the main difference being the condition replacing right-to-left continuity, which still implies that L_k would be eliminated by ECP.

11.1.6. W_p-whip and B_p-braid resolution theories; the W_pW and B_pB ratings

For each $1 \leq p \leq \infty$, one can define an increasing sequence (W_pW_n, $n \geq 0$) of resolution theories based on W_p-whips:

– $W_pW_0 = BRST$,
– $W_pW_1 = W_pW_0 \cup \{$rules for W_p-whips of length 1$\} = W_1$,
– ...
– $W_pW_n = W_pW_{n-1} \cup \{$rules for W_p-whips of length n$\}$,
– $W_pW_\infty = \cup_{n \geq 0} W_pW_n$.

One has obvious similar definitions for (B_pB_n, $n \geq 0$).

And, for each $1 \leq p \leq \infty$, one can also define in the usual way the W_pW [respectively B_pB] rating associated with the increasing sequence (W_pW_n, $n \geq 0$) [resp. B_pB_n, $n \geq 0$] of resolution theories.

One can also define the WW and BB ratings as being equal to $W_\infty W$ and $B_\infty B$, respectively, when no restriction is put *a priori* on the lengths of the inner whips [resp. braids] (of course, in each W-whip [resp. B-braid], they can only be smaller than its global length).

Remarks:

– it was important to properly define the length of a W_p-whip [or B_p-braid] in a way that takes into account the lengths of all its elements, because some W_p-whips will be equivalent to whips, g-whips, S_p-whips or gS_p-whips; for consistency of the ratings, they must be given the same size, independently of how they are considered;

– with the confluence property of all the B_pB_n resolution theories (see section 11.2), the B_pB ratings have the same good properties as those mentioned for previous generalised braid theories; however, non-anticipativeness is no longer true; it is replaced by a restricted form of look-ahead, controlled by the size p of the inner braids;

– as an obvious corollary to theorem 11.5 below, the BB rating is finite for any instance of a CSP that can be solved by T&E(2). *In Sudoku, this entails that all the known minimal puzzles have a finite BB rating – a rating that is obviously invariant under symmetry and supersymmetry*.

11.1.7. $gS_pW_n+W_pW_n$ and $gS_pW_n+B_pB_n$ theories; associated ratings

Allowing gS_p-subsets or whips[p] as right-linking patterns in different whips, one can hope to get still more powerful resolution theories. For each $1 \leq p \leq \infty$, one can define an increasing sequence $gS_pW_n+W_pW_n$, $0 \leq n \leq \infty$, of resolution theories:

– $gS_pW_0+W_pW_0 = $ BRST,

– $gS_pW_1+W_pW_1 = gS_pW_0+W_pW_0 \cup gS_pW_1 \cup W_pW_1 = W_1$,

– ...

– $gS_pW_n+W_pW_n = gS_pW_{n-1}+W_pW_{n-1} \cup gS_pW_n \cup W_pW_n$,

– ...

– $gS_pW_\infty+W_pW_\infty = \cup_{n \geq 0} gS_pW_n+W_pW_n$.

One can introduce obvious similar definitions for $gS_pB_n +B_pB_n$, $0 \leq n \leq \infty$.

And, for each $1 \leq p \leq \infty$, one can define in the usual way the gS_pW+W_pW [respectively gS_pB+W_pB] rating associated with the increasing sequence $gS_pW_n+W_pW_n$, $n \geq 0$, [resp. $gS_pB_n+B_pB_n$, $n \geq 0$] of resolution theories.

One can also define the gSW+WW and gSB+BB ratings in the usual way.

It is a straightforward corollary to lemma 4.1 and theorems 10.15 and 11.3 (below) that all the $gS_pB_n+B_pB_n$ resolution theories are stable for confluence and have the confluence property. A "simplest first" strategy can therefore be defined. Or rather several "simplest first" strategies: the question is, for each n, do we give precedence to gS_p-braids[n] or to B_p-braids[n]? As a result, these definitions leave us with a choice between Subsets and whips. Moreover, the (probably limited) increased resolution power of these combined theories (with respect to the B_p-braids) is probably not worth its cost in terms of computational complexity.

11.1.8. gSW-whip and gSB-braid theories; associated ratings

Going one step further, one can allow both gS_p-subsets and whips[p] as right-linking patterns in the same whips, in the hope of getting the most powerful theories. For each $1 \leq p \leq \infty$, one can define an increasing sequence $(gS_pW_p)W_n$, $0 \leq n \leq \infty$, of resolution theories:

– $(gS_pW_p)W_0 = BRST$,

– $(gS_pW_p)W_1 = W_1$,

– ...

– $(gS_pW_p)W_n = (gS_pW_p)W_{n-1} \cup \{$rules for whips of total length n, with inner gS_p-subsets and W_p-whips$\}$,

– ...

– $(gS_pW_p)W_\infty = \cup_{n \geq 0} (gS_pW_p)W_n$.

One can introduce obvious similar definitions for $(gS_pB_p)B_n$, $0 \leq n \leq \infty$.

And, for each $1 \leq p \leq \infty$, one can define in the usual way the $(gS_pW_p)W$ [respectively $(gS_pB_p)B$] rating associated with the increasing sequence $(gS_pW_p)W_n$, $n \geq 0$, [resp. $(gS_pB_p)B_n$, $n \geq 0$] of resolution theories.

One can also define the $(gSW)W$ and $(gSB)B$ ratings in the usual way.

Contrary to the previous case, the confluence property of the $(gS_pB_p)B_n$ resolution theories must now be proven directly; this can be done by combining the proofs for the gS_pB_n and the B_pB_n theories. A "simplest first" strategy can therefore be defined, or rather several "simplest first" strategies, each providing all the $(gS_pB_p)B$ ratings with good properties. But, as in the previous case, the (probably limited) increased resolution power is probably not worth the computational cost of so complex braids.

11.2. The confluence property of the B_pB_n resolution theories

We now prove the main property of B_p-braid resolution theories.

11.2.1. Proof of the confluence property

Theorem 11.3: each of the B_pB_n resolution theories ($1 \leq p \leq \infty$, $0 \leq n \leq \infty$) is stable for confluence; therefore, it has the confluence property.

Proof: in order to keep the same notations as in the proofs for the g-braids (section 7.5) and the S_r-braids (section 9.4), we prove the result for B_rB_n, r and n fixed.

We must show that, if an elimination of a candidate Z could have been done in a resolution state RS$_1$ by a B$_r$-braid B of length n' $\leq$ n and with target Z, it will always still be possible, starting from any further state RS$_2$ obtained from RS$_1$ by consistency preserving assertions and eliminations, if we use a sequence of rules from B$_r$B$_n$. Let B be: $\{L_1\ R_1\} - \{L_2\ R_2\} - \ldots - \{L_p\ R_p\} - \{L_{p+1}\ R_{p+1}\} - \ldots - \{L_m\ .\}$, with target Z, where the R$_k$'s are candidates or braids in B$_r$ modulo Z and the previous R$_i$'s. For inner braids, we use the notations in the definition (section 11.1).

The proof follows the same general lines as that for g-braids and S$_r$-braids. Indeed, it is remarkably close to that for S$_r$-braids, with transversal sets replaced by the sets of candidates linked to some right-linked object (see the similarities in Figure 11.3 and discussion in section 11.2.2). For technical reasons, we keep a separate case for inner braids of length 1, i.e. g-whips.

Consider first the state RS$_3$ obtained from RS$_2$ by applying repeatedly the rules in BRT until quiescence. As BRT has the confluence property by theorem 5.6, this state is uniquely defined. (Notice that, thanks to theorem 5.6 and the inclusion B$_n$ $\subset$ B$_r$B$_n$, we could use B$_n$ instead of BRT, but, apart from dispensing us of introducing marks, it does not seem to make the proof simpler.)

If, in RS$_3$, target Z has been eliminated, the proof is finished. If target Z has been asserted, then the instance of the CSP is contradictory; if not yet detected in RS$_3$, this contradiction can be detected by CD in a state posterior to RS$_3$, reached by a series of applications of rules from B$_r$, following the B$_r$-braid structure of B.

Otherwise, we must consider all the elementary events related to B that can have happened between RS$_3$ as well as those we must provoke in posterior resolution states RS. For this, we start from B' = what remains of B in RS$_3$ and we let RS = RS$_3$. At this point, B' may not be an S$_r$-braid in RS. We progressively update RS and B' by repeating the following procedure, for p = 1 to p = m, until it produces a new (possibly shorter) B$_r$-braid B' with target Z in resolution state RS – a situation that is bound to happen. We return from this procedure as soon as B' is a B$_r$-braid in RS. All the references below are to the current RS and B'.

a) If, in RS, any candidate that had negative valence in B – i.e. the left-linking candidate, or any t- or z- candidate, of CSP variable V$_p$, or any global or local t- or z- candidate of R$_p$ in case R$_p$ is an inner braid – has been asserted (this can only be between RS$_1$ and RS$_3$), then all the candidates linked to it have been eliminated by relevant rules from BRT in RS$_3$, in particular: Z and/or all the candidate(s) R$_k$ (k<p) to which it is linked, and/or all the elements of the g-candidate(s) R$_k$ (k<p) to which it is g-linked, and/or all the inner candidates to which it is linked of the inner R$_k$ braids (k<p) to which it is B-linked (by the definition of a B$_r$-braid); if Z is among them, there remains nothing to prove; otherwise, the procedure has already either been successfully terminated by case f1 or f2α and/or dealt with by case d2 of the previous such k's for which R$_k$ is an inner braid of length q$_k$ $\geq$ 2.

b) If, in RS_3, left-linking candidate L_p has been eliminated (but not asserted), it can no longer be used as a left-linking candidate in a B_r-braid. Suppose that either CSP variable V_p still has a z- or a t- candidate C_p, or that R_p is an inner braid of length $q_p \geq 2$ and there is another CSP variable V_p' in its W_p such that V_p' still has a z- or a t- candidate C_p; then, in B', replace L_p by C_p and (in the latter case) V_p by V_p'. Now, up to C_p, B' is a partial B_r-braid in RS with target Z. Notice that, even if L_p was linked or g-linked or B-linked to R_{p-1} (e.g. if B was a B_r-whip) this may not be the case for C_p; therefore trying to prove along the same lines a similar theorem for B_r-whips would fail here.

c) If, in RS, any t- or z- candidate of V_p or of the inner braid R_p (if R_p is an inner braid) has been eliminated (but not asserted), this has not changed the basic structure of B (at stage p). Continue with the same B'.

d) Consider now assertions occurring in right-linking objects of the global B_r-braid. There are two cases instead of one for g-braids: assertions occurring in a right-linking candidate or g-candidate (case d1) and assertions occurring anywhere in an inner braid R_p of length $q_p \geq 2$ (case d2).

d1) If, in RS, right-linking candidate R_p or a candidate R_p' in right-linking g-candidate R_p has been asserted (p can therefore not be the last index of B'), R_p can no longer be used as an element of a B_r-braid, because it is no longer a candidate or a g-candidate. As in the proof for S_r-braids, and only because of this d1 case, we cannot be sure that this assertion occurred in RS_3. We must palliate this. First eliminate by ECP or W_1 any left-linking or t- candidate for any CSP variable of B' after p, including those in the inner braids, that is incompatible with R_p, i.e. linked or g-linked to it, if it is still present in RS. Now, considering the B_r-braid structure of B upwards from p, more eliminations and assertions can be done by rules from B_r. (Notice that, as in the S_r-braids case, we are not trying to do more eliminations or assertions than needed to get a B_r-braid in RS; in particular, we continue to consider R_p, not R_p'; in any case, it will be excised from B'; but, most of all, we do not have to find the shortest possible B_r-braid!)

Let q be the smallest number strictly greater than p such that CSP variable V_q or some CSP variable V_q' in W_q still has a global left-linking, t- or z- candidate C_q that is not linked, g-linked or B-linked to any of the R_i for $p \leq i < q$. (For index q, there is thus a V_q' in W_q and a candidate C_q for V_q' such that C_q is linked, g-linked or B-linked to Z or to some R_i with $i < p$.)

Apply the following rules from B_r (if they have not yet been applied between RS_2 and RS) for each of the CSP variables V_u (and all the $V_{u,i}$ in W_u if R_u is an inner braid) with index (or first index) u increasing from p+1 to q-1 included:
- eliminate by ECP or W_1 or some $B_{r'}$ ($r' \leq r$) any candidate for any CSP variable in W_u that is incompatible with R_{u-1};

- at this stage, CSP variable V_u has no left-linking, z- or t- candidate and there remains no global t- or z- candidate in W_u if R_u is an inner braid;
- if R_u is a candidate, assert it by S and eliminate by ECP all the candidates for CSP variables after u, including those in the inner braids, that are incompatible with R_u in the current RS;
- if R_u is a g-candidate, it cannot be asserted; eliminate by W_1 all the candidates for CSP variables after u, including those in the inner braids, that are incompatible with R_u in the current RS;
- if R_u is an inner braid in Bq_u, it cannot be asserted by Bq_u; eliminate by Bq_u all the candidates for CSP variables after u, including those in the inner braids, that are incompatible with R_u in the current RS (this includes the target of R_u).

In the new RS thus obtained, excise from B' the part related to CSP variables and inner braids p to q-1 (included); if L_q has been eliminated in the passage from RS_2 to RS, replace it by C_q (and, if necessary, replace V_q by V_q'); for each integer $s \geq p$, decrease by q-p the index of CSP variable V_s, of its candidates and inner right-linking pattern (g-candidate or braid) and of the set W_s, in the new B'. In RS, B' is now, up to p (the ex q), a partial B_r-braid in $B_r B_n$ with target Z.

d2) If, in RS, a candidate C_p in a right-linking braid R_p with $q_p \geq 2$ has been asserted or eliminated or marked in a previous step, R_p can no longer be used as such as a right-linking inner braid of a B_r-braid, because it may no longer be an inner braid. Moreover, there may be several such candidates in R_p; consider them all at once. Notice that candidates can only have been asserted as values in the transition from RS_1 to RS_3 (the candidates asserted in case d1 are all excised from B') and that all the candidates for their CSP variables and all the (global or local) t-candidates they justified in B have also been eliminated in this transition.

Delete from R_p the CSP variables and the local t-candidates corresponding to these asserted candidates. Call R_p' what remains of R_p and replace R_p by R_p' in B'. A few more questions must be dealt with:
- is there still a candidate for one of the CSP variables of R_p' that could play the role of a left-linking candidate for R_p'? If not, R_p' has already become an autonomous braid in RS_3; excise it from B', together with a whole part of B' after it, along the same lines as in case d1;
- is R_p' still linked to the next part of B'? If not, excise it from B', together with a whole part of B' after it, as in the previous case;
- R_p' may be degenerated (modulo Z and the previous R_k's); this can easily be fixed by replacing R_p' with a sequence of right-linking candidates and/or smaller inner braids (modulo Z and the previous R_k's);
- R_p' or the sequence of right-linking candidates and/or smaller inner braids replacing it may have more targets than R_p; if any of these is a right-linking candidate or an element of a right-linking g-candidate or of an inner B_r-braid of B'

for some index after p, then mark it so that the information can be used in cases d2, f1, f2 or f3 of later steps.

In RS, B' is now, up to p (the ex q), a partial B_r-braid in $B_r B_n$ with target Z.

e) If, in RS, a left-linking candidate L_p has been eliminated (but not asserted) and CSP variable V_p has no t- or z- candidate in RS_2 (complementary to case b), we now have to consider three cases instead of the two we had for g-braids.

e1) If R_p is a candidate, then V_p has only one possible value, namely R_p; if R_p has not yet been asserted by S somewhere between RS_2 and RS, do it now; this case is now reducible to case d1 (because the assertion of R_p also entails the elimination of L_p); go back to case d1 for the same value of p (in order to prevent an infinite loop, mark this case as already dealt with for the current step).

e2) If R_p is a g-candidate, then R_p cannot be asserted by S; however, it can still be used, for any CSP variable after p, to eliminate by W_1 any of its t-candidates that is g-linked to R_p. Let q be the smallest number strictly greater than p such that, in RS, CSP variable V_q still has a global left-linking, t- or z- candidate C_q that is not linked or g-linked or B-linked to any of the R_i for $p \leq i < q$. Replace RS by the state obtained after all the assertions and eliminations similar to those in case d1 above have been done. Then, in RS, excise the part of B' related to CSP variables p to q-1 (included), replace L_q by C_q (if L_q has been eliminated in the passage from RS_2 to RS) and re-number the posterior elements of B', as in case d1. In RS, B' is now, up to p (the ex q), a partial B_r-braid in $B_r B_n$ with target Z.

e3) If R_p is an inner braid, then R_p is no longer linked via L_p to a previous right-linking element of the braid. If none of the CSP variables V_p' in W_p has a z- or t-candidate C_p that can be linked, g-linked or B-linked to Z or to a previous R_i, (situation complementary to case b), it means that the elimination of L_p has turned R_p into an unconditional braid. Let q be the smallest number strictly greater than p such that, in RS, CSP variable V_q has a global left-linking, t- or z- candidate C_q that is not linked or g-linked or B-linked to any of the R_i for $p \leq i < q$. Replace RS by the state obtained after all the assertions and eliminations similar to those in case d1 above have been done. Then, in RS, excise the part of B' related to CSP variables p to q-1 (included), replace L_q by C_q (if L_q has been eliminated in the passage from RS_2 to RS) and re-number the posterior elements of B', as in case d1. In RS, B' is now, up to p (the ex q), a partial B_r-braid in $B_r B_n$ with target Z.

f) Finally, consider eliminations occurring in a right-linking object R_p. This implies that p cannot be the last index of B'. There are three cases.

f1) If, in RS, right-linking candidate R_p of B has been eliminated (but not asserted) or marked, then replace B' by its initial part:
$\{L_1 R_1\} - \{L_2 R_2\} - .... - \{L_p .\}$. At this stage, B' is in RS a (possibly shorter) B_r-braid with target Z. Return B' and stop.

f2) If, in RS, a candidate in right-linking g-candidate R_p has been eliminated (but not asserted) or marked, then:

f2α) either there remains no unmarked candidate of R_p in RS; then replace B' by its initial part: $\{L_1 R_1\} - \{L_2 R_2\} - .... - \{L_p .\}$; at this stage, B' is in RS a (possibly shorter) B_r-braid with target Z; return B' and stop;

f2β) or the remaining unmarked candidates of R_p in RS still make a g-candidate and B' does not have to be changed;

f2γ) or there remains only one unmarked candidate C_p of R_p; replace R_p by C_p in B'. We must also prepare the next steps by putting marks. Any t-candidate of B that was g-linked to R_p, if it is still present in RS, can still be considered as a t-candidate in B', where it is now linked to C_p instead of g-linked to R_p; this does not raise any problem. However, this substitution may entail that candidates that were not t-candidates in B become t-candidates in B'; if they are left-linking candidates of B', this is not a problem either; but if any of them is a right-linking candidate or an element of a right-linking g-candidate or of an inner braid of B', then mark it so that the same procedure (i.e. f1, f2 or f3) can be applied to it in a later step.

f3) If, in RS, a candidate C_p in right-linking braid R_p of length $q_p \geq 2$ has been eliminated (but not asserted) or marked, this has been dealt with in case d2.

Notice that this proof works only because the notions of being linked and g-linked do not depend on the resolution state (they are structural) and the notion of being B-linked is persistent.

11.2.2. Similarities between Subsets and whips

As suggested by the above proof of confluence, there is a remarkable similarity between Subsets and whips/braids of same size p. The definitions of both concepts involve p different CSP variables and p sets of candidates for these variables:

– for S_p-subsets: p transversal sets of candidates, defined by p fixed constraints;

– for whips/braids[p]: p sets consisting of candidates linked (by any constraint) to the target or to one of the previous right-linking candidates (a total of p also!).

These similarities can be represented symbolically as in Figure 11.3 (for p = 4). Horizontal black lines represent CSP variables (V_1, V_2, V_3, V_4). In a Subset (leftmost part of the Figure), each vertical grey line represents a fixed constraint in (c_1, c_2, c_3, c_4). In a whip or a braid (rightmost part of the Figure), each of these lines represents the existence of a link (along *any* constraint) with the target or with a determined element in the sequence of p-1 right-linking candidates. In horizontal lines, candidates may exist only at the intersections with vertical lines; in the whip/braid case, an intersection may represent several candidates (in the same g-label). In spite of their deep conceptual differences, the ideas represented by "vertical lines" can be used in much the same ways in several proofs, such as confluence and the following "$T\&E(B_p)$ vs B_p-braids" theorem.

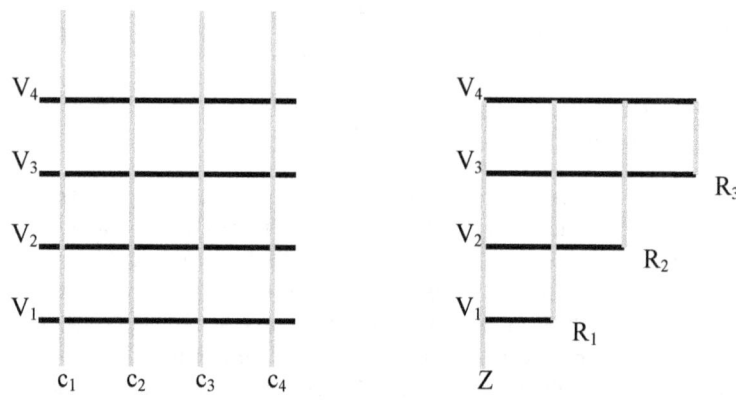

Figure 11.3. A symbolic representation of the similarities between a Subset and a whip.

For whips, the rightmost part of this Figure is an alternative view to that of Figure 11.1. The latter stressed the various links the target or a right-linking candidate can have with z- and/or t- candidates for various posterior CSP variables. The present view abstracts from these differences, considering that only the existence of a link is important. We insist that, contrary to the Subsets case in the leftmost part of the Figure and contrary to what these vertical lines may intuitively suggest, candidates in a vertical line do not have to be pairwise linked (and, in general, they are not).

11.3. The "$T\&E(B_p)$ vs B_p-braids" and "$T\&E(2)$ vs B-braids" theorems

For B_p-braids, for any $p \geq 1$, we are now prepared to expect some extension of the "T&E vs braids" theorem, a "$T\&E(B_p)$ vs B_p-braids"; it will be theorem 11.4.

But, the really new result (with respect to our above-mentioned expectations) is, if p is infinite, there will also appear a new kind of extension, the "$T\&E(2)$ vs B-braids" theorem (theorem 11.5), associated with the iteration of T&E at depth 2.

11.3.1. The "T&E(B_p) vs B_p-braids" theorem

As the T&E(T, Z, RS) procedure can been defined for any resolution theory T with the confluence property (see section 5.6.1), T&E(B_p, Z, RS) can be defined for every p. It is obvious that an elimination done by a B_p-braid can be done by T&E(B_p). The converse is true:

Theorem 11.4: for any $p\geq1$, any elimination done by T&E(B_p) can be done by a B_p-braid. As a result, any puzzle solvable by T&E(B_p) can be solved by B_p-braids.

Proof: it is an easy adaptation of that for g-braids (which are the case p=1 of B_p-braids). As the above proof of confluence, it is also remarkably close to the proof for S_p-braids, with transversal sets replaced by the sets of candidates linked to some right-linked object (see the similarities in Figure 11.3).

Let RS be a resolution state and let Z be a candidate eliminated by T&E(B_p, Z, RS), using some auxiliary resolution state RS'. Following the successive applications of rules from resolution theory B_p in RS', we progressively build a B_p-braid in RS with target Z. First, remember that B_p contains only four types of rules: ECP (which eliminates candidates), $B_{p'}$ (which eliminates targets of $B_{p'}$-braids, p'≤p), S (which asserts a value for a CSP variable) and CD (which detects a contradiction on a CSP variable).

Consider the sequence $(P_1, P_2, ..., P_k, ...P_m)$ of rule applications in RS' based on rules from B_p different from ECP and suppose that P_m is the first occurrence of CD (there must be at least one occurrence of CD if Z is eliminated by T&E(B_p, Z, RS)). We first define the R_k, V_k, W_k and q_k sequences, for k < m:
- if P_k is of type S, then it asserts a value R_k for some CSP variable V_k; let W_k = {V_k} and q_k=1;
- if P_k is of type $B_{p'}$, then define R_k as the non degenerated $B_{p'}$-braid used by the condition part of P_k, as it appears at the time when P_k is applied; let W_k be the sequence of CSP variables of R_k and q_k=p'; in this case, V_k will be defined later.

We shall build a B_p-braid[n] in RS with target Z, with the R_k's as its sequence of right-linking candidates or Bq_k-braids, with the W_k's as its sequence of sequences of CSP variables, with the q_k's as its sequence of sizes and with n= $\sum_{1\leq k\leq m} q_k$ (setting q_m=1). We only have to define properly the L_k's, q_k's and V_k's with $V_k \in W_k$. We do this by recursion, successively for k = 1 to k = m. As the proofs for k = 1 and for the passage from k to k+1 are almost identical, we skip the case k = 1. Suppose we have done it until k and consider the set W_{k+1} of CSP variables.

Whatever rule P_{k+1} is (S or Sq_{k+1}), the fact that it can be applied means that, apart from R_{k+1} (if it is a candidate) or the labels contained in R_{k+1} (if it is an Sq_{k+1}-braid), all the other labels for all the CSP variables in W_{k+1} that were still candidates in RS (and there must be at least one, say L_{k+1}, for some CSP variable V_{k+1} of W_{k+1}) have been eliminated in RS' by the assertion of Z and the previous rule applications. But these previous eliminations can only result from being linked or B-linked to Z or to some R_i, i≤k. The data L_{k+1}, R_{k+1} and $V_{k+1} \in W_{k+1}$ therefore define a legitimate extension for our partial B_p-braid.

End of the procedure: at step m, a contradiction is obtained by CD for a CSP variable V_m. It means that all the candidates for V_m that were still candidates for V_m in RS (and there must be at least one, say L_m) have been eliminated in RS' by the assertion of Z and the previous rule applications. But these previous eliminations can only result from being linked or B-linked to Z or to some R_i, i<m. L_m is thus the last left-linking candidate of the B_p-braid we were looking for in RS and we can take $W_m=\{V_m\}$. qed.

Here again (as in the proof of confluence), this proof works only because the notions of being linked and g-linked are structural and the notion of being B-linked is persistent. It is also again very unlikely that following the T&E(B_p) procedure to produce a B_p-braid, as in the construction in this proof, would produce the shortest available one in resolution state RS.

11.3.2. Definition of the T&E(T, P, n) procedure

In section 5.6.1, we defined the procedure T&E(T, Z, RS) and T&E(T, RS) for any resolution theory T with the confluence property, any candidate Z and any resolution state RS. We can now define the iterated versions of these procedures.

Definition: given a resolution theory T with the confluence property, a resolution state RS and an integer n, the two procedures *Trial and Error based on T at depth n for Z in RS* and *Trial and Error based on T at depth n in RS* [respectively *T&E(T, Z, RS, n)* and *T&E(T, RS, n)*] are defined by mutual recursion as follows:

T&E(T, Z, RS, 1) = T&E(T, Z, RS) and T&E(T, RS, 1) = T&E(T, RS), where the right-hand sides are defined in section 5.6.1.

For n>1, T&E(T, Z, RS, n) is defined as follows:
- make a copy RS_1 of RS; in RS_1, delete Z as a candidate and assert it as a value;
- apply T&E(T, RS_1, n-1);
- if RS_1 has become a contradictory state (detected by CD), then delete Z from RS (*sic*: RS, not RS_1); otherwise, do nothing (in particular if a solution is obtained in RS_1, merely forget it);
- return the (possibly) modified RS state.

For n>1, T&E(T, RS, n) is defined as follows:
a) in RS, apply the rules in T until quiescence; if the resulting RS is a solution or a contradictory state, then return it and stop;

b) mark all the candidates remaining in RS as "not-tried";
c) choose some "not-tried" candidate Z, un-mark it and apply T&E(T, Z, RS, n);
d) if Z has been eliminated from RS by this procedure,

 then goto a
 else if there remains at least one "not-tried" candidate in RS
   then goto c else return RS and stop.

Notice that every time a candidate is eliminated by step d of T&E(T, RS, n), all the other candidates (remaining after step a) are re-marked as "not-tried" by step b. Thus, the same candidate can be tried several times in different resolution states.

Definition: given a resolution theory T with the confluence property and an instance P with initial resolution state RS_P, we define *T&E(T, P, n)* as T&E(T, RS_P, n).

11.3.3. The "T&E(2) vs B-braids" theorem

In the previous definition, taking T = BRT(CSP) and n = 2, and forgetting the reference to BRT as usual, we get procedures T&E(Z, RS, 2), T&E(RS, 2) and T&E(P, 2). We write simply T&E(2) when the context is clear. It is obvious that an elimination done by a B_p-braid of any length can be done by T&E(2). The converse is more interesting:

Theorem 11.5: any elimination done by T&E(2) can be done by a B_p-braid[n] for some p and some n. As a result, any puzzle solvable by T&E(2) can be solved by B-braids.

The proof is a mere iteration of the previous proof. Let RS be a resolution state and let Z be a candidate eliminated by T&E(Z, RS, 2), using some auxiliary resolution state RS'. Following the successive events in RS', we progressively build a B_p-braid in RS with target Z. First, notice that there are only four types of such events: three are applications of rules from BRT [ECP (which eliminates candidates), S (which asserts a value for a CSP variable) and CD (which detects a contradiction on a CSP variable)] and the fourth is a call to some T&E(Z_{k+1}, RS_k), where RS_k is the resolution state reached after the k-th event.

Consider the sequence (P_1, P_2, ..., P_k, ...P_m) of such events in RS', forgetting those associated with rule ECP, and suppose that P_m is the first occurrence of CD (there must be at least one occurrence of CD if Z is eliminated by T&E(B_p, Z, RS)). We first replace the P_k sequence by a sequence of rule applications:
- if P_k is of type S, then we keep it unchanged;
- if P_k is a call to T&E(Z_{k+1}, RS_k), then, applying theorem 5.7, we replace it in RS_k by a braid[q_k] with target Z_{k+1} for some $q_k \geq 1$. We only have to notice that such a braid in RS_k is the same thing as a Bq_k-braid with target Z_{k+1} in RS, modulo Z and the previous right-linking candidates of the global B-braid under construction.

The rest of the proof is as in theorem 11.4. We skip it.

11.3.4. Application of the "T&E(2) vs B-braids" theorem to Sudoku

As the T&E(n) procedure is easy to code in very efficient ways, it is also easy to check that **all the known[11] minimal Sudoku puzzles can be solved by T&E(2)**; therefore **they can all be solved by B-braids and they all have a finite BB rating**.

As there is a finite number of minimal puzzles, it entails that **there is some p (possibly large) such all the known minimal Sudoku puzzles have a finite B_pB rating**. Two questions remain open: can all the minimal puzzles (not only the known ones) be solved by T&E(2) [we have strong reasons to conjecture that this is true] and what is the value of the smallest such p?

Knowing the smallest p would be interesting, because it would define the maximum look-ahead necessary when one tries to find a solution by structured search with only one hypothesis at a time and with no guessing. Whatever its actual value, it is also clear that such a p would provide a universal rating for Sudoku, in the restricted sense that it would ensure a finite rating to every puzzle (which the BB rating already does, but without a predefined value of p).

However, these universal ratings (BB or this B_pB) cannot be considered as universal in the non-technical sense that they would be associated with the "simplest" solution. As we have seen, although all the whip, braid and generalised whip or braid ratings we have introduced are largely mutually compatible (they give a puzzle different ratings in only rare cases), the rare cases where they differ also prove that it is not possible to have a single formal definition of simplicity.

11.4. Bi-whips/bi-braids and an alternative view of W-whips/B-braids

This section introduces another view of W-whips and B-braids.

11.4.1. Definition of bi-whips and bi-braids

Definition: in a resolution state RS, for any $n \geq 1$, a bi-whip[n] is a structured list $(Z_1, Z_2, (V_1, L_1, R_1), ..., (V_{n-1}, L_{n-1}, R_{n-1}), (V_n, L_n))$, such that:

---

[11] This includes the hardest ones recently generated by "Eleven", as mentioned in section 9.6. As we had previously checked that all the published "hardest" puzzles could be solved by T&E(2), after he announced his results, we conjectured that it was also true of his puzzles; Eleven kindly checked this with his program and provided a positive answer; later, when a sublist of his 26,370 hardest became available, we also checked them positively for this property with our independent program.

– for any $1 \leq k \leq n$, V_k is a CSP variable;

– Z_1, Z_2, all the L_k's and all the R_k's are candidates in RS;

– in the sequence $(L_1, R_1, ..., L_{n-1}, R_{n-1}, L_n)$, any two consecutive elements are different;

– Z_1 and Z_2 are different and they do not belong to $\{L_1, R_1, L_2, R_2, ... L_n\}$;

– L_1 is linked to Z_1 or Z_2;

– right-to-left continuity: for any $1 < k \leq n$, L_k is linked to R_{k-1};

– strong left-to-right continuity: for any $1 \leq k < n$, L_k and R_k are candidates for V_k;

– L_n is a candidate for V_n;

– Z_1 and Z_2 are not labels for V_n;

– for any $1 \leq k < n$: R_k is the only candidate for V_k compatible with Z_1, Z_2 and all the previous R_i ($i < k$);

– V_m has no candidate compatible with Z_1, Z_2 and all the previous R_i ($i < n$); (but V_m has more than one candidate – the usual non-degeneracy condition of the global structure being defined);

– $(Z_1, (V_1, L_1, R_1), ..., (V_{n-1}, L_{n-1}, R_{n-1}), (V_n, L_n))$ is not a whip[n];

– $(Z_2, (V_1, L_1, R_1), ..., (V_{n-1}, L_{n-1}, R_{n-1}), (V_n, L_n))$ is not a whip[n].

Definition: a bi-B_n-braid is a structured list as above, with the right-to-left continuity condition and the last two conditions replaced by:

– for any $1 < k \leq n$, L_k is linked to Z_1 or Z_2 or a previous R_i;

– $(Z_1, (V_1, L_1, R_1), ..., (V_{n-1}, L_{n-1}, R_{n-1}), (V_n, L_n))$ is not a braid[n];

– $(Z_2, (V_1, L_1, R_1), ..., (V_{n-1}, L_{n-1}, R_{n-1}), (V_n, L_n))$ is not a braid[n].

Definition: given a resolution state RS, two candidates Z_1 and Z_2 are said *bi-whip[n]* (respectively *bi-braid[n])* incompatible in RS if there exists in RS some bi-whip[n] (resp. some bi-braid[n]) with Z_1 and Z_2 as its first two elements.
Z_1 and Z_2 are said *bi-whip* (respectively *bi-braid*) incompatible in RS if there is some n such that, in RS, they are bi-whip[n] (resp. bi-braid[n]) incompatible.

11.4.2. Another view of W-whips and B-braids

We leave this as an easy exercise: show that the following definitions are equivalent to those of a W-whip and a B-braid, where neither the global length nor the maximum length of the inner whips or braids is specified.

Definition: in any resolution state RS of any CSP, a *W-whip* is a structured list $(Z, (V_1, L_1, R_1), ..., (V_{m-1}, L_{m-1}, R_{m-1}), (V_m, L_m))$ that satisfies the following conditions:

– for any $1 \leq k \leq m$, V_k is a CSP variable;

– Z, all the L_k's and all the R_k's are candidates;

– in the sequence of labels $(L_1, R_1, ..., L_{n-1}, R_{n-1}, L_n)$, any two consecutive elements are different;

– Z does not belong to $\{L_1, R_1, L_2, R_2, ... L_n\}$;

– L_1 is linked to Z;

– right-to-left continuity: for any $1 < k \leq m$, L_k is linked to R_{k-1};

– strong left-to-right continuity: for any $1 \leq k < n$, L_k and R_k are candidates for V_k;

– Z is not a label for V_m;

– for any $1 \leq k < m$: R_k is the only candidate for V_k compatible and not bi-whip incompatible *in RS* with Z and with all the previous right-linking candidates R_i;

– V_m has no candidate compatible and not bi-whip incompatible *in RS* with the target Z and with all the previous right-linking candidates (but V_m has more than one candidate – the usual non-degeneracy condition of the global W-whip).

Definition: a *B-braid* is a structured list as above, with "bi-whip compatible" replaced by "bi-braid compatible" and with the structural right-to-left continuity condition replaced by:

– for any $1 \leq k \leq m$, L_k is linked to Z_1 or Z_2 or a previous R_i.

These alternative definitions may give a better intuitive idea of what a W-whip and a B-braid are. In particular, *bi-whip[n] and bi-braid[n] incompatibility between Z1 and Z2 are restricted constructive forms of the abstract logical nand$_2$ predicate defined by: nand$_2$(Z1, Z2) $\equiv \neg(Z1 \wedge Z2)$.* This gives a more concrete logical content to the two complementary intuitive ideas that: 1) W-whips [respectively B-braids] are based on pairwise contradictions whereas ordinary whips [resp. braids] are based on contradictions arising from a single candidate; 2) but the ways a contradiction may be found are constructive and similar in both cases.

One could also imagine a resolution technique starting by pre-computing all the bi-whip or bi-braid incompatibility relations and then using them in the same way as ordinary links are used in ordinary whips and braids. Unfortunately, as these relations are not structural, they would have to be updated, with probably new instances created every time a candidate is deleted. Moreover, these new definitions do not allow to define ratings compatible with the previously defined ones: they reduce to the same complexity (zero) all the bi-whip[d] or bi-braid[d] pairwise contradictions, for all d. They are equivalent to a hidden level of T&E.

11.4.3. Generalisation: T&E(d), B^{d-1}-braids and the B-nand$_d$ predicate

All the definitions and results in this chapter can be generalised to T&E(d).

Exercise: define W^d-whips and B^d-braids (d levels of inner braids), prove the confluence property of the associated resolution theories and prove the "T&E(d) vs B^{d-1}-braids" theorem.

Exercise: define d-whips and d-braids in the same way as bi-whips and bi-braids in section 11.4.1. Redefine W^d-whips and B^d-braids in the same way as W-whips and W-braids in section 11.4.2. Define the $nand_d(Z_1, Z_2, ..., Z_d)$ predicate and show that W^{d-1}-whips and B^{d-1}-braids can be characterised by the use of such restricted constructive forms of $nand_{d'}$ contradictions between d' candidates, $d' \leq d$.

11.5. Depth of T&E and backdoor size

When speaking of T&E(n), the notion of a backdoor inevitably comes to mind and a question immediately arises: is there a relationship between the backdoor-size of an instance and the depth of T&E necessay to solve it? Le us first give the classical definition of backdoors, but in a formulation consistent with our approach.

Definition[12]: given an instance P of a CSP, *the backdoor-size of P, b(P)*, is the smallest integer n≥0 such that there exists a set B of n labels (a backdoor set) that, when added to the givens of P, allows to solve P within BRT.

We shall also define *the T&E-depth of P, d(P)*, as the smallest n≥0 such that P can be solved by T&E(n), with the convention that T&E(0) = BRT.

Now, given a puzzle P, one can associate with it two intrinsic constants: its backdoor size b(P) and its T&E-depth d(p). And our initial question gets formalised as: is there a relationship between d(p) and b(p)? The answer is not obvious because the backdoor-size b(P) is based on guessing b values (it is therefore largely incompatible with our approach and with the usual requirements of Sudoku players), whereas the T&E-depth d(P) is based on proving that some sets of d (or fewer) hypotheses are contradictory. In particular, none of the relations d ≤ b or b ≤ d or of their negations is obvious in the general CSP.

Let us therefore consider the Sudoku example again. As mentioned in section 11.3.4, all the known Sudoku puzzles can be solved by T&E(d) with d ≤ 2. Moreover:
d(P) = 0 ⇔ no T&E is necessary ⇔ P is solvable by BRT;

---

[12] More generally, one can also define the backdoor size b(P, T) of P for any resolution theory T as the smallest n such that there exists a set B of n labels (the backdoor set) that, when added to the givens of P, allows to solve P in T. And one can ask about the relationship between b(P, T) and d(P, T). For simplicity, we shall consider here only T = BRT, but more on this topic can be found on our website. The notion of a strong T backdoor is also introduced and an application to the famous EasterMonster puzzle is available.

d(P) = 1 ⇔ only one T&E hypothesis need be considered at a time ⇔ P is solvable by braids;

d(P) = 2 ⇔ only two or fewer T&E hypotheses need be considered at a time ⇔ P is solvable by T&E(B) ⇔ P is solvable by B-braids.

Consider first the question "d(P) ≤ b(P)?". If a puzzle can be solved by T&E at depth d, it does not mean that one can choose d fixed hypotheses to generate all the auxiliary grids necessary to the T&E procedure : indeed, this procedure may make hypotheses on any sets of d candidates. But we currently have no explicit counter-example to d(P) ≤ b(P). Notice that, if we consider gsf's lists related to backdoors [gsf www], either his FN-1 list of 1,183 puzzles P with b(P) = 1 or his FN-2 list of 28,948 puzzles P with b(P) = 2, all of them can be solved by ordinary T&E, i.e. they have d(P) = 1, thus satisfying d(P) ≤ b(P).

For the question "b(P) ≤ d(P)?", it is easy to find counter-examples. If we consider again gsf's FN-2 list of 28,948 puzzles P with b(P) = 2, all of them can be solved by ordinary T&E, i.e. they satisfy 1 = d(P) < b(P) = 2.

One can even find counter-examples to the question "b(P) ≤ d(P) + 1?". If we consider gsf's list of 14 puzzles P with b(P) = 3, with their names and their SER:

#1   1......2.9.4...5...6...7...5.9.3.......7......85..4.7....6...3...9.8...2.....1 Easter-Monster SER= 11.6
#2   9......5.4.3...6...2...1...8.74.......2......8.6.7.1....9...3...7.4....5.....2 tarek-ultra-.3.. SER= 11.3
#3   7......4.2.6...1...5...8...3.91........5.....2.3.9.8....7...6...9.2...4.....5 tarek-ultra-.3.1 SER= 11.3
#4   1......89.....91.2......4....76......3..4...9....2..5..4.7....5....8.1..6.3..... tarek-4/.8 SER= 11.5
#5   1.......2..34...5..6.....7.....89..4...3.6.....9.4.....2...1..7........6..5.8..3. jpf-.4/14/.8 SER= 11.2
#6   5......3.2.6...1...8...9...4.7.1......3.......42..7.9.....5...1...7.2...3.....8 tarek-ultra-.3.2 SER= 11.2
#7   1.......2..34...5..6.....7.....5..4...3.1....894.....2...1..7........6..5.9..3. jpf-.4-1. SER= 11.2
#8   1......6.2.5...4...3...7...4.85........1........24.8...7...3...5...9.2.6.......1 coloin SER= 11.3
#9   1......6.2.5...4...3...7...4.89........2.4.......15.8...7...3...5...9.2.6.......1 coloin-.5/11/.1 SER= 11.4
#10 ..1...2...3.....4.5...3...6...1.7....4.....8...9.2...3......8.6..5..3...2...7.. ocean-2..7-.5-29-1 SER= 9.4
#11 3.......2.....54.......6.....1.2..3...........648..........9.7...5.......1.8.5..6.... gfroyle-2..7-.5-3.-4 SER= 3.6
#12 .8..9.....3......6...3...4.....1...5..2...9....7..8..65.....1.....2.7........4.... gfroyle-2..7-.5-3.-3 SER= 4.2
#13 ...2.58..4.3......1..........6...715.....2....4.....2...59.........3.67........ gfroyle-2..7-.5-3.-2 SER= 5.7
#14 ...5....13.8......4....3.......61...9....8.....5....6.7..2..1...3.........49. gfroyle-2..7-.5-3.-1 SER= 6.6

then four of them (#10, #11, #12 and #14) can be solved by ordinary T&E(1), i.e. they satisfy 1 = d(P) < b(P) - 1 = 2; [the remaining 10 can be solved by T&E(2), i.e. they satisfy d(P) = 2 and therefore d(P) = b(P) - 1].

As a result, it does not seem that the notion of backdoor size (intrinsically based on guessing) can shed much light on classifications of puzzles, like those based on the ratings defined in this book, that reject *a priori* any form of guessing.

11.6. The scope of B$_p$-braids

Table 11.1, relative to gsf's list, is the analogue for Bp-braids of Table 9.1 for Sp-braids. For easier comparison of their resolution power, small figures recall the values obtained in Table 9.1. All the puzzles in the list are solved by B$_6$-braids (and all but 4 are solved by B$_5$-braids).

Resolution theory → ↓ slice of puzzles	gB$_\infty$	B$_2$B$_\infty$	B$_3$B$_\infty$	B$_4$B$_\infty$	B$_5$B$_\infty$	B$_6$B$_\infty$
1-500	187	369_{336}	457_{414}	482_{443}	496	500
500-1000	178	364_{335}	462_{415}	496_{460}	500	
1001-1500	163	421_{382}	492_{451}	500_{486}		
1501-2000	168	437_{397}	499_{476}	500_{490}		
2001-2500	135	412_{367}	498_{434}	500_{474}		
2501-3000	116	386_{334}	495_{443}	500_{479}		
3001-3500	120	389_{335}	496_{424}	500_{473}		
3501-4000	113	372_{325}	493_{426}	500_{472}		
4001-4500	104	345_{298}	475_{395}	499_{448}	500	
4501-5000	231	433_{399}	493_{450}	500_{482}		
5001-5500	348	495_{487}	500_{500}			
5501-6000	490	500_{500}				
Total solved /6000	2353	4923 4495	5860 5328	5977+ 5707	5996	6000
Total unsolved /6000	3647	1077 1505	140 672	23 293	4	0

Table 11.1. *Number of puzzles solved by B$_p$-braids, for each slice of 500 puzzles in gsf's list (restricted to the first 6000). The first column here (for gB$_\infty$) corresponds to the second in Table 9.1.*

12. Final remarks

In these final retrospective remarks, which are intended neither as a summary nor as a conclusion, we shall first highlight and comment some overlapping facets of what has been achieved for the general CSP (with a few open questions). Considering that a third of this book is an illustration of the general theory with the Sudoku CSP, the second section will be a quick review, mainly for the readers of *HLS*, of what is new with respect to *HLS2*.

12.1. About our approach of the finite CSP

12.1.1. *About the general distinctive features of our approach*

There are five main inter-related reasons why this book diverges radically from the current literature on the finite CSP[13]:

– almost everything in our approach, in particular all our definitions and theorems, is formulated in terms of *pure logic*, independently of any algorithmic implementation; (apart from the obvious logical re-formulation of a CSP, the current literature on CSPs is mainly about algorithms for solving them and comparisons of such algorithms);

– we systematically use redundant sets of CSP variables;

– we fix the main parameter defining the "size" of a CSP and we are not concerned with the usual theoretical perspectives of complexity, such as NP-completeness of a CSP with respect to its size;

– we nevertheless tackle questions of complexity, in terms of the statistical distribution of the *minimal instances* of a fixed size CSP; although all our rules are valid for all the instances of a CSP, without any kind of restriction, we grant minimal instances a major role in all our statistical analyses and classification results; the thin layer of instances they define in the whole forest of possible instances (see chapter 6 for this view) allows to discard secondary problems that

---

[13] We are not suggesting that our approach is better than the usual ones; we are aware that our purposes are non-standard and they may be irrelevant when speed of resolution is the main criterion.

multi-solution or over-constrained instances would raise; (the notion of minimality is almost unknown in the CSP world);

– last but not least, our *purposes* lie much beyond the usual ones of finding a solution or finding the fastest algorithm for it. Here, *instead of the solution as a result, we are interested in the solution as a proof of the result, i.e. in the resolution path*. Accordingly, we have concentrated on finding *understandable, meaningful, pure logic* resolution paths – though these words do not have a clear pre-assigned meaning.

We have taken this purpose into account in Part I by interpreting the "pure logic" requirement literally – i.e. as a solution completely defined in terms of mathematical logic (with no reference to any algorithmic notions). Thus, we have introduced a general resolution paradigm based on progressive candidate elimination (which amounts to progressive domain restriction, a classical idea in the CSP community), but in which each of these eliminations is justified by a single, well defined *resolution rule* of a given *resolution theory* and is interpreted in modal (non algorithmic) terms. We have established a clear logical (intuitionistic) status for the notion of a candidate (a notion that does not pertain to the CSP Theory). Moreover, we have shown that the modal operator that naturally appears in any formal definition of a candidate can be "forgotten" when we state resolution rules, provided that we work with intuitionistic (or constructive) instead of classical logic (which is not a restriction in practice).

Once this logical framework is set, a more precise purpose can be examined, not completely independent from the original vague "understandable" and "meaningful" ones: one may want the *simplest* pure logic solution. As is generally understood without saying when one speaks of the simplest solution to a mathematical problem, we mean neither easiest to discover for a human being nor computationally cheapest, but simplest to understand for the reader. Even with such precisions, we have shown that "simplest" may still have many different, all logically grounded, meanings, associated with different (purely logical) ratings of instances.

Taking for granted that hard minimal instances of most fixed size CSPs cannot be solved by elementary rules but they require some kind of chain rules (with the classical xy-chains of Sudoku as our initial inspiration), we have refined our general paradigm by defining families of resolution rules of increasing (logical and computational) complexity, valid for any CSP: some reversible (Bivalue Chains, Reversible Subset Chains, Reversible g-Subset Chains) and some oriented, much more powerful ones (whips, g-whips, S_p-whips, gS_p-whips, W_p-whips and similar braid families).

The different resolution paths obtained with each of these families correspond to different legitimate meanings of "simplest" (when they lead to a solution) and, in spite of strong subsumption relationships, we have shown that none of them can be completely reduced to another. Said otherwise: there does not seem to be any universal notion of simplicity for the resolution of a CSP; this is exemplified by the various ratings ascribed to some CSP instances in several chapters.

12.1.2. About our resolution rules (whips, braids, ...)

Regarding these new families of chain rules, now reversing the history of our theoretical developments, four main points should be recalled:

– We have introduced a formal definition of Trial-and-Error (T&E), a procedure that, in noticeable contrast with the well known structured search algorithms (breadth-first, depth-first, ...) and all their CSP specific variants implementing some form of constraint propagation (arc-consistency, path-consistency, MAC, ...), allows no "guessing", in the sense that it accepts no solution found by sheer chance during the search process: a value for a CSP variable is accepted only if all its other possible values have been tested and each of them has been proven to lead to a contradiction.

– With the "T&E vs braids" theorem and its "T&E(T) vs T-braids" extensions to various resolution theories T, we have proven that a solution obtained by the T&E(T) procedure can always be replaced by a "pure logic" solution based on T-braids, i.e. on sequential patterns accepting simpler patterns taken from the rules in T as their building blocks.

– Because its importance could not be over-estimated, we have proven in great detail that all our generalised braid resolution theories (braids, g-braids, S_p-braids, gS_p-braids, B_p-braids, ...) have the confluence property. Thanks to it, we have justified the idea that these types of logical theories can be supplemented by a "simplest first" strategy, defined by ascribing in a natural way a different priority to each of their rules. When one tries to compute the rating of an instance and to find the simplest, pure logic solution for it, in the sense that it has a resolution path with the shortest possible braids in the family (which the T&E procedure alone is unable to provide), this strategy allows to consider only one resolution path; without this property, all of them should *a priori* be examined, which would add an exponential factor to computational complexity. Even if the goal of maximum simplicity is not retained, the property of stability for confluence of these T-braid resolution theories remains useful, because it guarantees that valid eliminations and assertions occasionally found by any other consistent opportunistic solving methods cannot introduce any risk of missing a solution based on T-braids.

– With the statistical results in chapter 6, we have also shown that, in spite of a major structural difference between whips and braids (the "continuity" condition), whips (even if restricted to no-loop ones) are a very good approximation of braids (at least in the Sudoku case), in the double sense that: 1) the associated W and B ratings are rarely different when the W rating is finite and 2) the same "simplest first" strategy, *a priori* justified for braids but not for whips, can be applied to whips, with the result that a good approximation of the W rating is obtained after considering only one resolution path (i.e. the concrete effects of non confluence of the whip resolution theories appear only rarely). This is the best situation one can desire for a restriction: it reduces structural (and computational) complexity but it entails little difference in classification results.[14] Of course, much work remains to be done to check whether this proximity of whips and braids is true for CSPs other than Sudoku and for all the types of extended whips and braids defined in this book.

12.1.3. About human solving based on these rules

The four above-mentioned points have their correlates regarding a human trying to solve an instance of a CSP "manually" (or should we say "neuronally"?), as may be the "standard" situation for some CSPs, such as games:

– It should first be noted that T&E seems to be the most natural resolution method for a human who is unaware of more complex possibilities and who does not accept guessing. This was initially only a vague intuition. But, with time, it has received some concrete support from our experience in the Sudoku micro-world (with friends, students, contacts, or from questions of newcomers on forums), considering the way new players spontaneously re-invent it without even having to think of it. Indeed, it does not seem that they reject guessing *a priori*; they start by using it and they feel unsatisfied about it after some time; "no-guessing" appears as an additional *a posteriori* requirement.

– The "T&E vs braids" theorem means that the most natural T&E solving technique, in spite of being strongly anathemised by some Sudoku experts, is not so far from being compatible with the abstract "pure logic" requirement. Moreover, its proof shows that a human solver can always modify a T&E solution in order to present it as a braid solution. Thanks to the subsumption theorems or to the more general "T&E(T) vs T-braids" theorem, this remains true when he learns more elaborate techniques (such as Subset or g-Subset rules) and starts to combine them with T&E.

– Finding the shortest braid solution is a much harder goal than finding any solution based on braids and this is where the main divergence with a solution obtained by mere T&E occurs. For the human solver who started with T&E, it is nevertheless a natural step to try to find a shorter (even if not the shortest) solution.

---

[14] By contrast, the "reversibility" condition often imposed on chains (never clearly formulated before *HLS* but widely accepted in some Sudoku circles) is very restrictive and it leads some players to reject solutions based on non-reversible (or "oriented") chains (such as whips and braids) and to the (in our opinion, hopeless for really hard instances) search for extremely complex patterns (such as all kinds of what we would call extended g-Fish patterns: finned, sashimi, chains of g-Fish, …). This said, we acknowledge that Reversible Subset Chains (Nice Loops, AICs) may have some appeal for moderately complex instances.

An obvious possibility consists of excising the useless branches; but one can also look for alternative braids, either for the same elimination or for a different one.

– As for the fourth point, a human solver is very likely to have spontaneously the idea of using the continuity condition to guide his search for a contradiction on some target Z: it means giving a preference to the last result obtained.

Finally, for a human solver, the transition from the spontaneous T&E procedure to the search for whips can be considered as a very natural process. Looking for Subsets and then for g-Subsets can also be considered as a natural, though different, evolution. And the two can be combined. Once more, there is no unique way of defining what "the best solution" may mean.

Of course, a human player can follow a very different learning path, starting with xy-chains and progressively trying to spot patterns from the ascending sequence of more complex rules following our discovery path in *HLS*.

12.1.4. About the similarity between Subsets and whips/braids of same size

We have noticed a remarkable formal similarity between Subsets and whips/braids of same size (see Figure 11.3 and comments there). It has appeared in very explicit ways in the proofs of confluence and of the generalised "T&E vs braids" theorems for the S_p-braids and B_p-braids.

But whips/braids have a much greater resolution power than Subsets of same size, at least in the Sudoku CSP, as shown by the general subsumption theorems in section 8.7 and the specific statistical results in Table 8.1. As mentioned in section 8.7.3, these results indicate that the definition of Subsets is much more restrictive than the definition of whips. In Subsets, transversal sets are defined by a single constraint. In whips, the fact of being linked to the target or to a given previous right-linking candidate plays a role very similar to each of these transversal sets. But being linked to a candidate is much less restrictive than being linked to it via a pre-assigned constraint; in this respect, the three elementary examples for whips of length 2 in sections 8.7.1.1 and 8.8.1 are illuminating. As shown by the subsumption and almost-subsumption results in section 8.7, the few cases of Subsets not covered by whips because of the restrictions related to sequentiality are too rarely met in practice to be able to compensate for this.

For the above reasons, we conjecture that, in any CSP, whips/braids have a much greater resolution power than Subsets of same length p, for small values of p; for larger values of p, it is less clear because there may be an increasing number of cases of non-subsumption but there may also be more ways of being linked to a candidate. Probably, much depends on how many different constraints a given candidate can participate in. This is an area where much more work is necessary.

12.1.5. About minimal instances and uniqueness

Considering that, most of the time, we restrict our attention to minimal instances that (by definition) have a unique solution, one may wonder why we do not introduce any "axiom" of uniqueness. Indeed, there are many reasons:

– it is true that we restrict all our statistical analyses of resolution rules to minimal instances, for reasons that have been explained in the Introduction; but it does not entail that validity of resolution rules should *per se* be limited to minimal instances; on the contrary, they should apply to any instance; in a few examples in this book, our rules have even been used to prove non-uniqueness or non-existence of solutions;

– as mentioned in the Introduction, from the point of view of Logic, uniqueness cannot be an *axiom*, at least not an axiom that could impose uniqueness of a solution; it can only be an *assumption*; moreover, when incorrectly applied to a multi-solution instance, the *assumption* of uniqueness can lead, via a vicious circle, to the erroneous *conclusion* that an instance has a unique solution; we have given an example in *HLS1*, section XXII.3.1 (section 3.1 of chapter "Miscellanea" in *HLS2*);

– uniqueness is not a constraint the CSP solver (be he human or machine) is expected or can choose to satisfy; in some CSPs or some situations (such as for statistical analyses or for games like Sudoku), uniqueness may be a requirement to the provider of instances (he should provide only "well formed", i.e. minimal, instances); the CSP solver can then decide to trust his provider or not; if he does and he uses rules based on it in his resolution paths, then uniqueness can best be described as an oracle; for this reason, in all the solutions we have given, uniqueness is never assumed, but it is proven constructively from the givens;

– the fact is, there is no known way of exploiting the assumption of uniqueness for writing any general resolution rule for uniqueness; and we can take no inspiration in the Sudoku case, because all the known rules for uniqueness are based on Sudoku specific constraints;

– in the Sudoku case, the known rules of uniqueness, when added to a resolution theory with the confluence property, destroy confluence (see *HLS* for an example);

– still in the Sudoku case, it does not seem that the known rules for uniqueness have much resolution power; there is no known example that could be solved if they are used but that could not without them.

Of course, we are not trying to deter anyone from using uniqueness, if they like it, in CSPs for which it allows to formulate specific rules, such as Sudoku (where it has always been a very controversial topic, but it has also led to the definition of smart rules); in some rare cases, it can simplify the resolution paths. We are only explaining why we chose not to use it in our theoretical approach.

12.1.6. About minimal instances vs density and tightness of constraints

Two global parameters of a CSP, its "density of constraints" and its "tightness", have been identified in the classical CSP literature. Their influence on the behaviour of general-purpose CSP solving algorithms has been studied extensively and they have also been used to compare such algorithms. (As far as we know, these studies have been about unrestricted CSP instances; we have been unable to find any reference to the notion of a minimal instance in constraint satisfaction.)

Definitions: the *density of constraints* of a CSP is the ratio between the number of label pairs linked by a constraint (supposing that all the constraints are binary) and the total number of label pairs; the *tightness* of a CSP is the ratio between the number of label pairs linked by a "strong" constraint (i.e. a constraint due to a CSP variable) and the number of label pairs linked by a constraint.

Density reflects the intuitive idea that the nodes of a graph (here, the graph of labels) can be more or less tightly linked by the edges (here the binary contradictions); it also evokes a few general theorems relating the density and the diameter of a random graph (a topic that has recently become very attractive because of communication networks). Tightness evokes the difference we have mentioned between Sudoku (tightness 100%) and n-Queens (tightness ~ 50%, depending on n).

In the context of this book, relevant questions related to these parameters should be about their influence on the scope of the various types of resolution rules with respect to the set of minimal instances of the CSP. However, how these two parameters should be defined in this context is less obvious than it may seem at first sight. The question is, should one compute these parameters using all the labels of the CSP or only the actual candidates?

Taking the 9×9 Sudoku example (again!), the computation is easy for labels: there are 729 labels (all the nrc triplets) and each label is linked to 8 different labels on each of the n, r, c axes, plus 4 remaining labels on the b axis. Each label is thus linked to the same number (28) of other labels and one gets a density equal to $28/728 = 3.846\%$. More generally, for n×n Sudoku with $n = m^2$, density is: $(4m^2-2m-2)/(m^6-1)$; it tends rapidly to zero (as fast as $4/n^2$) as the grid size n increases.

However, considering the first line of each resolution path in this book, one can check that for a minimal puzzle, after the initial Elementary Constraint Propagation rules have been applied (i.e. after the straightforward initial domain restrictions), the number of candidates remaining in the resolution state RS_P of an instance P is much smaller. As all that happens in a resolution path depends only on RS_P, a definition of density based on the candidates in RS_P can be expected to be more relevant. But, the analysis of the first series of 21,375 puzzles produced by the controlled-bias generator, leads to the following conclusions, showing that neither the number of candidates in RS_P nor the density of constraints in RS_P have any significant correlation with the difficulty of a puzzle P (measured by its W rating):

– the number of candidates in RS_P has mean 206.1 (far less than the 729 labels) and standard deviation 10.9; it has correlation coefficient -0.20 with the W rating;

– the density of constraints in RS_P has mean 1.58% (much less than when computed on all the labels) and standard deviation 0.05%; its has correlation coefficients -0.16 with the number of candidates in RS_P and -0.06 with the W rating.

Can tightness give better or different insights? This parameter plays a major role in the left to right extension steps of the partial chains of all the types defined in this book. In n×n Sudoku or n×n LatinSquare, tightness is 100%, whatever the value of n; these examples may therefore not be used to investigate this parameter. If there are few CSP variables, there may be few chains. However, from the millions of Sudoku puzzles we have solved, problems that appear for the hardest ones solvable by whips or g-whips arise from two opposite causes: not only because there are too few partial whips or g-whips (and no complete ones), but also because there are too many useless partial whips or g-whips (eventually leading to memory overflow problems).

One idea that needs be explored in more detail is that the possible effects of initial density or tightness of constraints are minimised by considering only the thin layer of minimal puzzles (as is the case for the number of givens).

12.1.7. About a strategic level

We have used the confluence property to justify the definition of a "simplest first strategy" for all the braid (and, by extension, the whip) resolution theories. This strategy fits the goals of finding the simplest solution (keeping the above comments on "simplest" in mind) and of rating an instance.

Other systematic strategies can also be imagined. One of them is considering subsets of CSP variables of "same type" and restricting all the rules to such subsets. This is what we have done for Sudoku in *HLS1*, with the 2D rules. It is easy to see that, as the "2D" rules are the projections of the "3D" ones presented here, all the 2D-braid theories (in the four rc, rn, cn and bn spaces) are stable for confluence and have the confluence property; it is therefore also true of their union. In *HLS1*, we have shown that 97% of the Sudogen0 puzzles can be solved by such 2D rules (the real percentage may be a little less for an unbiased sample). We still consider these rules as interesting special cases that have an obvious place in the "simplest first" strategy and that may be easier to find and/or to understand for a human player.

Now, it is very unlikely that any human solver would proceed in such a systematic way as described in the above strategies. He may prefer to concentrate on some aspect of the puzzle and try to eliminate a candidate from a chosen cell (or group of cells). As soon as he has found a pattern justifying an elimination, he applies it. This could be called the opportunistic "first-found-first-applied strategy". And, thanks to stability for confluence, it is justified in all the generalised braid resolution theories defined in this book (there can be no "bad" move blocking the way to the solution).

What may be missing however in our approach is more general "strategic" knowledge[15] for orienting the search: when should one look for such or such pattern? But the fact is, we have no idea of which criteria could constitute a basis for such knowledge. Moreover, even in the most studied Sudoku CSP, whereas there is a plethora of literature on resolution techniques (sometimes misleadingly called strategies), nothing has ever been written on the ways they should be used, i.e. on what might legitimately be called strategies. In particular, one common prejudice is that one should first try to eliminate bivalue/bilocal candidates (i.e., in our vocabulary, candidates in bivalue rc, rn, cn or bn cells). Whereas this may work for simple puzzles, it is almost never possible for complex ones. This can easily be seen by examining some of the examples of this book, with the long sequences of whip eliminations necessary before a Single is found: if any of these eliminations had occurred for a candidate from a bivalue/bilocal cell, then it would have been immediately followed by a Single.

12.1.8. About ratings and the requirement for the "simplest" solution

Our initial motivations included a "pure logic solution" and a "simplest solution". If the first goal has been reached in Part I, one may wonder what the second goal has become.

For any instance P of any CSP, several ratings of P have been introduced. All of them are defined in pure logic terms and are intrinsic properties of P; moreover, they have been shown to be largely mutually consistent, i.e. "most of the time" they assign the same ratings to P (at least for Sudoku). Moreover, for any CSP whose minimal instances can be solved with at most two levels of Trial-and-Error, the BB rating, a rating that can thus be considered as universal, has even been defined.

But what the multiplicity of these logically grounded ratings shows is that there is one thing all our formal analyses cannot do in our stead: choosing what should be considered as "simplest". And this can only depend on one's specific goals. Let us illustrate this with the Sudoku CSP. Whether these remarks would apply to other CSPs, or how they should be modified, remains an open question.

---

[15] [Laurière 1978] presents a different perspective, based on general-purpose heuristics.

If one is interested in the simplest solution for all the minimal puzzles, then, considering the statistical results of chapter 6, a whip solution would certainly be the simplest one, *statistically*; a g-whip solution would be a good alternative. "Statistically" means that, in rare cases, a better solution based on Subsets or g-Subsets or Reversible Subset Chains could be found.

If one is interested in providing examples of some particular set of techniques or promoting them, then a solution considered as the simplest must (tautologically) use only these techniques; the job will then be to provide nice examples of such puzzles; this is the approach implicitly taken by most puzzle providers and most databases of "typical examples" associated with solvers. Unfortunately, apart from those here and in *HLS*, we lack both formal studies of such sets of techniques and statistical analyses of their scopes.

If one is interested in the "hardest" puzzles, then the first thing should be to specify what is meant by "hardest" (in particular with respect to which rating); puzzles harder than the "hardest" known ones (wrt the SER rating) keep being discovered; one can consider that Part III of this book (apart from chapter 8) is dedicated to resolution rules for the hardest puzzles; it shows that different techniques of increasing complexity can be used to deal with them. Much seems to depend on two parameters: the maximal depth d of Trial-and-Error necessary to solve these instances and the maximal look-ahead l necessary to solve them at depth d-1. (Even for Sudoku, although we can reasonably conjecture that d=2, we have no formal proof of this; and we have no estimate for l, except that $l \geq 6$).) However, for the very hardest puzzles, it may happen that the whole requirement of simplicity becomes merely meaningless: the existence of extremely rare but very hard instances that cannot be solved by any "simple" rules is a fact that cannot be ignored.

12.2. About the Sudoku CSP, beyond *HLS2* [for the readers of *HLS*]

This section is a quick review of the main points related to the Sudoku CSP that are new with respect to *HLS*.

12.2.1. About the general framework

The general formal logic framework introduced in this book (all of Part I), when applied to Sudoku, is globally the same as in *HLS*. Only two slight differences appear: we now use Gentzen's "natural logic" instead of Hilbert's axioms; and, when dealing with the resolution states (formerly called "knowledge states" in *HLS*), we refer to the logic of necessitation instead of epistemic logic (or logic of knowledge). None of these changes has any practical consequences; in particular, resolution theories should still be understood as theories in intuitionistic (constructive) logic. Our definition of a resolution theory was slightly less precise in *HLS*.

But the main difference with *HLS* is that Sudoku is now considered as a CSP in a more systematic way than it was there; in particular:

– the cells in the Extended Sudoku Board are now explicitly interpreted as representations of CSP variables;

– the nrc notation introduced in *HLS2* as a convenient way of representing chains, now appears as an obvious special case of a natural notation for any CSP;

– basic interactions ("pointing" and "claiming") are systematically written as instances of whip[1];

– Subsets rules are first formulated in a general way, meaningful for any CSP, in terms of CSP variables and transversal sets, before being re-expressed for Sudoku in the usual terms of numbers, rows, columns and blocks; Naked, Hidden and Super-Hidden (Fish) Subsets are not only related by symmetry as they were in *HLS*; from the CSP point of view, they are now the very same rule, because the symmetries have been used at a higher level to introduce new CSP variables; (of course, this does not change anything for all practical purposes);

– the main change brought by this new perspective of Subsets is, it allows to introduce their "grouped" version (g-Subsets) as a natural generalisation (in the same way as several chain patterns have a grouped version) and to re-interpret the well-known Franken and Mutant Fish as g-Subsets.

12.2.2. New resolution rules and new ratings

For the parts specifically dedicated to the Sudoku CSP (about a third of this book), they are very far from constituting a third edition of *HLS* (none of the examples, specialised versions of resolution rules or independence theorems present in *HLS2* has been reproduced here). Indeed, these parts start where *HLS2* ended.

We first introduced the following patterns, resolution rules and/or topics (that were not in *HLS2*) on the Sudoku Player's Forum in 2008; but in this book, most of them are now presented in the more general CSP framework (with a somewhat different, simpler terminology):

– whips (which are a more synthetic view of both nrczt- chains and lassoes);

– braids, with a detailed proof of the confluence property of braid resolution theories;

– definition of the Trial-and-Error procedure (T&E) and proof of the "T&E vs braids" theorem;

– definition of the controlled bias generator; comparison of various kinds of generators; unbiased statistics for the W rating for a much larger sample than in

HLS; [even though it was already possible to generate large random samples of minimal puzzles, before *HLS1* the literature had concentrated on isolated examples illustrating specific rules; systematic studies of large collections of puzzles had been lacking; to palliate this deficiency, detailed numerical results about the number of puzzles solved by each type of rule had been given in *HLS*, but they were still based on the biased samples produced by the currently available generators];

– g-whips and g-braids; proofs of the associated confluence property and "gT&E vs g-braids" theorem;

– detailed subsumption theorems for whips and braids, showing that they capture "almost all" but not all the cases of Subset rules;

– Reversible S_p-chains, obtained by allowing the insertion of Subsets as right-linking objects in bivalue chains; proof that these chains are the same thing as grouped AICs or Nice Loops (but our definition does not involve the unnecessarily complex notion of a "restricted common"); proof of the confluence property for the associated resolution theories;

– S_p-whips and S_p-braids, obtained by allowing the insertion of Subsets as right-linking objects in whips and braids; proofs of the associated confluence property and "T&E(S_p) vs S_p-braids" theorem; analysis of their scope;

– gS_p-Subsets, the "grouped" version of Subsets, a generalisation allowed by considering Subset patterns (S_p-subsets) from the general CSP point of view; they provide a new view of Franken and Mutant Fish;

– Reversible gS_p-chains, gS_p-whips and gS_p-braids, obtained by allowing the insertion of g-Subsets as right-linking objects in bivalue chains, whips and braids;

– W_p-whips and B_p-braids; proofs of the associated confluence property and "T&E(B_p) vs B_p-braids" theorem; relation of B-braids with T&E(2), providing a finite BB rating for all the known minimal puzzles [and indications that B_6B could also be universal].

References

[Angus 2005-2007]: ANGUS J., Simple Sudoku, http://www.angusj.com/sudoku/, 2005-2007.

[Armstrong 2000-2007]: ARMSTRONG S., Sadman Software Sudoku, Solving Techniques, http://www. sadmansoftware.com/sudoku/techniques.htm, 2000-2007.

[Barcan 1946a]: BARCAN M., A Functional Calculus of First Order Based on Strict Implication, Journal of Symbolic Logic, Vol. 11 n°1, pp. 1-16, 1946.

[Barcan 1946b]: BARCAN M., The Deduction Theorem in a Functional Calculus of First Order Based on Strict Implication, Journal of Symbolic Logic, Vol. 12 n°4, pp. 115-118, 1946.

[Barker 2006]: BARKER M., Sudoku Players Forum, Advanced solving techniques, post 362, *in* http://www.sudoku.com/forums/viewtopic.php?t=3315

[Berthier 2007a]: BERTHIER D., *The Hidden Logic of Sudoku*, First Edition, Lulu.com Publishers, May 2007.

[Berthier 2007b]: BERTHIER D., *The Hidden Logic of Sudoku*, Second Edition, Lulu.com Publishers, November 2007.

[Berthier 2008a]: BERTHIER D., From Constraints to Resolution Rules, Part I: Conceptual Framework, *International Joint Conferences on Computer, Information, Systems Sciences and Engineering (CISSE 08)*, December 5-13, 2008, Springer. Published as a chapter of *Advanced Techniques in Computing Sciences and Software Engineering*, Khaled Elleithy Editor, pp. 165-170, Springer, 2010.

[Berthier 2008b]: BERTHIER D., From Constraints to Resolution Rules, Part II: chains, braids, confluence and T&E, *International Joint Conferences on Computer, Information, Systems Sciences and Engineering (CISSE 08)*, December 5-13, 2008, Springer. Published as a chapter of *Advanced Techniques in Computing Sciences and Software Engineering*, Khaled Elleithy Editor, pp. 171-176, Springer, 2010.

[Berthier 2009]: BERTHIER D., Unbiased Statistics of a CSP - A Controlled-Bias Generator, *International Joint Conferences on Computer, Information, Systems Sciences and Engineering (CISSE 09)*, December 4-12, 2009, Springer. Published as a chapter of *Innovations in Computing Sciences and Software Engineering*, Khaled Elleithy Editor, pp. 11-17, Springer, 2010.

[Berthier www]: BERTHIER's website: http://www.carva.org/denis.berthier (permanent URL)

[Bridges et al. 2006]: BRIDGES D. & VITA L., *Techniques of Constructive Analysis*, Springer, 2006.

[Brouwer 2006]: BROUWER A., Solving Sudokus, http://homepages.cwi.nl/~aeb/games/ sudoku/, 2006.

[Davis 2006]: DAVIS T., The Mathematics of Sudoku, www.geometer.org/mathcircles/ sudoku.pdf, 2006.

[Felgenhauer et al. 2005]: FELGENHAUER B. & JARVIS F., Enumerating possible Sudoku grids, http://www.afjarvis.staff.shef.ac.uk/sudoku/sudgroup.html, 2005.

[Feys 1965]: FEYS R., *Modal Logics*, Fondation Universitaire de Belgique, 1965.

[Fitting 1969]: FITTING M., *Intuitionistic Logic, Model Theory and Forcing*, North Holland, 1969.

[Fitting et al. 1999]: FITTING M. & MENDELSOHN R., *First-Order Modal Logic*, Kluwer Academic Press, 1999.

[Garson 2003]: GARSON J., Modal Logic, *Stanford Encyclopedia of Philosophy*, http://plato.stan ford. edu/entries/logic-modal/, 2003.

[Gary et al. 1979]: GARY M. & JOHNSON D., *Computers and Intractability: A Guide to the Theory of NP-Completeness*, Freeman, 1979.

[Gentzen 1934], GENTZEN G., Untersuchungen über das logische Schließen I, *Mathematische Zeitschrift*, vol. 39, pp. 176-210, 1935.

[gsf www]: Glenn FOWLER's website: http://www2.research.att.com/~gsf/sudoku

[Guesguen et al. 1992]: GUESGUEN H.W. & HETZBERG J., *A Perspective of Constraint-Based Reasoning*, Lecture Notes in Artificial Intelligence, Springer, 1992.

[Hanssen 20xx]: HANSSEN V., http://www.menneske.no/sudoku/eng/index.html

[Hendricks et al. 2006]: HENDRICKS V. & SYMONS J., Modal Logic, *Stanford Encyclopedia of Philosophy*, http://plato.stan ford. edu/entries/logic-modal/, 2006.

[Hintikka 1962]: HINTIKKA J., *Knowledge and Belief: An Introduction to the Logic of the Two Notions*, Cornell University Press, 1962.

[HLS1, HLS2, HLS]: respectively, abbreviations for [Berthier 2007a], [Berthier 2007b] or for any of the two.

[Jarvis 2006]: JARVIS F., Sudoku enumeration problems, http://www.afjarvis.staff.shef.ac. uk/ sudoku/, 2006.

[Kripke 1963]: KRIPKE S., Semantical Analysis of Modal Logic, *Zeitchrift für Matematische Logik und Grundlagen der Matematik*, Vol. 9, pp. 67-96, 1963.

[Kumar 1992]: KUMAR V., Algorithms for Constraint Satisfaction Problems: a Survey, *AI Magazine*, Vol. 13 n° 1, pp. 32-44, 1992.

[Laurière 1978]: LAURIERE J.L., A language and a program for stating and solving combinatorial problems, *Artificial Intelligence*, Vol. 10, pp. 29-117, 1978.

[Lemmon et al. 1977]: LEMMON E. & SCOTT D., *An introduction to Modal Logic*, Blackwell, 1977.

[Meinke et al. 1993]: MEINKE K. & TUCKER J., eds., *Many-Sorted Logic and its Applications*, Wiley, 1993.

[Mepham xx]: MEPHAM M., http://www.sudoku.org.uk/

[Moschovakis 2007]: MOSCHOVAKIS J., Intuitionistic Logic, Stanford Encyclopedia of Philosophy, http://plato.stan ford.edu/entries/logic-intuitionistic/, 2006.

[Newell 1982]: NEWELL A., The Knowledge Level, *Artificial Intelligence*, Vol. 59, pp 87-127, 1982.

[Riley 2008]: RILEY G., *CLIPS online documentation*, http://clipsrules.sourceforge.net/OnlineDocs.html, 2008.

[Rossi et al. 2006]: ROSSI F., VAN BEEK P. & WALSH T., *Handbook of Constraint Programming*, Foundations of Artificial Intelligence, Elsevier, 2006.

[Royle 20xx]: ROYLE G., Minimum Sudoku, http://www.csse.uwa.edu.au/~gordon/sudokumin.php

[Russell et al. 2005]: RUSSELL E. & JARVIS F., There are 5472730538 essentially different Sudoku grids ... and the Sudoku symmetry group, http://www.afjarvis.staff.shef.ac.uk/sudoku/sudgroup.html, 2005.

[SPlF]: Sudoku Player's Forums, http://www.sudoku.com/forums/index.php

[SPrF]: Sudoku Programmers Forums, http://www.setbb.com/sudoku/index.php?mforum= sudoku

[Sterten xx]: STERTEN, http://magictour.free.fr/sudoku.htm

[Sterten 2005]: STERTEN (alias DUKUSO), *suexg*, http://www.setbb.com/phpbb/viewtopic. php?t=206&mforum= sudoku, 2005.

[Stuart 2006]: STUART A., http://www.scanraid.com, 2006.

[Stuart 2007]: STUART A., *The Logic of Sudoku*, Michael Mepham Publishing, 2007.

[Sudopedia]: Sudopedia, http://www.sudopedia.org/wiki/Main_Page

[Werf 2005-2007]: van der WERF R., Sudocue, Sudoku Solving Guide, http://www.sudocue. net/guide.php, 2005-2007.

[Yato et al. 2002]: YATO T. & SETA T., Complexity and completeness of finding another solution and its application to puzzles, IPSG SIG Notes 2002-AL-87-2, http://www-imai.is.s.u-tokyo.ac.jp/~yato/data2/SIGAL87-2.pdf, 2002.